Basic

Plasma

Physics

Beryl Browning

For W. E.

First published 2008

Published by Lulu

ISBN 978-1-4092-1919-4

About the author:
Beryl Browning obtained her PhD in Plasma Physics from
the University of Manchester Institute of Science &
Technology.   She is now the senior partner of a
management training consultancy.

Contents

Introduction 1
Symbols and constants 2

Chapter
1 Strange glows, striations and dark spaces (1675-1900) 3
2 The development of plasma theory (1900-1950) 23
3 Statistical mechanics and the kinetic theory of gases 43
  3.1   Introduction 43
  3.2   The gas laws 43
  3.3   Kinetic theory 45
      3.3.1   The link between pressure and kinetic energy 45
      3.3.2   A definition of temperature 46
  3.4   Maxwell-Boltzmann velocity distribution 48
  3.5   Plasma temperature 51
      3.5.1   Temperature as an energy 52
  3.6   Population distributions — The Boltzmann factor 53
      3.6.1   The Saha equation 54
  3.7   Summary 57
4 Motion of charged particles in electric and magnetic fields 59
  4.1   Introduction 59
  4.2   The equation of motion: the Lorentz force 59
  4.3   Maxwell's equations 60
      4.3.1   Gauss's law for electric fields 61
      4.3.2   Faraday's law of electromagnetic induction 61
      4.3.3   Gauss's law for magnetic fields 61
      4.3.4   Ampère's law with displacement current 61
  4.4   Motion in a uniform electric field 62
      4.4.1   Drift velocity 62
      4.4.2   The influence of ionisation 63
  4.5   Motion in a uniform magnetic field 63
      4.5.1   Magnetic moment, M 65
  4.6   Motion in combined electric and magnetic fields 66
  4.7   Effect of an external force field 69
      4.7.1   Gravitational drift 69
  4.8   Diamagnetic drift — the effect of temperature or density gradients .. 70
  4.9   Non-uniform electric and magnetic fields 71

	4.9.1	Motion in non-uniform magnetic fields:	..	..	..	71
		magnetic mirror, "grad B" drift, curvature drift				
	4.9.2	Time-varying spatially uniform magnetic field	..	..	76
	4.9.3	Time-varying electric field	..	..	..	..	..	78
4.10		Summary	..	..	..	..	..	..	..	..	78
5	**Plasma characteristics**	..	..	..	..	..	..	..	..	81
5.1		What is plasma?	..	..	..	..	..	..	..	81
5.2		Examples of plasmas	..	..	..	..	..	..	82
5.3		Collective behaviour	..	..	..	..	..	..	83
5.4		Debye length, λ_D	..	..	..	..	..	..	84
5.5		Sheaths	..	..	..	..	..	..	..	85
	5.5.1	Double layers	..	..	..	..	..	..	87
5.6		The plasma parameter	..	..	..	..	..	88
5.7		Collisions	..	..	..	..	..	..	..	88
	5.7.1	Coulomb collisions	..	..	..	..	..	89
	5.7.2	Collision cross section, σ_C	..	..	..	..	90
	5.7.3	Diffusion in a plasma	..	..	..	..	..	91
	5.7.4	Collision frequency, ν	..	..	..	..	..	92
	5.7.5	"Collisionless" plasmas	..	..	..	..	93
5.8		Plasma conductivity, σ, and resistivity, η	..	..	..	94
5.9		Summary	..	..	..	..	..	..	..	96
6	**Plasma waves and oscillations**	..	..	..	..	..	99
6.1		Introduction	..	..	..	..	..	..	..	99
6.2		Simple harmonic motion and dispersion	..	..	..	99
6.3		Plasma frequency, ω_p	..	..	..	..	..	..	102
	6.3.1	Langmuir waves	..	..	..	..	..	..	104
	6.3.2	Significance of ω_p	..	..	..	..	..	104
6.4		Damping mechanisms	..	..	..	..	..	..	106
	6.4.1	Collision damping	..	..	..	..	..	106
	6.4.2	Landau damping	..	..	..	..	..	107
6.5		Ion oscillations	..	..	..	..	..	..	107
	6.5.1	Ion acoustic waves	..	..	..	..	..	108
6.6		The influence of magnetic fields	..	..	..	..	109
	6.6.1	High-frequency wave propagation parallel to B: polarisation	..	110
	6.6.2	Very low frequency wave propagation parallel to B: Alfvén waves	114
	6.6.3	High-frequency wave propagation perpendicular to B	..	115
	6.6.4	Low-frequency wave propagation — magnetosonic waves	..	116
6.7		Summary	..	..	..	..	..	..	..	117
7	**Plasma radiation**	..	..	..	..	..	..	..	119
7.1		Introduction	..	..	..	..	..	..	..	119
7.2		Ionisation	..	..	..	..	..	..	..	119

7.3	The effect of impurities	.. ..	120
	7.3.1	Impurity production	.. ..	121
7.4	Radiation processes	.. ..	122
	7.4.1	Bremsstrahlung	.. ..	123
	7.4.2	Cyclotron radiation	.. ..	124
	7.4.3	Recombination radiation	.. ..	125
7.5	Line radiation	.. ..	126
	7.5.1	Effect of external fields	.. ..	127
	7.5.2	Spectral line broadening	.. ..	127
7.6	The effect of plasma on radiation	.. ..	129
7.7	Information obtainable from radiation	.. ..	129
	7.7.1	Limitations to the accuracy of measurement	.. .. ..	131
7.8	Summary	.. ..	132
8	**Mathematical models**	.. ..	133
8.1	Introduction	.. ..	133
8.2	The single-fluid model — magnetohydrodynamics (MHD)	.. ..	133
	8.2.1	Momentum transfer equation	.. ..	134
	8.2.2	Continuity equation	.. ..	136
	8.2.3	The MHD equations	.. ..	137
8.3	Magnetic pressure	.. ..	138
	8.3.1	Plasma beta	.. ..	140
8.4	Plasma instabilities	.. ..	141
	8.4.1	"Ideal" instabilities	.. ..	143
	8.4.2	Resistive instabilities	.. ..	145
8.5	Kinetic theory	.. ..	145
	8.5.1	Phase space	.. ..	146
8.6	Boltzmann equation	.. ..	147
	8.6.1	Collisionless Boltzmann, or Vlasov equation	.. .. ..	148
	8.6.2	The effect of collisions: Fokker-Planck equation	.. ..	149
8.7	Summary	.. ..	150
9	**Gas discharges**	.. ..	151
9.1	Introduction	.. ..	151
9.2	What is a gas discharge?	.. ..	151
9.3	Breakdown	.. ..	152
9.4	Types of discharge	.. ..	156
	9.4.1	Dark discharge or Townsend discharge	.. .. ..	157
	9.4.2	Corona discharge	.. ..	157
	9.4.3	D.c. low-pressure glow discharge	.. ..	157
	9.4.4	Arc discharges	.. ..	160
	9.4.5	Radio frequency (r.f.) discharges	.. ..	162
	9.4.6	Microwave gas discharges	.. ..	164

	9.4.7	Electron cyclotron resonance (ECR) discharges	164
10	**Industrial and environmental applications of gas discharges**		167
	10.1	Background	167
	10.2	Arc applications	168
	10.3	Gas discharges as light sources	169
	10.4	Synthesis of chemical compounds	170
	10.5	Sputtering: surface cleaning and thin-film deposition	171
	10.6	Surface modification	172
		10.6.1 Surface modification by ion implantation	173
	10.7	paint spraying	174
	10.8	Xerographic processes	174
	10.9	Removal of compounds from flue gases: electrostatic precipitators	175
	10.10	Plasma etching	175
	10.11	Gas laser	176
	10.12	MHD power generators	177
	10.13	Particle accelerators	177
	10.14	Environmental applications and waste processing	178
	10.15	Space propulsion	180
11	**Nuclear fusion**		183
	11.1	"A lot of nonsense"	183
	11.2	The binding energy of the nucleus	186
	11.3	Nuclear fusion	188
		11.3.1 The raw materials of fusion	188
		11.3.2 Fusion reactions	189
		11.3.3 Fusion cross section, σ_f	191
		11.3.4 Reaction rates	192
12	**Controlled nuclear fusion: the confinement problem**		195
	12.1	Introduction	195
	12.2	Historical background	195
	12.3	Energy confinement — Lawson's criterion ($n\tau_E$)	200
	12.4	Magnetic confinement and plasma β	201
		12.4.1 The pinch effect	202
	12.5	Magnetic confinement systems	203
		12.5.1 Open systems	203
		12.5.2 Closed systems	204
		12.5.3 Safety factor, q	207
	12.6	Confinement geometries	209
		12.6.1 Externally-generated magnetic confinement: tokamak, stellarator	209
		12.6.2 Self-generated magnetic confinement: RFP, spheromak	211
		12.6.3 Spherical tokamaks — the latest design	213
	12.7	Heating the plasma	215

12.8	Turbulence	216
12.9	The prospect for fusion	216
	12.9.1 ITER	217
12.10	Inertial confinement	217
12.11	Summary	220
13	**Aurorae and sunspots**	223
14	**Astrophysical, space and atmospheric plasmas**	237
14.1	Introduction	237
14.2	The interstellar medium	237
14.3	Stellar plasmas — the Sun	239
	14.3.1 Structure of the Sun	240
	14.3.2 The solar magnetic field	242
14.4	The Earth's magnetosphere	244
	14.4.1 Bow shock	246
	14.4.2 Magnetopause	247
	14.4.3 Magnetotail	247
	14.4.4 Magnetic storms	248
	14.4.5 Van Allen belts	248
	14.4.6 The ring current	250
	14.4.7 Aurorae	251
14.5	Earth's ionosphere	252
	14.5.1 Wave reflection and absorption in the ionosphere	254
	14.5.2 Whistlers	256
	14.5.3 Lightning	257
14.6	Other planets	258
14.7	Summary	261
15	**Plasma classification**	263
Suggested further reading		271
Appendix A: A mathematical diversion on ω_p		272
Appendix B: Solution of circularly-polarised and X-wave cut-offs		274
Historical development of plasma physics and related fields		276
Glossary		281
SI units		283
Index		285

Introduction

Plasma physics is central to the development of nuclear fusion as a clean renewable source of energy. It is fundamental to our understanding of our environment in space. A knowledge of the plasmas surrounding the Earth is essential for modelling and predicting disturbances which can affect ground-based communications and sensitive instrumentation orbiting the Earth, as well as the safety of spacecraft and astronauts in future interplanetary missions. Closer to home, plasma technology and processing are rapidly-expanding fields with ever-increasing industrial, commercial and environmental applications: thin-film deposition, semiconductor production and toxic waste treatment, to name but three.

This book is intended as a basic introduction to the various aspects of plasma physics, its history and development. In providing an overview of the theory and its applications, I have tried to draw together the underlying principles of physics and the more specialised theories and terminology of plasma. I have deliberately kept the mathematics to a minimum and the equations as simple as possible. An understanding of vector calculus is assumed but a knowledge of tensors is not required. I make no apologies if the end result is a more descriptive and qualitative treatment of the subject than is usual. I have opted for breadth rather than depth, and plasma physics is now a very wide-ranging field. There are many other books currently available which examine specific aspects of plasma physics in much greater detail. It is nevertheless important that students of one branch of the subject should understand where their topic of interest fits into the broader picture. So, whilst this book is intended for use by first- and second-year undergraduates, it will also provide a useful introduction for postgraduate students for whom plasma physics is a new subject.

Until the 1950s, plasma was investigated as part of the development of electrical discharges in gases and, to a lesser extent, through the study of its atmospheric effects, in particular magnetic storms, the aurorae and the ionosphere. In the last 50 years, serious attempts have been made to understand plasma and its interactions with electromagnetic fields, the driving forces being those of space physics, fusion research and industrial applications. The first two chapters of this book therefore deal with the history of plasma physics up to 1950. The subsequent development of specific topics is then outlined in the relevant chapters.

I would like to thank the staff at the British Library for allowing me freedom of access to their archived journals, without which, much of this work would have been impossible.

Symbols and Constants

a = acceleration

a_0 = Bohr radius ($\approx 5.3 \times 10^{-11}$ m)

A = atomic mass number; area

b = impact parameter

$\boldsymbol{B}$ = magnetic flux density; (magnetic field)

B_θ = poloidal magnetic field

B_ϕ = toroidal magnetic field

c = speed of light in vacuum

e = electron charge (1.602×10^{-19} C)

E = energy

$\boldsymbol{E}$ = electric field

f = frequency (Hz)

$\boldsymbol{F}$ = force

g = gravity, plasma parameter

h = Planck's constant (6.626×10^{-34} J s)

I = current

J = current density

k = wave number

K = kelvin (temperature)

k_B = Boltzmann constant (1.38066×10^{-23} J K^{-1})

L = length

M = magnetic moment

m_e = electron rest mass (9.1095×10^{-31} kg)

m_n = neutron rest mass (1.675×10^{-27} kg)

m_p = proton rest mass (1.6726×10^{-27} kg)

n = particle number density (m^{-3}); neutron; energy level in Bohr atom

N = index of refraction

N_D = particles in Debye sphere

p = pressure

P = power

q = charged particle; safety factor

R_E = Earth radius (≈ 6378 km.)

R_g = gyroradius

T_e = electron temperature

v_A = Alfvén speed

$\langle v_e \rangle$ = mean thermal speed of electrons

v_g = group velocity

v_p = phase velocity

v_s = sound speed

V = volume

Z = ionic charge; atomic number

Z_{eff} = effective ionic charge

α = radiative recombination coefficient

β = beta factor

γ = ratio of specific heats

ε_0 = permittivity of a vacuum (8.854×10^{-12} F m^{-1} ($= $ C^2 N^{-1} m^{-2}))

η = resistivity

λ = wavelength

λ_D = Debye length

λ_{mfp} = mean free path for collisions

μ = permeability of plasma

μ_0 = permeability of free space (1.2566×10^{-6} H m^{-1} ($= $ N A^{-2}))

υ_e = electron collisional frequency

ρ = charge density

ρ_m = plasma mass density

σ = electric conductivity

σ_c = momentum transfer or collision cross section

σ_f = fusion cross section

Σ = sum of

τ = gyro-period

τ_e = confinement time

ω = angular frequency (rad s^{-1})

ω_c = cyclotron frequency

ω_{ce} = electron cyclotron frequency

ω_p = plasma frequency

ω_{pe} = electron plasma frequency

$\ln \Lambda$ = Coulomb logarithm

Temperature conversion:-

1 eV $= e$ J $= 1.602 \times 10^{-19}$ J

1 eV $= 1.16 \times 10^4$ K

1 K $= k_B$ J $= 1.38066 \times 10^{-23}$ J

1 K $= 8.62 \times 10^{-5}$ eV

To convert eV to K, multiply by e/k_B

Chapter 1

Strange glows, striations and dark spaces (1675-1900)

In November 1991, in a laboratory in the English countryside near Oxford, an international group of scientists demonstrated that electricity can be obtained from processes similar to those occurring in the Sun. Their controlled thermonuclear fusion experiment lasted just two seconds but produced over a megawatt of power. It was the culmination of three hundred years of research by some of the greatest names in physics; research which united the once-separate fields of rarefied gases, electricity and magnetism to produce what is now known as plasma physics.

The story begins in 1675. French astronomer Jean Picard (1620-82), returning home one night from his Paris Observatory, noticed a strange flickering light in the barometer he was carrying: when the mercury moved, an eerie glow appeared in the vacuum of the tube. It was rather like the greenish glow produced by the recently-discovered (but not understood) phosphorus, and many people thought that the flickering in the tube might be caused by a "mercurial phosphorus".

Evangelista Torricelli (1608-47) had created the first mercury barometer in 1643 when he filled a long, sealed glass tube with mercury, covered the open end with his finger, and immersed the inverted tube in a bowl of the metal. To the amazement of his audience, the level of mercury in the tube dropped until it was about 76 cm. above that in the bowl. No air had entered the tube, so the space above the column must be empty. Torricelli's barometer, with its "Torricellian vacuum" at the top of the tube, shattered the long-held belief that a vacuum could not exist, and soon, other ways of producing the effect were developed.

The first air-extraction pump (a spherical flask labouriously evacuated using a leather-covered wooden piston and two stop-cocks) was built seven years later by the German engineer Otto von Guericke (1602-86). Robert Boyle (1627-91) was working in Oxford in 1657 when he read about von Guericke's pump and, with his assistant Robert Hooke (1635-1703), set out to improve the design. As a result of his efforts, Boyle established the relationship (discovered independently in 1676 by French physicist Edme Mariotte (1620-84)) between the pressure, volume and temperature of a given quantity of air which would become known as *Boyle's law*. In the Netherlands, Christian Huygens (1629-95) replaced the flask-and-stopcock air-pump used by von Guericke and Boyle with the simpler bell-jar and plate which soon became essential equipment for vacuum experiments.

The 1600s saw much experimentation, but little understanding of what was observed, particularly in the fields of electricity and magnetism. The early Greeks had known that amber will attract lightweight matter when rubbed with cloth or fur — the Greek word for amber is *elektron*. They also knew that magnetite — an oxide of iron (Fe_3O_4) then called lodestone — would cling to iron tools; but little progress was made, despite the arrival from China (around 1100 AD) of the magnetic compass in the form of a magnetised needle floating in water. William Gilbert (1540-1603) experimented with electricity and magnetism in the late 1500s, introducing the terms "electric forces" and "electric attraction". He compared the Earth to a large spherical lodestone and developed the idea of a "magnetic pole". French philosopher René Descartes (1596-1650) later demonstrated the patterns made by iron filings sprinkled around a magnet.

By the mid-1600s anything activated by rubbing was believed to have an amber quality, or *electricity*, and the idea grew that the electric charge given to an object was a quantity. Around 1672, von Guericke built the first friction machine to collect electric charge: a winch turned a large ball of sulphur which quickly built up large amounts of charge when spun between cupped hands. By 1700, many amateur scientists were using these early electrostatic generators to study the properties and behaviour of electric charges at rest.

In 1705, Francis Hauksbee (d.1713?) replaced von Guericke's sulphur ball with a large glass cylinder and set about investigating the strange glow of Picard's barometer. It was still widely thought that "mercurial phosphorus" was to blame, and Hauksbee reported to the Royal Society in London that a partial vacuum did indeed make phosphorus glow brighter.[1] Turning his attention to mercury, he covered a bowl of the metal with a bell jar, passed a tube through the top of the jar into the mercury, then removed as much air as possible from the jar. When air was allowed to return via the tube and through the liquid, it blew the mercury against the sides of the jar in an "abundance of glowing globules". Shaking mercury in air just produced tiny pinheads of light, but in rarefied air it became "luminous all round, not as little bright sparks but as a continued circle of light", sometimes producing flashes of light at the top of the jar, which he compared to lightning. Movement and a partial vacuum were therefore needed to produce the glow. He noticed that particles of mercury were attracted to the sides of the jar and concluded that electricity, produced by friction, was probably responsible for the effect.[2] It would be more than a hundred years before the knowledge and the equipment were available to investigate further.

Electrostatics continued to fascinate experimenters throughout the 1700s. They repeated Hauksbee's work on glowing mercury but gained no understanding and merely marvelled at the effects. The electric friction machine became an essential piece of laboratory equipment, but its main drawback was its inability to store the charge generated and deliver it in a single discharge — all it produced was a constant trickle. A solution came in 1745 with the invention of the Leyden jar. Although it was hailed as a great scientific advance, it was essentially a modified friction machine. Credit for its invention is shared by two men, Ewald Georg von Kleist (d.1748) in Poland and Dutch physicist Pieter van Musschenbroek (1692-1761) in Leyden. Both were independently experimenting with electricity, and, perhaps trying to collect "electric fluid", they inserted the wire or nail conductor being charged into a hand-held glass jar partly filled with water, unwittingly building the first capacitor. The glass jar (an insulator) separated the conductor connected to the generating machine from another, grounded, conductor (the sweaty hand). Charge could now be collected and stored in significant quantities for delivery in a single pulse.

In America, Benjamin Franklin (1706-90) watched some electrical demonstrations in Boston and began to experiment, using the Leyden jar. In July 1746 he put forward his theory of electricity. Franklin imagined a single electric fluid permeating space and existing within all matter. If an object acquired more than its normal share it was called "plus"; if less, it was called "minus". To account for the flow of charge, he suggested a flow of electricity in which the charge carriers were positive. In London, the Royal Society thought his ideas derisory and only a brief notice of his research was published.

Franklin then turned to thunder and lightning, at that time thought to be caused by exploding gases. In 1752, he sent a kite into the interior of a thundercloud and obtained sparks from the string, confirming that lightning was an electrical phenomenon like electric sparks. He then erected an iron rod to draw the lightning down into his house so that he could experiment on it. He concluded that thunderclouds are generally "in a negative state of electricity", and therefore, "for the most part, in thunder strokes it is the earth that strikes into the clouds, not the clouds that strike the earth".[3] Franklin's announcement of his results caused a surge of interest in lightning studies.

Major progress in understanding the glow in Picard's barometer came in February 1752 when William Watson (1715-87) informed the Royal Society of his work on electricity under vacuum conditions. Watson had begun studying electrical discharges in 1747, soon after the invention of the Leyden jar. His experiments, together with those of Abbé Nollet (1700-70) in France with whom he corresponded, are the first recorded on the subject. Watson had designed what he claimed was a considerable improvement on Boyle's air-extraction pump and used it to discover whether the "electrical fluid" would appear in a vacuum and whether it would travel further than in air.[4]

Using a long glass tube (Fig. 1.1, overleaf) fitted with two slender brass rods, one of which was connected to an electric friction machine, Watson found that "brushes of electrical fire" passed from one rod to the other. In air, this effect lasted as long as the gap between the rods was less than five inches (125 mm). In vacuum, it continued even when they were withdrawn to their maximum extent. It was, he reported, "a most delightful spectacle, when the room was darkened, to see the electricity in its passage. The coruscations were of the

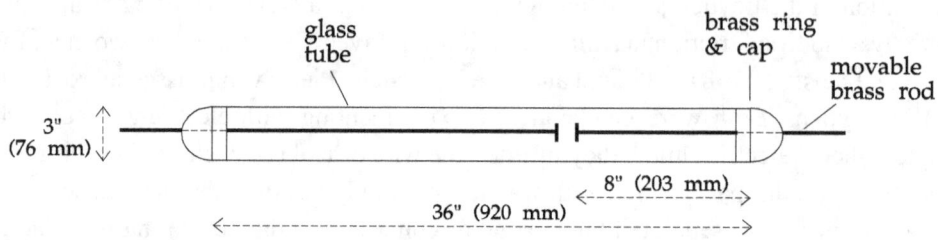

Fig. 1.1   Sketch of Watson's discharge tube

whole length of the tube between the plates, of a bright silver hue, and resembled very much the most lively coruscations of the aurora borealis." Under the best vacuum conditions electricity passed between the brass plates in one continuous stream. He repeated the experiment using a Leyden jar: at the instant that the jar was discharged, Watson observed a "mass of very bright embodied fire" jump from one brass plate to the other when the distance between them was less than ten inches. At greater separation, the "fire" was weaker and began to diverge.

Some 20 years later, William Henly published his own electrical experiment involving a discharge through vacuum.[5] Using an exhausted glass tube 18 inches (457mm) long and two inches (51mm) in diameter (A), with brass balls cemented to each end (B and C), and supported on columns of sealing wax (G), he had tried to determine the direction of the electricity passing through the tube. The fine-pointed wire (F) attached to ball C was positioned close to the glass globe of an electric friction machine to collect the charge. A chain hung from ball B to the table. When he operated the friction machine, the small brass ball at D was surrounded by a "dense white atmosphere of electricity" (Fig. 1.2). Bringing

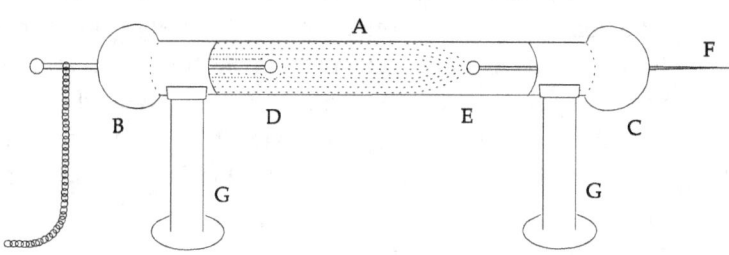

Fig. 1.2   Henly's evacuated tube

a negatively-charged Leyden jar in contact with ball C made the cloud envelop the small brass ball at E. Henly suggested that the power supply drove the "particles of electricity" through the brass balls and wires with great velocity, but once inside the vacuum, the tube filled with light as the particles expanded away from each other due to their mutual repulsion.

The theory was beginning to evolve. By the late 1780s, it had been established that electric charge can be positive or negative, with the charge carriers in any flow of charge then believed to be positive. However, what actually caused lightning and the discharges in evacuated tubes remained a mystery, and there was, as yet, no way of measuring the amount of charge produced other than by the strength of the shock received. Research was still confined to electrostatics, the small amount of current delivered by friction machines and Leyden jars making steady-state experiments difficult. That situation changed in March 1800 when Italian physicist, Count Alessandro Volta (1745-1827) announced to the Royal Society that an electric current can be generated by placing a salt solution or damp material between two different metals, zinc and copper. This *voltaic cell*, as it became known, rapidly shifted attention from electrostatics to electric currents and increasingly-large collections of cells (batteries) appeared, despite initial concerns that electricity from batteries was not the same as that produced by friction machines.

Links were being found between previously-unrelated fields of research and the idea grew that electricity and magnetism might be connected, since amber attracts lightweight matter when rubbed, and magnets attract iron filings. The evidence had, in fact, been around for many years: In 1676, a report of a lightning strike on one of two Barbados-bound ships described how the affected vessel turned around and headed homeward, pursued by the other. Investigation of the damaged ship's compasses revealed that all the north and south points had changed positions. The lightning was (correctly) blamed, but how it had changed the magnetism was never considered. Some sixty years later, lightning struck a tradesman's house, damaging a box of cutlery. Many of the knives and forks melted and when the tradesman emptied the box onto his counter, he was amazed to find that the cutlery picked up nails and other iron objects lying nearby.[6] Once again, the cause was not investigated

A chance discovery, in 1820 by Hans Christian Oersted (1777-1851), that an electric current will deflect a magnetic compass needle provided the catalyst. A wire attached to a battery lay over a compass on Oersted's bench, almost parallel to the needle. When the circuit was closed, the compass needle swung perpendicular to the current-carrying wire. In Paris, André Marie Ampère (1775-1836) deduced that a compass needle close to a current-carrying wire experiences a magnetic force which acts at right angles to the current in a series of concentric circles. This meant that electricity could now be measured using the deflection of a compass needle instead of relying on a subjective estimate of the strength of shock received.

The obvious question was then asked: does magnetism produce electricity? It was assumed that since a steady current generates a steady magnetic field, the reverse would be true. Michael Faraday (1791-1867) disproved this in 1831 when he showed that an electric current is generated by a *changing* magnetic field.[7] Joseph Henry (1797-1878) in New York had performed a similar experiment the previous year but, unlike Faraday, he had not published his results.

Following Nicholson and Carlysle's discovery of electrolysis while using a voltaic battery in May 1800, Sir Humphry Davy (1778-1829) began to investigate chemical decomposition. In 1808 he passed an electric current through a mercurial (Torricellian) vacuum containing pieces of charcoal. A brilliant purple flame appeared to come from the charcoal, forming a "conducting chain of light nearly an inch long". Davy later described it as "an arc, or column of electrical light between two burning electrodes".[8]

In December 1821 he published his work on electrical discharges under vacuum conditions. He had found that electricity and sparks could permeate even the best mercurial vacuum, contrary to earlier beliefs; while the shock from a Leyden jar made it luminous. The intensity depended on temperature: with a very hot tube, the light in the vapour was bright green and dense; as the temperature dropped, it dimmed; until at temperatures below freezing it was barely visible in a darkened room. Boiling mercury in an evacuated tube produced globules of mercury vapour. Friction between the globules and the glass tube produced electricity which discharged through the vapour with sparks visible in daylight (as noted by Hauksbee over a century earlier).[9] Later, he described a discharge between charcoal terminals which were gradually separated after contact — a *drawn arc*. In air, the discharge crossed a gap of at least four inches (100 mm). In vacuum, it would cross a half-inch space, but the addition of heat increased the distance to 6 or 7 inches.[10] Progress was being made as techniques improved, but understanding of the underlying processes was still minimal.

When Davy died in 1829, Faraday took over his work on the chemical changes produced by electrical discharges and went on to define the laws of electrolysis five years later.[11] He suggested that a flow of charged particles caused the passage of current across the solution, which he called *electrolyte*, and this separated the fluid into *ions*. The metal surfaces in contact with the electrolyte he named *electrodes* (*anode* and *cathode*).

Four years later, in 1838, Faraday presented a set of definitions for electrical discharges which is still in use today.[12] He defined a *disruptive discharge* as one which violently displaces particles of the dielectric across which it breaks. The discharge itself might be in the form of a spark, brushes or a glow. An electric *spark*, he said, occurs when an insulating dielectric, such as air, is placed between two oppositely-charged conducting surfaces. By increasing the power to the conductors or narrowing the gap between them a spark finally appears. In an early attempt to describe the *breakdown* of the dielectric, Faraday suggested that the moment of discharge was determined by the molecule of the dielectric which was under greatest strain: the resistance broke down suddenly, producing a disruptive discharge; its path perhaps depending on the potential difference experienced by the particles in its route.

Brushes he described as having "pale ramifications with a quivering motion and accompanied by a low dull chattering sound". Sir Charles Wheatstone (1802-75) had recently shown[13] that brushes are a succession of intermittent discharges. Using very fine wires, Faraday produced barely distinguishable brushes, but as long as sound was heard the discharge remained intermittent. When the sound ceased, the light became a continuous glow.

He decided that the brush was actually a discharge where one of the electrodes was a poor conductor, and that it usually occurred between a conductor and air. The quality of the power source and the size, shape and polarity of the conductor determined the point at which a spark changed to a brush. A spark discharge starting at the cathode became a brush much sooner and changed to a glow much later than one originating from the anode. Obtaining a glow was easier in rarefied air than at atmospheric pressure.

He described bringing two thin brass rods into contact in a partial vacuum, then separating them while an electrical discharge passed through them. At the moment of separation a continuous glow covered the end of the negative rod. As the distance was increased, a purple haze appeared on the end of the positive rod, extending towards the negative rod and elongating as the rod separation grew, but never joining the negative glow. There was always a short dark space, about 1.5 mm long, between them. Its length and position relative to the negative rod never changed. This dark area had been observed by others and called the "neutral point" in the belief that there were two "separate electricities" which combined and neutralised each other at that spot. Faraday decided this made no sense and described the dark area as a *dark discharge* — subsequently called the *Faraday dark space*. The discharge continued across the dark region at the same level as in the luminous part. The intervening gas particles were, he said, responsible for the different effects within the tube and therefore needed further investigation.

Faraday was self-taught. He had had no mathematical training and so could not develop the equations needed to support his ideas. In 1845, he used the image of a force field to illustrate his theories, defining a line of force, either electric or magnetic, as indicating the direction in which the interaction between charges or magnetic poles is transmitted, the density of lines representing the strength of the force.[14] The concept of a force field avoided the difficulties in understanding "action-at-a-distance": electric and magnetic forces acting on objects separated by relatively large distances, with no visible connection between them. By imagining lines of force between electrodes or magnetic poles, the action of electric and magnetic forces could now be seen as the force field acting directly on an object.

Faraday found that some materials behave differently from iron or lodestone when placed in a magnetic field. These he labelled *diamagnetic*.[a] He placed a piece of diamagnetic "heavy glass"[b] between the poles of a strong electromagnet and passed a polarised image of a flame through it. The rotation of the image increased with the intensity

---

[a]   Diamagnetism is associated with the orbital motion of electrons in the atom. An external magnetic field alters the angular momentum of the electrons, producing a field that opposes the applied magnetic field and results in a very small net decrease in the applied field within the diamagnetic material.

[b]   Silicated borate of lead

of the magnetic field. With no current to the electromagnet, both the magnetic field and the polarised image disappeared. When he passed an electric current around a beam of polarised light in a plane perpendicular to the beam, the plane of polarisation rotated in the same direction as the current. Faraday had confirmed his theory that there was a direct relation between light, magnetism and electricity. He had made what Maxwell later described as one of his most significant discoveries: that a magnetic field will rotate the plane of polarisation of light — an effect which became known as *Faraday rotation*.[15]

At the time there was no obvious reason for the link between light, electricity and magnetism. Faraday modestly credited Samuel Hunter Christie of Trinity College, Cambridge with discovering a direct relation between light and magnetism in the 1820s, when he had shown that solar rays have magnetic properties: the vibrations of a magnetised needle came to rest sooner when not exposed to sunlight.[16] In 1856, William Thomson (1824-1907), later Lord Kelvin, explained that the direction of motion of moving particles determined the effect of the magnetic field on light: when two identical circularly-polarised rays of light rotate in opposite directions, the ray rotating in the same direction as the electricity of the magnetising current will travel with greater velocity than the other ray.[17] Eight years later, James Clerk Maxwell proved the link between electricity, magnetism and light mathematically.

By 1840, Faraday's attention had turned elsewhere and it was left to others to pursue the investigation of electrical discharge phenomena. William Robert Grove (1811-96) became professor of experimental philosophy at the London Institution in 1840. During the next ten years he investigated disruptive discharges and arcs using voltaic batteries. Borrowing from earlier work on electrolysis, he used a voltaic arc — what would now be called a d.c. arc — with various gases, but nothing useful resulted because the intense heat of the arc affected the terminals and masked any effects on the gas. Grove needed a different power source.

While visiting Paris in 1851, he heard of a new invention by Heinrich Daniel Rühmkorff (1803-77): a thick copper wire wound round a soft iron core, with a secondary coil of very thin wire on the outside of it, and a vibrating hammer acting as an automatic contact breaker. A similar induction coil had been designed in the US some 13 years earlier by Charles Grafton Page (1812-68),[18] but it was unknown in Europe. Using Rühmkorff's coil, Grove discovered transverse dark bands, or *striations*, in an electrical discharge through a vacuum tube containing a small piece of phosphorus.[19] Unbeknown to Grove, striations had been observed nine years earlier in France by Abria, while experimenting with induction coils and evacuated tubes in Bordeaux. Abria had produced a brush discharge, in rarefied air, extending from a positive (spherical) terminal almost to the negative (pointed) terminal where there was a dark interval. He had seen alternate dark and luminous bands in the discharge, concave towards the ball when the point was near to it and convex when the point was some distance away.[20]

Several European researchers were investigating electrical discharges in the 1850s. Heinrich Geissler (1814-79), a glass-blower in Tübingen, Germany, found a way of fusing platinum electrodes directly into glass and began producing vacuum tubes. Improvements in evacuation techniques meant lower densities were achievable and a wider range of phenomena could be seen. In Paris in 1857, du Moncel discovered that when he touched the evacuated tube with his finger, the luminous discharge was attracted to the wall of the tube. The following year, French researchers Quet and Seguin discharged a Leyden jar across an evacuated tube to see if the Rühmkorff coil was responsible for striations. Only when the bottle was weakly charged did the striations appear.[21]

Many years earlier, Davy had noticed that a magnet would divert the arch of light which he formed between carbon points using a powerful battery. In Bonn in 1857, Julius Plücker (1801-68) tested the effect of a magnet on Geissler tubes containing various gases and was surprised to see the stream of light split in two: one part was undulating and flickering, the other became a brilliant fine point of light. Metal from the electrodes, especially the cathode, was deposited on the glass of the tube. He also noticed striations in the tubes and decided these could not be explained by the rapid interruptions to the current produced by Rühmkorff's coil, but must be caused by matter from the discharge itself collecting at definite parts of the tube. The stratified light was separated from the cathode by the dark area of the discharge identified by Faraday some twenty years earlier.[22]

In London, the scientific writer John Peter Gassiot (1797-1877) had been investigating sparks in air between the electrodes of various types of batteries[c] throughout the 1840s and now turned his attention to electrical discharges. In a lecture to the Royal Society in 1858, he recounted his attempts to observe striations in a Torricellian vacuum and called for further investigation of the effect of a magnet on both striations and the intense phosphorescent light.[23]

In the 1850s, the best vacuum obtainable was still the Torricellian vacuum, but no instruments could be inserted through the mercury into the vacuum nor could any substance be used which reacts with mercury. Although the air pump did not share these limitations, even the best could not match the Torricellian vacuum. With Faraday's help, Gassiot performed a series of experiments using voltaic batteries and vacuum tubes which were initially filled with air, hydrogen, oxygen or nitrogen before the mercury was introduced and the tube evacuated. There was therefore a mixture of mercury vapour and the residue of the air or gas in the evacuated tubes that he studied. Pure mercury vapour glowed greenish-blue, but the slightest trace of gas in the vapour changed the colour of the striations: air or nitrogen

---

[c]   One battery, described by Gassiot in a letter published in 1838 in *Philosophical Magazine*, consisted of "160 half-pint earthenware jars, the zinc elements being placed inside, the size permitting the use of brown paper instead of membrane, the exciting liquids being saturated solutions of sulphate of copper and common salt."

12

gave them a red colour, whilst hydrogen or oxygen turned them a bluish-grey (Fig.1.3a).[24] Gassiot found that striations were absent with both a poor and an almost perfect vacuum. Enclosing the negative terminal (the cathode) in glass tubing, with the open end of the tube extending 3 mm beyond the end of the terminal, destroyed the striations and made the negative discharge appear to shoot out of the tube. It could be deflected by a magnet and wherever it touched the side of the vacuum tube, a brilliant blue fluorescent spot appeared and the tube at that point quickly became hot (Fig.1.3b).

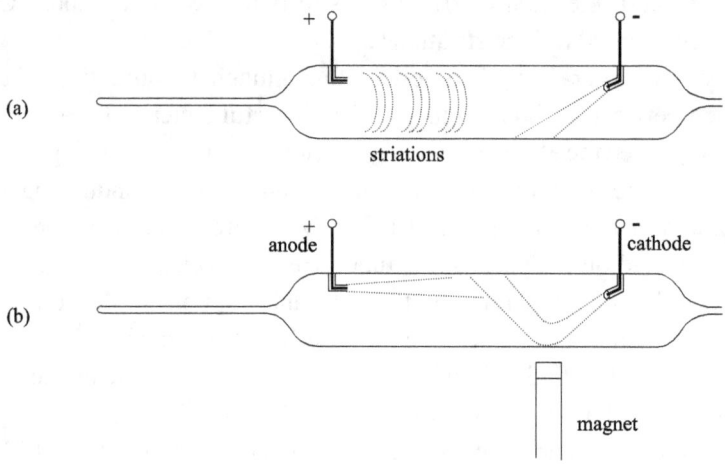

Fig. 1.3   Gassiot's discharge tube experiments

In 1859 Julius Plücker reported observing a green phosphorescence on the glass tube, near to the cathode, which he thought was due to rays travelling from the cathode to the walls of the tube and back again.[25] Plücker's attention then turned to the latest developments in spectral analysis and, with Johann Wilhelm Hittorf (1824-1914) of Münster, he began using electrical discharges to study the spectra of ignited gases and vapours.[26]

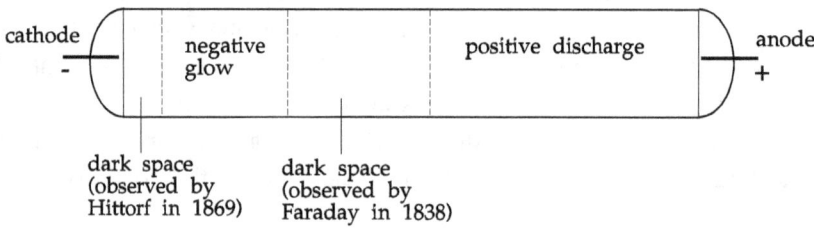

Fig. 1.4   Discharge zones identified by the 1870s

Hittorf reported, in 1869, that objects placed between a point-like electrode and the side of the vacuum tube cast shadows in the phosphorescence at the end of the tube opposite to the cathode, confirming Plücker's suggestion that some kind of radiation proceeded in straight lines from the cathode. He also identified a new dark space in the tube between the cathode and the negative glow (Fig. 1.4), noting that its width increased as the vacuum improved, ultimately filling the whole tube.[27]   Seven years later, Eugen Goldstein, in Berlin, gave the name *Kathodenstrahlen* (cathode rays) to the radiation producing the phosphorescence. It was now widely accepted that the spectrum of the gas near the cathode was different from that of the positive discharge on the other side of the dark space. Goldstein confirmed that changing the pressure in the tube also altered the colour of any striations, depending on the gas.[28]

In 1875, Warren de la Rue (1815-89), William Spottiswoode (1825-83) and Hugo W. Müller (1833-1915), who had left Germany in 1854 to work in London as a private assistant to de la Rue, began investigating what caused the striations in discharges.[29] Spottiswoode soon left to pursue his own research and in 1878 de la Rue and Müller presented their conclusions to the Royal Society. They had found that the type of gas, size of battery and the shape of the terminals — whether pointed, spherical or disk-shaped — all influenced the spark length. The appearance of the voltaic arc at atmospheric pressure also varied with the gas used and stratification was just visible under a microscope. The discharge in a vacuum tube was essentially the same as at atmospheric pressure, with the molecules of the gas acting as "carriers of electrification". Metal from the terminals was often deposited on the inside of the tube, leaving a permanent record of the spaces between the striations, which originated at the anode. The greatest heat was found in the vicinity of the strata. Varying the current often produced a total change in the colour of the strata — from cobalt blue to pink in the case of a hydrogen tube. Goldstein had noticed similar colour variations with changes in pressure. Changing the amount of current supplied to an irregular discharge with indistinct strata rendered them steady and distinct, but even then a pulsation could be detected in the current.[30]

De la Rue and Müller concluded that the electric arc and the striated discharge in vacuum are variations of the same phenomenon. For all gases, they said, there is a minimum pressure which offers the least resistance to the passage of an electrical discharge. After that minimum has been reached, resistance to the discharge rapidly increases as the pressure of the medium decreases. This concept was incorporated in Paschen's Law (1889). Despite much experimental work, and producing, in effect, a summary of what was then known about the workings of electrical discharges, de la Rue and Müller were no nearer to discovering the cause of the striations which had aroused so much interest.

While de la Rue and Müller were investigating striations, elsewhere in London, William Crookes (1832-1919) was working with low-density discharge tubes. The son of a tailor, he was largely self-taught, trying experiments and reading any book he could find on science. He entered the Royal College of Chemistry in London in 1847 and discovered

14

thallium (a toxic white lead-like metal) some fourteen years later using Bunsen and Kirchhoff's newly-developed spectral analysis techniques. The vacuum balance he had used aroused his interest in rarefied gases and, from 1873, he began investigating discharge tubes and the dark space separating the cathode from the negative glow, first observed by Hittorf in 1869.

Crookes presented his results in December 1878, confirming Hittorf's observation that the pressure in the tube affected the width of the dark space. He had also found width variations with cathode temperature and with the type of gas used: greatest in hydrogen and least in carbonic acid at the same pressure. The shape and size of the dark space was unaffected by the distance between the electrodes or the power supplied to the vacuum tube. At high power the dark space was obscured by the brilliance of other parts of the discharge; at low power, with only a very faint spark, it was barely visible; yet in both cases, careful inspection revealed no change in the size of this second dark space.

At high exhaustion, the dark space filled the tube and Crookes observed the phosphorescent spot on the glass seen by Gassiot twenty years earlier. Under close scrutiny, the dark violet focus at the cathode could be seen, while the rays issuing from it produced a sharply-defined spot of greenish-yellow light where they touched the glass tube. The difference in colour, from the blue seen by Gassiot, was due to the type of glass used. The English lead glass used by Gassiot gave a blue colour. Most of Crookes' apparatus was made of what he described as "soft German glass" — a soda glass which produced a greenish-yellow phosphorescence. He confirmed earlier observations that the rays producing the phosphorescence travelled in straight lines from the cathode, since objects inserted into their path cast sharp shadows in the glow at the opposite end of the tube (Fig. 1.5). Under similar conditions, he said, a surface which merely radiates light, such as a flame, would throw a scarcely visible penumbra, not a clearly-defined shadow.

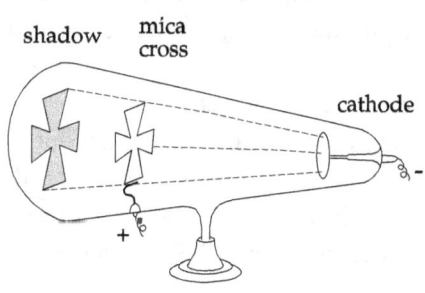

Fig. 1.5   Shadow of mica cross on wall of discharge tube

The width of the dark space at the cathode was, he announced, a measure of the mean free path between successive collisions of molecules of the residual gas. Molecules rebounding with increased velocity from the excited cathode would prevent slower-moving ones from reaching it. The conflict occurred at the boundary of the dark space, where the luminous area revealed the energy of the discharge. Crookes postulated that the phosphorescent spot was caused by the mean free path between collisions being longer, at high vacuum, than the distance between the source of the rays (the cathode) and the glass. The sudden loss of velocity when the stream of fast-moving particles hit the glass produced the phosphorescence.

His paper concluded by speculating on the state in which matter exists in highly exhausted discharge tubes. The (then) new ideas regarding the gaseous state of matter viewed a given space as containing millions of molecules moving rapidly in all directions and undergoing frequent collisions. The length of the mean free path of the molecules is therefore very small when compared with the dimensions of the vessel. It is these factors which determine the properties of an ordinary gas. Under conditions of high vacuum, however, the mean free path becomes long because the number of particles is reduced and so fewer collisions take place. Each molecule can therefore move without interference. If the mean free path is comparable in length to the dimensions of the vessel, Crookes explained, the properties which constitute a gas are reduced to a minimum producing "an ultra-gaseous state in which the properties now under investigation come into play. ... The phenomena in these exhausted tubes reveal to physical science a new world — a world where matter exists in a fourth state, where the corpuscular theory of light holds good, and where light does not always move in a straight line; but where we can never enter, and in which we must be content to observe and experiment from the outside."[31]

This idea of a fourth state of matter was treated with scorn both in the UK and abroad, and especially by the German scientists. In defending his theory, Crookes declared: "Between the third and fourth states there is no sharp line of demarcation any more than there is between the solid and liquid states or between the liquid and gaseous states". He pointed out that we know nothing of the absolute length of the free path or the absolute velocity of an individual molecule, but must limit ourselves to the *mean* free path and the *mean* velocity. All that those who claimed to disprove his theory had shown was that a few molecules can travel much further than the mean free path with perhaps much higher velocities than the mean before they are stopped by collisions.[32]

The dark region between the cathode and the negative glow became known as the Crookes or cathode dark space; in Germany, it was often called the Hittorf dark space. Crookes' paper, which summarised both his five years of research and the current state of British knowledge about electric discharges in vacuum tubes, laid the foundations of plasma physics. Previously, interest had focused on the positive electrode — the anode — and the striations which originated from it. Now it was the discharge from the negative electrode, the cathode, which would attract most attention. The rays producing the phosphorescence became known as *cathode rays*. The British view, led by Crookes, was that these were streams of negatively-charged particles, which, unless acted upon by magnetic forces, moved in straight lines and at high speed away from the cathode. The particles were presumed to have acquired this velocity under the influence of the strong electric field which exists in the neighbourhood of the cathode. The opposing German view, first proposed by Goldstein, held that the rays were transverse waves, like light, in the ether (a hypothetical fluid traditionally believed to fill space). J.J. Thomson wondered later if the phosphorescence, which was like that produced by ultra-violet light, had led Goldstein towards a wave theory. He pointed out that the path taken by the rays in a magnetic field was easily understood from the particle theory but no attempt was made to explain it in the ether theory, except for a vague idea that it was similar to the rotation of the plane of polarisation in the magnetic field.[33]

It was generally accepted that matter was made up of individual atoms, but there was still no theory of the structure of the atom. The molecule was often seen as the electrically-neutral particle, consisting of atoms which may carry positive or negative charge. Irish physicist George Johnstone Stoney (1826-1911) had pointed out, in August 1874, that Faraday's laws of electrolysis implied that a fixed quantity of electricity passes for each chemical bond broken. There must therefore be a fundamental unit quantity of electricity, which Stoney labelled e. He estimated that the amount of electricity corresponding to each chemical bond separated in hydrogen is 10^{-20} ampere and suggested that this quantity should become the basic unit of charge. In the 1870s, the ampere was used as the electromagnetic *quantity* of electricity. By the 1890s, it had become the unit *current*.[34] In SI units, the charge on the electron is 1.6×10^{-19} C. Stoney's estimate was too small by a factor of 16 but this was pioneering work in a field about which little was then known.

In 1881, Hermann von Helmholtz (1821-94) proposed that, if elementary substances are composed of atoms, electricity, positive as well as negative, must also be divided into definite elementary portions, which behave like atoms of electricity. Stoney returned to his idea in 1891, suggesting that there may be more than one chemical bond, with its associated charge of 10^{-20} ampere, in each atom. "These charges, which it will be convenient to call *electrons*, cannot be removed from the atom; but they become disguised when atoms chemically unite."[35] Although Stoney was suggesting that the atom may have a more complex structure than had earlier been thought, he still regarded it as an indivisible unit. Stoney's electrons were not particles — they had no mass, just charge.

The cause of striations continued to fascinate. In 1880, William Spottiswoode and J. Fletcher Moulton, investigating sparks and disruptive discharges, found that the luminous discharge in high vacuum was often too feeble to observe, with the phosphorescence on the tube wall providing the only indication of what was going on inside. They concluded that striations must be regions of higher density since they have a definite configuration and a higher temperature than that of the intervening dark spaces.[36] Four years later, German-born physicist Arthur Schuster (1851-1934), working in Manchester with mercury vapour discharges, reported that striations were seen best in compound gases. A mixture of air and mercury, for example, showed beautiful luminous bands but the effect disappeared as the vacuum improved. He also identified a thin luminous layer close to the cathode under high vacuum which was possibly due to impurities in the cathode surface.

The electrical conductivity of the gas in the discharge tube was still not understood. It was believed that gas molecules acquired a charge through contact with the electrodes and then moved under the influence of the electric field existing between them. Schuster argued that there was strong spectroscopic evidence for a dissociation similar to that in electrolysis, with the gas molecules splitting into two parts, charged positively and negatively, rather than the direct transfer of electricity from one molecule to another. The glow discharge generally revealed several distinct spectra within the tube: dissociation seemed strongest in the negative glow, while the characteristic spectrum of gases near the cathode indicated a complex molecule.[37]

Unknown to Schuster, his ideas on conductivity in gases were not new. Two years earlier, in 1882, a German scientist had put forward the same theory. In 1890, Schuster acknowledged: "I have only recently become acquainted with Mr. Giese's work. We both assume that each molecule of a gas contains atoms which carry equal and opposite charges and that these charges are the same as those carried by the ions in electrolytes." Both men believed that the current of electricity through a gas could only be maintained by the diffusion of the charged atoms within it. Schuster also noted that, in a magnetic field, the curved path of the particles emitted from the cathode depended on two, then unknown, quantities: particle velocity and how much electricity they carried. If particles of mass m and charge e move with velocity v at right angles to a magnetic field, B, said Schuster, the radius of curvature of the rays is determined by the equation $mv^2/r = Bve$, which gives a charge-to-mass ratio of $e/m = v/(Br)$.[38] The significance of that equation did not go unnoticed.

Also in 1890, Sir Joseph John (J. J.) Thomson (1856-1940), then director of the Cavendish Laboratory at Cambridge University, decided that a more up to date measurement was required of the speed with which the light accompanying electrical discharges travelled through the tube. Charles Wheatstone (1802-75) had used a rotating mirror and a vacuum tube six feet long in 1835, and had concluded that the speed of the flash through the tube was greater than $2 \times 10^5 \, m \, s^{-1}$. Thomson believed that knowing the speed of propagation would help him to understand the processes occurring in the discharge. He confirmed Plücker's 1859 suggestion that the discharge begins close to the anode and reported[39] that the luminosity of the positive discharge, by this time called the "positive column", travelled away from the anode with an approximate velocity of $1.6 \times 10^8 \, m \, s^{-1}$ — just over half the speed of light.

The following year, Thomson produced the first "electrodeless" ring discharge, in which the currents formed closed circuits in the gas without passing from the gas to metal or glass, as is the case when metal electrodes are placed outside the tube. The vessel containing the low-pressure gas was placed inside a solenoid through which passed rapidly-alternating currents produced by the discharge of Leyden jars. Many years later, Thomson showed that the circulating currents in the gas were of the same order of magnitude as those in the coil.[40] In 1891, it was merely a curiosity — one of several different designs of discharge tube. It would be a long time before its significance was recognised.

In 1892, German physicist Heinrich Hertz (1857-94) showed that cathode rays can pass through thin metal foil. His assistant, Czech-born Philipp Lenard (1862-1947) produced an evacuated cathode ray tube with a small window of aluminium foil at the opposite end of the tube to the cathode. When a beam of cathode rays passed from the tube through the window into air in a darkened room, a diffuse light appeared, spreading from the window into the air outside the tube.[41] The properties of the rays outside the tube resembled those of cathode rays, but for a time they were known as Lenard rays. Lenard found that they affected sensitive paper and photographic plates. Three years later, Wilhelm Conrad Röntgen (1845-1923) discovered X-rays while studying the luminescence.

Even in the mid 1890s, knowledge of physics was insufficient to explain the effects observed in the discharge tubes. It was known that cathode rays were easily deflected by a magnetic field and produced heat when they fell on matter. They needed a potential gradient to start them off, but once started they travelled in straight lines and therefore cast sharp shadows, although their path was independent of the anode's position. Some of the evidence suggested that the cathode rays were particles and the direction of deflection in a magnetic field indicated that they carried a negative charge; but they did not appear to be deflected by an electric field. There were those (mainly German scientists), such as Goldstein, Hertz and Lenard, who thought that if the cathode rays were not vibrations in the ether, they must be a form of short-wave radiation, but they were unable to explain the deflection of the rays by a magnetic field. Others, including Crookes and Thomson, believed them to be negatively-charged particles travelling at great speed. Schuster had talked of "particles of mass m and charge e" in 1890; the following year, George Johnstone Stoney had suggested the name "electron" for the electric charge associated with chemical bonds. It remained for Thomson to prove that cathode rays consisted not of molecules as Crookes had thought, but of sub-atomic particles — electrons.

At the Royal Institution on 30th April 1897, Thomson presented his results. Two years earlier, French physicist Jean Baptiste Perrin (1870-1942) had directed a narrow beam of cathode rays across an evacuated tube. When the beam was deflected by a magnet, making the rays strike a small metal object inside the tube, the object acquired a substantial negative charge. This charge, and the direction of the magnetic field used to deflect the rays, confirmed that cathode rays were negatively charged particles, not waves. Thomson observed that, while the wave theorists accepted that charged particles were emitted from the cathode, they believed that they had nothing to do with cathode rays.[42]

Previous attempts to deflect cathode rays using electric fields had failed, he said, because the pressure in the discharge tube was too high. The rays made the gas through which they passed a conductor and putting two electrodes across the discharge merely produced a current across it. By reducing the pressure in the tube until the cathode rays were only faintly visible, Thomson revealed that the rays *were* deflected by an electric field and in a direction consistent with their being negatively charged. Cathode rays were therefore not ether waves but particles of matter. He then passed the rays through uniform electric and magnetic fields (E and B), adjusted so that the forces were equal and opposite, and the beam passed through undeflected. The velocity of the particles could now be calculated using the equation: $v = E/B$. Using Schuster's 1890 equation, deflection in a magnetic field gave a value for the radius of curvature of the deflected path ($r = mv/(eB)$) and by eliminating v from the equation, he arrived at the charge-to-mass ratio:

$$\frac{e}{m} = \frac{E}{rB^2} \tag{1.1}$$

The right-hand side of this equation could be determined by measurement. Although the electric field could not be measured precisely, Thomson found that the particles were travelling at speeds of about one tenth that of light, a speed which until then was not associated with matter. He obtained a value for e/m_e of about 10^7 electrostatic units per gramme — in SI units: 10^{11} C kg^{-1}. The modern accepted value is 1.759×10^{11} C kg^{-1}. Thomson found that this ratio was the same for any gas and concluded that "we have in cathode rays, matter in a new state ... in which all matter is of the same kind; this matter being the substance from which all the chemical elements are built up."

George Francis FitzGerald (1851-1901) of Dublin, reviewing Thomson's lecture in *The Electrician* in May 1897, remarked that the idea of atoms being divisible into much smaller parts, which Thomson called "corpuscles", was essentially William Prout's (1785-1850) hypothesis that all matter is made up of corpuscles which are all of the same kind. FitzGerald suggested that, instead of corpuscles, the cathode rays consisted of free electrons.[43]

Lenard and Kaufmann confirmed Thomson's e/m results, and, the following year, Wilhelm Wien (1864-1928) announced he had proved that cathode rays still carry a negative charge after passing through an earthed aluminium window, and so were, indeed, charged particles. Positively-charged particles did not pass through the window, indicating they were larger than the negative particles.[44]

The mystery of the cathode rays and the electrical conductivity of the gases in the tubes had finally been solved. As J.J. Thomson explained in 1899, the conductivity results from the presence of charged ions. It is their motion in the electric field which constitutes the current. Measurements of the charge-to-mass ratio for the positive and negative ions in the tubes indicated that, under the same potential gradient, the ratio of their velocities would be very large. This accounted for most of the differences in appearance between the anode and cathode, since the greater electrical intensity close to the cathode results from the greater velocity of the electrons.[45] At the time, it was difficult to see what use the electron might have, but the way was now clear for the major developments that were to take place in physics in the 20th century.

20

References

1.　　Hauksbee, F.: *Phil. Trans.*, **24**, p.1865-6, (Feb. 1705)

2.　　Hauksbee, F.: *Phil. Trans.*, **24**, p.2129-35 (Aug. 1705)

3.　　Cajori, F.: *A History of Physics*, p. 132, (1933) MacMillan Co. (NY)

4.　　Watson, W.: *Phil. Trans.*, **47** p.362-76 (1752); *Dictionary of National Biography*

5.　　Henly, W.: *Phil. Trans.*, **64**, p.388-432 (1774)

6.　　*Phil. Trans.*, **11**, p.647-53 (1676);　Cookson, Dr.: *Phil. Trans.* **39**, p.74-5 (1735-36)

7.　　Faraday, M.: *Phil.Trans.*, **51** I, p.125-62 (1832)

8.　　Davy, H.: *Phil. Trans.* **99**, p.39-104 (1809); *Phil. Trans.*, **39**, p.425-39 (1821)

9.　　Davy, H.: *Phil. Trans.*, (1822) p.64-75

10.　　Faraday, M.: *Phil.Trans.* (1833), p.23-54

11.　　Faraday, M.: *Phil. Trans.*, (1834) I, p.77-122

12.　　Faraday, M.: *Phil. Trans.*, (1838) Pt.I, pp.83-124, 125-168

13.　　Wheatstone, Charles: *Phil. Trans.*, **53** I, p.583-9 (1834)

14.　　Faraday, M.: *Phil. Trans.*, **136**, p.1-20 (1846)

15.　　Maxwell, J.C.: *Nature*, **7**, p.323-5, 341-3 (1873); *Phil. Trans.*, **155**, p.459-512 (1865)

16.　　Christie, S.H.: *Phil. Trans.*, p.219-39 (1826); p.379-96 (1828).

17.　　Thomson, W.: *Proc. Roy. Soc.*, **8**, p.150-8 (1856)
　　　Maxwell, J.C.: *Nature*, **7**, p.323-5, 341-3 (1873)

18.　　Page, Charles G.: *Am.J.Sci.* **35**, p.252-68 (1839)

19.　　Grove, W.R.: *Phil. Trans.*, **142** I, p.87-101 (1852)

20.　　Abria: *Ann. de Chim.*, **7**, p.462-88 (1843)

21.　　de la Rue, W. & H. Müller: *Phil. Trans.*, **169** I, p.55-118 (1878)
　　　Quet & Seguin: *Compt. Rendus* **47**, p.964-7 (1858)

22.   Plücker, J.: *Phil. Mag.* 4th ser. **16**, pp.119-32, 408-18 (1858)

23.   Gassiot, J.P.: *Phil. Trans.*, **148** p.1-16 (1858). Bakerian Lecture.

24.   Gassiot, J.P.: *Phil. Trans.*, **149** I, p.137-60 (1859); *Roy. Soc. Proc.*, **9**, p.601-5 (1859); *Proc. Roy. Soc.*, **10** p.393-404 (1859-60)

25.   Plücker, J.: *Annalen der Phys. und Chem.*, **107**, p.77-113 (1859)
      Thomson, J.J.: *Discharge of electricity through gases* (1898) p.137

26.   Plücker, J. and J.W. Hittorf: *Proc. Roy. Soc.*, **13**, p.153-7 (3.3.1864)

27.   Crookes, W.: *Phil. Trans.*, **170**, p.135-64 (1879).
      Hittorf, J.W.: *Ann. Phys. & Chem.*, **136**, p.1-31 (1869)
      Thomson, J.J.: *Discharge of electricity through gases* (1898) p.137

28.   Goldstein, E.: *Phil. Mag.* 5th ser., **4**, p.353-63 (1877)
      Thomson, J.J.: *Discharge of electricity through gases* (1898) p.138

29.   de la Rue, W., H.W. Müller & W. Spottiswoode: *Proc. Roy. Soc.*, **23** p.356-61 (1875)

30.   de la Rue, W., & H.W. Müller: *Proc. Roy. Soc.*, **26**, pp.324-5, 519-23 (1877); *ibid.*, **27**, p.374-81 (1878); *Phil. Trans.*, **169**, p.55-118 (1878); *Proc. Roy. Soc.*, **29,** p.281-9 (1879)

31.   Crookes, W.: *Proc. Roy. Soc.*, **28**, p.103-11 (1878-9); *Phil. Trans.*, **170**, p.135-64 (1879); *Proc. Roy. Soc.*, **28,** p.477-82 (1878-9)

32.   Crookes, W.: *Proc. Roy. Soc.*, **30**, p.469-72 (1879-80); *Proc. Roy. Soc.*, **30**, p.446-58 (1880/1)

33.   Thomson, J.J.: *Discharge of electricity through gases* (1898) p.189-92

34.   Stoney, J.G., *Phil. Mag.*, **38**, p.418-20 (1894)

35.   Johnstone Stoney, G.: *Sci. Proc. Roy. Dublin Soc.* (16 Feb. 1881); *Phil. Mag.* **11**, p.381-90 (1881); *Sci. Trans. Roy. Dublin Soc.*, **4**, p.563-608 (1888-92)

36.   Spottiswoode, W., & J.F. Moulton: *Proc. Roy. Soc.*, **30**, p.302-4 (1879-80); *Proc. Roy. Soc.*, **32**, p.385-90 (1881)

37.   Schuster, A.: *Proc. Roy. Soc.*, **37**, p.317-39 (1884)

38.   Schuster, A.: *Proc. Roy. Soc.*, **47**, p.526-61 (1889-90)
      Giese, W.: *Ann. d. Phys. & Chem.*, **17**, p.519-50 (1882)

39.   Thomson, J.J.: *Proc. Roy. Soc.*, **49**, p.84-100 (1890-1)
      Plücker, J.: *Ann. d. Phys. & Chem.*, **107**, p.77-113 (1859)

40.   Thomson, J.J.: *Phil. Mag.*, ser.5, **32**, p.321-36, 445-64 (1891). *Proc. Phys. Soc.*, **40**, p.79-89 (1927-8); *Phil. Mag.*, **4**, p.1128-60 (1937)

41.   Thomson, J.J.: *Discharge of electricity through gases* (1898) p.182-5
      Lenard, P.: *Ann. d. Phys. & Chem.,* **51**, p.225-67 (1894)

42.   Perrin, J.B.: *Comptes Rendus*, **121**, p.1130-6 (1895)
      Thomson, J.J.: *Electrician*, **39**, p.104-9 (May 1897); *Phil. Mag.* 5th ser., **44**, p.293-316 (1897)

43.   FitzGerald, G.F.: *Electrician,* **39**, p.103-4 (1897)

44.   Kaufmann, W.: *Ann. Phys. Chem.* (Supp.) **61**, p.544-52 (1897)
      Wien, W.: *Berlin Phys. Gesell. Verh.* **16**, p.165-72 (1897) [Abstr.]

45.   Thomson, J.J.: On the theory of the conduction of electricity through gases by charged ions. *Phil. Mag.,* **47** (5th ser.) p.253-68 (1899)

Chapter 2

The development of plasma theory (1900-1950)

Work on discharge tubes in the 1890s had produced some exciting developments and almost every laboratory in the world was soon experimenting with them. Having discovered the nature of the cathode rays, scientists now had to determine the detail of what was happening inside the tubes. Until Niels Bohr published his model of the hydrogen atom in 1913, much of the theorising was just guesswork. To fully understand what produced the effects inside the tube required knowledge of atomic structure; collisional processes and ionisation; and particle size and velocity, all of which took many years to accumulate. Much of the information was acquired through work with electrical discharges, and the history of atomic physics in the late 19th and early 20th centuries is inextricably linked with that of discharges through gases.

J.J. Thomson established the basic collisional theory for ionised gases in 1900 when he explained why, when ultraviolet light falls on one of two metal plates, the current between them increases with their separation distance, especially at low pressure. This was, he said, because negative ions — there was still a reluctance to call them electrons — start from the illuminated plate and gain sufficient velocity (and therefore kinetic energy) from the electric field[a] as they travel through the gas to produce other ions in collisions with surrounding molecules.[1]

Irish-born John Sealy Edward Townsend (1868-1957) had worked for Thomson at the Cavendish Laboratory in Cambridge before moving to Oxford in 1900. He had noticed that electrons produced new ions when moving in an electric field which was too small to sustain a continuous discharge. They were therefore, he thought, responsible for maintaining the current. The slower-moving positive ions differed little in mass from ordinary molecules, and only made a noticeable contribution in a strong electric field. Ionisation by collision was, therefore, fundamentally important in developing large currents from comparatively small electric forces and could explain many of the phenomena observed in gas discharge tubes.[2]

He began to formulate the equations needed to explain both Thomson's photo-

---

[a]   In an electric field, electron velocities increase rapidly, their energy being almost equal to that imparted by the field.

24

ionisation ideas and his own observations. Although his experiments were largely conducted at or near atmospheric pressure at a time when ionisation processes were not fully understood, his equations, in modified form, are still in use today; as is his sparking criterion.

Townsend had seen that, when a potential difference is established between two electrodes, the gas acts as an insulator until the applied potential equals or exceeds a certain value. This he called the *sparking potential* or *breakdown voltage*. At this point, the insulation breaks down and a current, accompanied by a glow, passes through the gas. Between parallel-plate electrodes, the sparking potential is a function only of the gas pressure and the distance between the electrodes, a result obtained experimentally by de la Rue and Müller in 1879.[3] The relationship, now known as Paschen's Law, was formally described

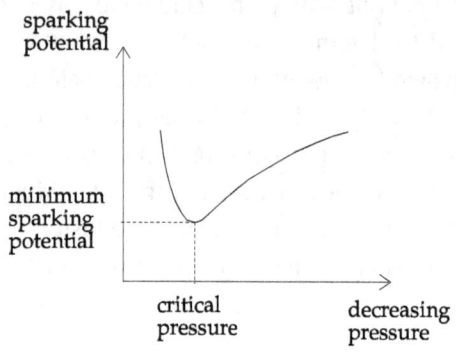

Fig.2.1 Variation in sparking potential with pressure

by Paschen in 1889.[4] He showed that, for parallel-plate electrodes a fixed distance apart, the resistance to the discharge, and hence the sparking potential, is large at high pressure and decreases with decreasing pressure to a minimum, before rising rapidly as the pressure continues to decrease (see Fig. 2.1).

Townsend calculated the breakdown voltage for various gases and found that its value depended on the minimum distance between the electrodes at which a continuous discharge takes place. The condition for a spark is then one in which the current becomes self-sustaining without the introduction of outside (photo-ionised) electrons. At low pressure, with a small gap between the plates, no large current will pass through the gas.[5]

By 1910 the basic principles of ionisation were understood, largely due to the work of Townsend and Thomson. It was known that, when the energy acquired by an ion exceeded "a certain definite value" an electron "escapes from the atom", producing a free electron and a positively-charged ion. The radiation emitted when the two recombined occurred as a series of pulses, each containing the same amount of energy.[6] It required Bohr's model of the atom to provide the detail and enable the actual ionisation processes to be worked out.

In 1916, Karl Taylor Compton (1887-1954) confirmed that Townsend's equations were correct in any theory of ionisation based exclusively on inelastic collisions. Compton's experiments indicated that the current through an ionised gas reaches a maximum value at a particular gas pressure. At very low pressures, the low density, and therefore lack of collisions, means that electrons acquire the high velocities needed for ionisation but because collisions are rare, there is no noticeable increase in current due to ionisation. At high pressures there are many collisions but little ionisation, because the electrons have insufficient velocity. Collisional frequency was not simply a function of the mean free path of a gas molecule, nor of its dimensions. It was also affected by the forces acting on the electron as it approaches a molecule.[7]

With the basic theory of ionisation established, investigations returned to the processes occurring in the discharge tubes themselves. In 1900, Johannes Stark (1874-1957) had observed that, unlike metallic conductors, the conductivity in gases carrying an electric discharge varies from point to point depending on the density and mobility of the ions. He suggested that ions of different charge might carry different amounts of current which could produce strata of positive and negative ions as in the striated discharge. Eight years later, J.W. Bispham observed that all types of striated glow discharges are intermittent in character.[8] By 1909, Thomson had decided that striations were caused by fluctuations in electric force affecting ionisation in the gas: the electric force being negative on the cathode side of each bright band, causing an accumulation of ions. The resulting increase in recombination, he thought, produced the bright bands.[9]

The cathode dark space, meanwhile, was still occupying William Crookes' attention. In 1907, at the aged of 74, he suggested that the luminous layer on the cathode might result from ionisation of residual gas molecules and the subsequent recombination of positive ions with slow-moving electrons released from the metal electrode. The cathode dark space perhaps indicated the mean free path of high-velocity electrons which start close to the cathode. Under normal discharge pressures, the discharge tube dimensions had no effect on the size of the dark space, but as exhaustion continued, the space increased in a large tube and disappeared altogether in a small one. The neighbouring negative glow also varied in size with pressure and was, he suggested, the scene of collisions between free electrons and positive ions.[10]

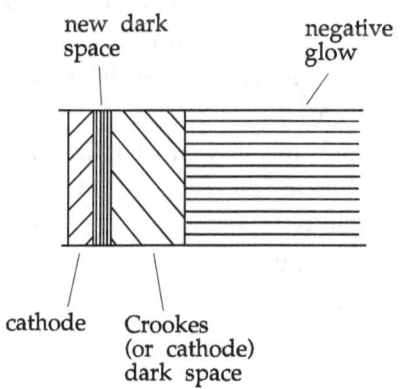

Fig. 2.2   **Aston's dark space (helium)**

In Birmingham, Francis William Aston (1877-1945) was also investigating the size of the cathode dark space under varying conditions in different gases. In helium, an additional dark space appeared very close to the cathode, between it and the cathode dark space (Fig. 2.2), with very different properties from the latter. It was practically independent of pressure and was not detected in other gases. At 1 mm pressure, the new space was intensely black and about 0.5 mm thick, followed by a region of moderate greenish light 1 cm long (the cathode dark space), terminating in a brilliant bluish-green negative glow.

Building on Crookes' ideas, Aston proposed that electrons leaving the cathode with insufficient energy to ionise the gas will only acquire that energy by falling through a definite potential. There will therefore be a space in front of the cathode in which no ionisation will occur, and where no light would be expected. The newly-discovered dark space could therefore be regarded as the distance through which the electrons fall in order to gain the necessary ionisation energy. He predicted that, if this distance d was less than the mean free path of the electrons, a region of maximum ionisation (and therefore light) could be expected in the Crookes dark space just beyond d. At a point $2d$ from the cathode, the second generation of electrons, formed by collisions at d, will have reached ionisation velocity and a second maximum of light just beyond $2d$ should be observed: the Crookes dark space should therefore be striated near the cathode. Only the first maximum was found in hydrogen, but in helium, the striation was clearly visible.[11]

By 1910 Aston was at Cambridge, continuing the work on the electric force in the cathode dark space and negative glow begun by Schuster in the 1880s. While previous researchers had made their measurements using probes inserted into the discharge, Aston fired a beam of cathode rays across it, taking their deflection as a measure of the electric force at that point. He found that the force in the dark space was much greater than in the negative glow and was almost linearly proportional to the distance from it, while the total fall in potential inside the dark space agreed with that observed across the electrodes. He concluded that the Crookes dark space is a region of uniform positive space charge which ceases abruptly in the negative glow.[12]

Other research continued to confirm the long-held belief that the different types of discharge were all variations of the same thing, but, despite worldwide investigation, the mechanism involved in producing the various discharges and their associated phenomena remained largely a mystery. Mrs. Hertha Ayrton[b] (1854-1923) in a paper presented to the Royal Society in June 1901, suggested that the arc was simply a gap in a circuit, providing its own conductor through the evaporation of its own material. All solid surfaces are irregular, she explained, and as the carbon tips are moved apart, the heat caused by the resistance of the surfaces still in contact with each other evaporates the carbon at these points. By the time contact between the tips is completely broken, the small gap between them is full of carbon vapour, which would be sufficient to maintain the arc.[13]

The outbreak of war in 1914 diverted most research towards war applications. Discharge tubes were needed for use in radio work, etc., and their development made rapid progress. As life returned to normal in the 1920s, attention focused once again on the origins of discharges. Aston, writing in 1923, admitted that the nature of the negative glow and the mechanism by which the current is carried across the Crookes dark space at moderately low

---

[b]   Professor Ayrton is noted for her work on arc lamps and searchlight technology used in anti-aircraft work in both the 1st and 2nd World Wars. In 1898 she was elected to the Institution of Electrical Engineers and was their only woman member. Her nomination for fellowship of the Royal Society in 1902 was rejected because she was a married woman.

pressures were still matters of speculation. It was generally accepted that positive ions passing through a gas "do not retain their identity, but by capturing and losing electrons may lose and gain their charge from time to time."[14]

In Princeton, New Jersey, Karl Compton believed that the electric arc was simply the negative glow region of the glow discharge. Thomson had earlier suggested that the arc current consisted of electron emission from a very hot cathode plus ionisation of the gas by these electrons. Compton questioned whether thermionic emission alone could account for the primary arc currents. He thought the cathode fall in potential, occurring over a distance equivalent to the electron mean free path in the gas, was equally important.[15]

At the Cavendish Laboratory in Cambridge in 1923, Appleton and West detected electric waves in a striated glow discharge, confirming Bispham's 1908 observation that striated discharges are intermittent.[16] The oscillation frequency was independent of the external circuitry but increased with pressure and with anode potential. The production of such waves indicated that the discharge was actually pulsating. "The oscillations," they declared, "are therefore of a new type, being ionic in character and origin." In 1906, Lord Rayleigh had reported similar oscillations of electrons disturbed from equilibrium.[17] Their cause was unknown.

In America, Compton, Turner and McCurdy had also been studying striated discharges. When striations occurred, they said in 1924, both electric field and ion concentration revealed periodic variations, as Thomson had observed 15 years earlier. The electron velocity distribution was Maxwellian[c] everywhere, except between the striations, indicating that it might be produced by regions of ionisation. Light emission in the discharge seemed to be linked to excitation of neutral atoms by electron impact rather than ionisation and recombination. If recombination was solely responsible, the light from the negative glow, where most recombination occurred, would be significantly brighter than that from the positive column, which it was not. They therefore proposed that most of the light in the much brighter positive column came from readjustments within excited atoms. Atoms excited by electron impacts were to be found in the striations but not in the regions between, and seemed unable to diffuse between the layers.[18]

Gas discharge theory was beginning to take shape. In May 1925, Prof. R. Whiddington summarised current knowledge in a series of three lectures at the Royal Institution in London.[19] Whiddington identified three regions in the discharge tube: the cathode, with its softly-luminous cathode glow, Crookes' dark space and negative glow; a central region, containing the continuously-glowing, sometimes striated positive column, separated from the cathode region by the Faraday dark space and extending right up to the anode; and the anode itself, consisting of a very thin layer of light over the anode surface (Fig. 2.3).

---

[c]   See Chapter 3

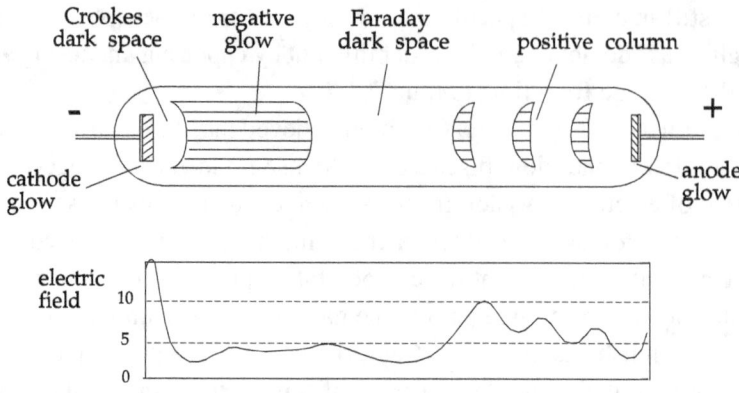

Fig. 2.3   Whiddington's discharge tube

The immediate neighbourhood of the cathode was still a mystery. Positive ions, said Whiddington, travel from anode to cathode and eject electrons from it by some (then) unknown process. These electrons are soon accelerated by the electric field and ionise the residual gas by collision, producing the negative glow, which widens as the gas pressure falls. The positive ion bombardment results in a negative space charge and consequent lowering of the electric field close to the cathode. The sharply defined edge of the Crookes dark space was thought to be the region where ionisation by cathode rays occurs, together with recombination resulting in light emission. It had long been known that objects placed in the dark space cast a shadow in both directions. This indicated that most of the electrons from the cathode were somehow produced by the arrival of positive ions originating at the edge of the negative glow. Since ion velocity is much less than electron velocity, there was presumed to be a concentration of positive space charge in the Crookes dark space.

The anode, like the cathode, is covered by a film of luminescence, in which the drop in potential is as abrupt as at the cathode. The anode fall, said Whiddington, depends only on the gas and the current. As the current decreases, the anode fall rises to a constant value approximately equal to the ionisation potential of the gas. The cloud of electrons near the anode was thought to produce the fall, and at low currents, single-impact ionisation perhaps occurred in a film of gas on the anode surface.

The positive column, occupying the central section of the tube, was the most puzzling region of all. It generally formed a uniform column of light, independent of the length or shape of the tube, in which striations may appear when the gas contained impurities. Conductivity was high and the electric force small. What caused the light emission, and the appearance and behaviour of striations was still a mystery.

Professor Whiddington's survey of the state of knowledge in 1925 shows that considerable progress had been made since the turn of the century, but much was still not understood. That situation was about to change dramatically.

The theory of plasma physics grew largely from the work of one man, American chemist and physicist Irving Langmuir (1881-1957), and his co-workers at the General Electric Company's research laboratory in Schenectady, New York. It was here in 1909 that Langmuir started investigating electron behaviour in low-pressure, strongly-ionised gases.

Townsend had conducted his experiments on pre-breakdown or "dark" discharges where the primary electrons were produced externally, often photoelectrically. While he had explained the avalanche process that started the discharge, no one, in 1909, understood the glowing gas which followed breakdown. Langmuir began by looking at the space charge distribution in the tubes and thermionic emission of electrons from a hot cathode. By the early 1920s he was beginning to understand what was happening. In 1923, he announced that, since gradients in electrostatic fields are associated with regions of space charge, and the electric field in the positive column was constant, the space charge in the positive column must be practically zero. Positively- and negatively-charged particles were therefore present in equal numbers in any uniformly-ionised gas.[20]

By placing a small auxiliary electrode, or *probe*, in the path of an electrical discharge and determining the volt-ampere characteristic of the electrode, he was able to measure electron and ion current densities, the distribution of electron velocities, and study the distribution of potential along the discharge. The idea of using a sounding electrode was not new — Johannes Stark had used a similar device in 1905 to investigate space potentials and the cathode drop in an arc[21] — but the auxiliary electrode became known as a *Langmuir probe* and would later be a frequently-used tool in plasma research.

A Langmuir probe is a metallic electrode, often made of tungsten wire, which is insulated except for the tip. Langmuir's measurements established the potential assumed by the electrode when it was electrically insulated from other parts of the discharge tube, and therefore taking no current: it was "floating". By applying an external variable potential, V, to the probe and measuring the current through it as a function of the external potential, Langmuir obtained what is now called the *I-V characteristic* (Fig. 2.4) of the probe. A floating probe will rapidly acquire a negative charge because electrons in the ionised gas move faster than the heavier positive ions. Eventually, the electron and ion current densities to the probe equalise as more electrons are repelled and the current through the probe falls to zero. The probe at this point assumes a negative potential, called the *floating potential* (V_f in Fig. 2.4). At

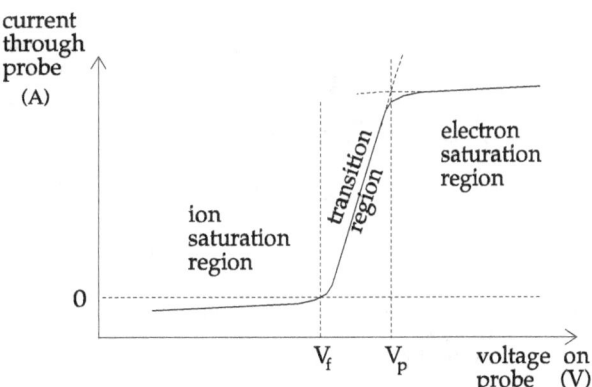

Fig. 2.4   I-V characteristic of a Langmuir probe

potentials below V_f electrons are repelled. Only positive ions are collected by the probe and the *ion saturation current* can be measured. At potentials above V_f (the transition region), the probe becomes increasingly less negative with respect to the ionised gas and more electrons are collected, until the current through the probe equals the electron current. All electrons arriving close to the probe tip are collected by it. This point, at which the *electron saturation current* can be measured, is now known as the *plasma potential* (V_p in Fig. 2.4). Any electrode inserted in an ionised gas disturbs the electric field and hence the particle density and energy. This disturbance is minimal at V_p.

Langmuir found that, when a cylindrical or spherical electrode is immersed in an ionised gas, it becomes surrounded by a symmetrical space charge region of (usually) positive ions. A negatively-charged electrode repels electrons from its neighbourhood and attracts positive ions, so that a positive ion *sheath*, whose thickness he calculated to be about 0.1 mm, forms around the electrode. Electrons are reflected from the outer surface of the sheath while all positive ions reaching the sheath pass through it to the electrode. When he changed the negative voltage of the electrode from 10 to 100 volts the sheath thickness only increased by about 0.15 mm, a distance which was small compared to the mean free path of the electrons or ions and the dimensions of the tube, so no change occurred in the positive ion current reaching the electrode. The sheath thus acts as an electrostatic screen preventing the field due to the negative charge on the electrode from extending beyond the edge of the sheath. The electrode's potential does not affect what happens in the arc or discharge, or the current flowing to the electrode. The whole drop in potential between the ionised gas and the probe, said Langmuir, is concentrated within the sheath, with the positive space charge of the ions in the sheath neutralising the effect of the negative charge on the electrode (see Fig. 2.5). The number of positive ions taken up by the probe is thus limited to the number reaching the outer edge of the sheath through their normal random motion, and the probe itself has no influence on that. With a positively-charged electrode and low gas pressure, he found that an electron sheath formed in a similar way.[22]

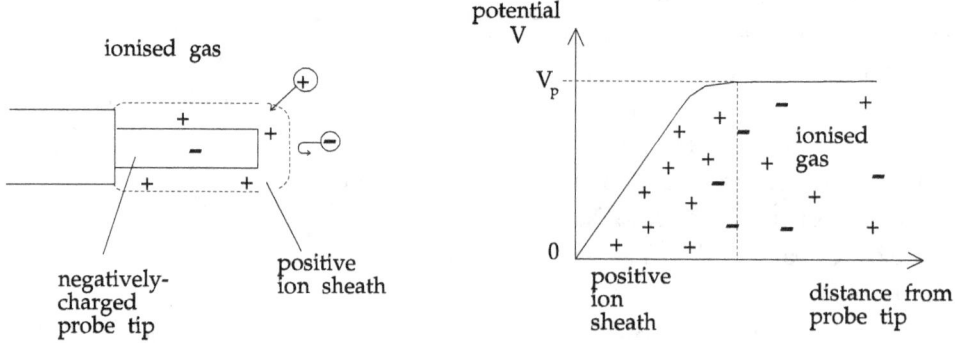

Fig. 2.5   Space charge in the sheath

It had been thought that, when a wire was introduced into a discharge, the voltage being measured was the actual voltage at that point in space. Langmuir's discovery of a sheath revealed that the electrode disturbed the conditions in its vicinity and, instead of being negative as was generally found, the space potential in the body of the discharge was usually positive relative to the walls of the tube.

In 1923, Langmuir began to assemble his results. While developing a set of space charge equations in 1913, he had shown that, in high vacuum in a space devoid of positive ions, the mutual repulsion of electrons (as space charge) limits the flow of current from a hot cathode to a cold anode. In 1920 he had noticed that, when a metal electrode is heated to a high temperature in high vacuum, electrons are emitted from its surface. Their quantity was determined by the cathode material and the condition of its surface; their initial velocities depended on the cathode temperature. Experiments also showed that most of the electrons in an ionised gas have a Maxwellian velocity distribution (see Chapter 3), so, from kinetic theory, electron velocities can be defined in terms of an electron temperature, T_e. The ion velocity distribution was less obviously Maxwellian and was harder to determine.[23]

Adding these observations to his theory of sheaths, Langmuir noted that electrons emitted from a heated cathode will flow out through the positive ion sheath.[24] If this electron current is large enough, it will neutralise the positive ion space charge close to the cathode. Eventually, the current is limited by space charge and no further increase in electron current is possible. The cathode is then covered by a *double layer*, or double sheath, with an inner negative space charge and an equal outer positive charge, the field being zero at the cathode and at the sheath edge. He calculated that the ratio of the electron current to ion current was equal to $(m_i/m_e)^{1/2}$ where m is the mass of the particle (positive ion or electron). The anode normally has a positive ion sheath, but with small anodes a detached double sheath may exist at the boundary of the anode glow. Later, in 1930, E.C. Childs suggested that the dark space between the cathode and the Crookes dark space, first observed by Aston in 1907, was possibly a sheath of space charge, since it was too thin for a current-carrier to collide with another particle in its journey across.[25]

In 1925, while Whiddington was summarising current knowledge of discharge tube phenomena in his Royal Institution lectures, Langmuir embarked on perhaps the most significant stage of his investigations. Projecting electrons from a hot cathode with uniform velocity into the ionised vapour of a low pressure mercury arc, he found that many of them quickly gained velocities whose voltage-equivalent was greater than the total drop across the tube. An even larger number had less than the expected energy. As a group, therefore, the electrons had gained no energy. Instead, a random Maxwellian velocity distribution had been imposed on the uniformly-moving stream by some as yet unidentified scattering mechanism. The velocity distribution corresponded to temperatures ranging from 5000 K to 60,000 K despite the mean free path of the electrons being so long that collisions alone could not have produced it. Gas pressure was so low that only a small percentage of the scattered electrons could have collided with atoms of the gas.

The speed with which the velocity distribution was imposed indicated that some kind of randomly-directed impulse was delivered to the electrons. Electrical oscillations, with an intrinsic natural period which was independent of the current passing through the tube, had previously been reported[d] in gas discharge tubes and it was suggested that electrical oscillations in the arc might have caused the scattering by subjecting the electrons to rapidly-changing electronic fields.[26] In 1926, Arthur Dittmer suggested that oscillations at frequencies above 10^8 Hz were probably responsible. Penning, in Eindhoven, reported that radio-frequency oscillations, between 3×10^8 and 6×10^8 Hz, and electron scattering generally occurred together in low-pressure discharges through mercury vapour but, like Dittmer, he could not explain them.[27]

Langmuir and his co-worker Lewi Tonks confirmed Penning's observations, reporting frequencies up to 1.2×10^9 Hz. These waves, wrote Langmuir in August 1928, had a well-defined high frequency and short wavelength, and seemed to depend on the presence of low-velocity electrons. "These oscillations must be regarded as compressional electric waves somewhat analogous to sound waves. Except near the electrodes, where there are sheaths containing very few electrons, the ionised gas contains ions and electrons in about equal numbers, so that the resultant space charge is very small. We shall use the name *plasma* to describe this region containing balanced charges of ions and electrons."[28]

This is the first appearance, in print, of the word "plasma" to describe the ionised gas of the discharge tube. The October issue of *Nature* provided a definition: "Dr. Langmuir has published a theoretical analysis of the possible modes of vibration of what he refers to as a 'plasma', a highly ionised gaseous medium at low pressure which contains, when undisturbed, equal numbers of positive ions and of electrons and appears to have accounted for the majority of these hitherto unexplained observations. Waves in the component electron gas should be of high frequency, with a zero group velocity, and so be incapable of transmitting energy. Similar vibrations of lower frequency should theoretically also occur in a beam of electrons, and have in fact been detected, whilst the electrical analogue of sound waves has been found in a vibration of the heavier positive ions and tentatively identified with the type of ionic oscillations which is supposed to be associated with moving striations."[29]

No explanation was given for the choice of the name "plasma", other than the comment in a paper by Tonks and Langmuir: "When the electrons oscillate, the positive ions behave like a rigid jelly with uniform density of positive charge ne. Imbedded in this jelly and free to move there is an initially uniform electron distribution of charge density $-ne$."[30] It has been pointed out by later writers that "plasma" is Greek for "something mouldable, like jelly" — they compare the oscillations in the discharge tubes to a bowl of trembling jelly or

---

[d]   E.g. Rayleigh in 1906; Appleton & West in 1923

to the seething movement in the protoplasm of living cells, and suggest that one of these images might have inspired Langmuir. Lewi Tonks later firmly denied this: "Quite definitely," he wrote in 1967, "neither the oscillatory characteristics of plasmas nor the 'seething movement in living cells' was relevant. Most certainly Langmuir did not look upon his plasma as an analogue of 'the cellular jelly, protoplasm'." According to Tonks, the two men were investigating room temperature arcs when Langmuir asked him what they should call the main part of the discharge. Tonks promised to think about it. "The next day, Langmuir breezed in and announced, 'I know what we'll call it! We'll call it the "plasma".' The image of blood plasma immediately came to mind; I think Langmuir even mentioned blood."[31]

Harold Mott-Smith confirmed this in September 1971 in a letter to the editor of *Nature*. Irrespective of whether they were studying mercury-arc columns, glow discharges or gas-filled thermionic tubes, he recalled, the team noticed that many of the discharge phenomena were markedly similar. Langmuir saw the significance of this and everyone searched for a name. They all knew that credit for a discovery goes to the person naming it and not necessarily to the one who actually discovers it. According to Mott-Smith, Langmuir noticed that what they called the equilibrium or uniform part of the discharge "acted as a sort of sub-stratum carrying particles of special kinds, like high-velocity electrons from thermionic filaments, molecules and ions of gas impurities. This reminded him of the way blood plasma carries round red and white corpuscles and germs. So he proposed to call our 'uniform discharge' a 'plasma'. For a long time, we were pestered by requests from medical journals for reprints of our articles."[32]

It took a while for the name "plasma" to gain acceptance. Physicists and chemists regarded it with suspicion, perhaps because it was already used in another field, while the engineering world saw it as a trade name of General Electric. The confusion with blood plasma would continue for many years.

Tonks and Langmuir identified two types of oscillations: electron oscillations which are too rapid for the heavier positive ions to follow; and ion oscillations which are so slow that the electrons can maintain their density distribution. With the former, displaced electrons oscillate around their original position with simple harmonic motion, the frequency of oscillation being

$$f = 8.98 n^{1/2} \text{ Hz.} \tag{2.1}$$

This is a natural frequency, now known as the *plasma frequency*[e] and given by

$$\omega_p = \left(\frac{n_e e^2}{m_e \varepsilon_0} \right)^{\frac{1}{2}} \quad rad \ s^{-1} \tag{2.2}$$

where n_e is the electron density. Apart from secondary effects from the resulting electric

---

[e]   See Chapter 6

fields, no energy is transmitted. As Langmuir explained, these oscillations do not propagate through the plasma but are simply the effect that changes in electron density have on the electric field.[33]

One property of this natural frequency had already been observed. T.L. Eckersley had reported in 1927 that electromagnetic waves originating in a localised pulse were affected by the state of ionisation. The presence of free ions or electrons gave a dispersive character to the medium through which the waves passed, causing the various frequency components of the pulse to separate out. The higher frequencies arrived at a distant point before the lower frequencies, producing a disturbance of constantly lowering pitch. This, Eckersley found, did not continue to zero frequency but was cut off at a critical frequency, given by equation (2.2), and with a phase velocity which depended on the state of the medium. Waves of lower frequency than the critical frequency could not travel through the medium. Eckersley explained this as: "the electrons in a medium rob the wave of its momentum, and at the critical frequency they steal all the momentum so that the wave can travel no further."[34]

The oscillations of positive ions, said Tonks and Langmuir, vary in type according to their wavelength. Shorter-wavelength oscillations were comparable to the electron vibrations, approaching a natural frequency

$$\omega_{pi} = \omega_{pe}\left(\frac{m_e}{m_p}\right)^{\frac{1}{2}} \qquad (2.3)$$

as the group velocity approaches zero. Longer wavelength oscillations were identified as electric sound waves. Webb & Pardue had recently found[35] ion oscillations at frequencies between 1 to 240 kHz in low pressure discharges in air. Langmuir's ion sound waves fell within this range.

Langmuir's theory of plasma oscillations drew on recent work by Peter Joseph Wilhelm Debye (1884-1966) in the field of electrolysis. Debye and his German co-worker, Eric Hückel had shown that, in low concentrations of electrolytes, an ion under the influence of an external electric field travels with a velocity at which the friction balances the field. As the concentration increases, Coulomb forces between the ions have greater effect and the presence of other ions increases the friction. Both factors reduce the conductivity of the fluid.[36] Langmuir adapted this argument to ionised gases: Debye and Hückel's work indicated that under equilibrium conditions in a plasma the average potential near a charged plane varies according to the equation $V = V_0 \, exp(-x/\lambda_D)$ where x is the distance from the plane and λ_D is what he termed the *Debye distance*. When x is large compared to λ_D, the average potential and the average electric field become zero. Comparing this to charged particles, Langmuir suggested that the field due to electrons should disappear at distances greater than λ_D. If some transient influence changes the electron concentration in a volume of plasma with dimensions which are large compared to λ_D, the resulting electric fields will act in the direction necessary to restore the balance. This converts the potential energy of these fields into kinetic energy of the electrons, producing the observed plasma oscillations and electric waves.

In the space of twelve months, Langmuir and Tonks had drawn up the framework for the new field of plasma physics, providing a set of equations which, with some modifications, are still in use today. Langmuir received the Nobel Prize for Chemistry in 1932 for his work on vacuum phenomena, and published his theory in the same year. Many visible features of gas discharges result from differences between the sheaths, which cover the electrodes and walls and contain strong electric fields, and the relatively field-free region of the plasma. To illustrate this, he used a glass tube containing mercury vapour. When a current of a few milliamperes flows from a hot cathode and the tube is filled with the characteristic green-blue glow of the mercury discharge, a dark space separates the glow from the walls. This space is a positive ion sheath, while the glow is a typical plasma. Ions from the plasma arrive at the sheath with energies of less than one volt. Once inside the sheath, strong fields accelerate them towards the walls. As the current increases, the dark space becomes thinner. "It is the scarcity of electrons," said Langmuir, "which causes the lack of luminosity in the sheath, for the luminosity of the plasma is due to the excitation of mercury atoms by electrons."

To create a plasma, he said, ionisation must be sufficient to produce a potential maximum within the tube which is higher than at both electrodes. Low velocity electrons from the ionisation become trapped in this region until their concentration approximates that of the ions. This, he believed, was a fundamental characteristic of plasma.[37]

It was widely believed that knowledge of discharge tube phenomena was virtually complete and there was little further to be gained from their study — there were more exciting developments in other branches of physics to attract the attention of researchers. Calls for a standard terminology[38] went unheeded. Electrical engineers, spectroscopists and gas discharge researchers all used words such as "arc" and "spark" to define discharge phenomena, but the meaning varied from one group to another. To the discharge researcher, "sparking potential" involved a faint, almost invisible, spark between cold electrodes, while the electrical engineer's spark resulted from the discharge of a large condenser, producing intense light and a large current. To the engineer, an "arc" was a stream of hot gases carrying an electric current across a gap between two electrodes. To the discharge researcher it was a discharge in which the current is carried by the metal vapour from the hot, partially-vaporised cathode, as well as the gas. To the spectroscopist, the terms "arc" and "spark" had been adopted before line spectra production was understood and indicated whether the spectral lines originated from neutral or ionised atoms: certain lines in the complete spectrum of an element were absent, for example, when the substance was excited by an arc discharge.

An increasing variety of discharges were being investigated and several attempts had been made to determine when and why one type of discharge became another. Differences in gas pressure, cathode temperature and current strength from one experiment to another added to the confusion and although interest in discharge tubes had waned, there were still many questions to be answered.

It had long been known that there is a time lag in the production of the discharge after the voltage across the electrodes reaches the sparking potential, but what initiated the spark, and the subsequent glow discharge, was still not understood, largely because of the short time involved — some 10^{-8} sec. — and the lack of suitable equipment. Despite its difficulty in explaining various breakdown effects, the widely accepted theory was still that of Townsend, now modified to include the effect of space charges which develop just before breakdown occurs. In 1927, James Taylor suggested that ionisation initially produces a very small "dark" current. A space charge develops and at a certain current density the normal glow or spark discharge sets in. This interim stage is rarely visible, said Taylor, because the external circuit usually supplies sufficient energy to rapidly initiate the spark or glow.[39] A later idea was that an electron avalanche produced the space charge needed for breakdown. An avalanche occurs when ionising collisions of electrons with other particles create free electrons which themselves undergo ionising collisions with other particles.

Compton, Turner and McCurdy had concluded in 1924 that electrode spacing explained the difference between the glow and the arc discharge: a small gap produced an arc — it was merely the negative glow of the longer glow discharge. Others disagreed, claiming that a hot cathode was essential for an arc but not for the glow discharge.[40] Karl Compton continued to develop his theory of the electric arc and in 1927 he reported that only the fall in potential at the cathode had a definite value, characteristic of the gas. Falls elsewhere in the discharge — at the anode, along the positive column and in the region of the negative glow — varied or disappeared with changes in the current, pressure or geometry of the arc path. He therefore defined the arc as a "discharge of electricity, between electrodes in a gas or vapour, which has a voltage drop at the cathode of the order of the minimum ionising or minimum exciting potential of the gas or vapour."[41]

Using a generalised curve of the gas discharge characteristic (Fig. 2.6), Compton described the transition from a glow to an arc discharge. Between points a and b, the "normal glow" phase, current density was constant, cathode fall potential remained steady at around 300 volts and the glow only completely covered the cathode when the current had increased to point b. A further increase in current produced an increase in current density and a

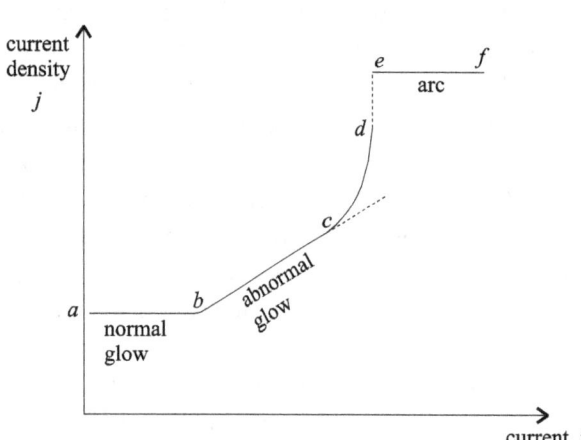

Fig. 2.6   Transition from glow to arc

large increase in the cathode drop (up to several thousand volts). This, he labelled the "abnormal glow". At *c* the cathode glow began to concentrate into a hot spot, becoming more so with further increase in current. By point *d* the cathode drop had fallen below that of the original normal glow, producing instability, and the discharge changed abruptly, at *e*, into a true arc. The glow became a hot spot with a cathode drop in the region of 10 volts.

In 1907, E.F. Northrup had passed an alternating current along a trough of mercury and noticed that the metal contracted in cross-section as the current increased, until the depression reached the bottom of the trough and broke the circuit. He described this as a *pinch effect*.[42] In Ohio in 1933, Willard H. Bennett suggested that a cylindrically-symmetric stream of high-velocity electrons flowing along the axis of a discharge tube might produce a magnetic self-focusing effect if the current momentarily increased at one particular point (Fig. 2.7). This focusing effect had been discussed in the 1920s with reference to gases in cathode ray oscillographs, but Bennett thought the magnetic effects might also be important. The increase in magnetic field would force all the charged particles, both ions and electrons, closer to the axis, probably in a very small interval of time — of the order of 10^{-4} s. Such focusing, he said, could explain some of the phenomena observed in high voltage tubes. Constriction of an arc under its own magnetic field — the pinch effect — would later be regarded both as a means of containing a laboratory plasma, and as a cause of its instability.

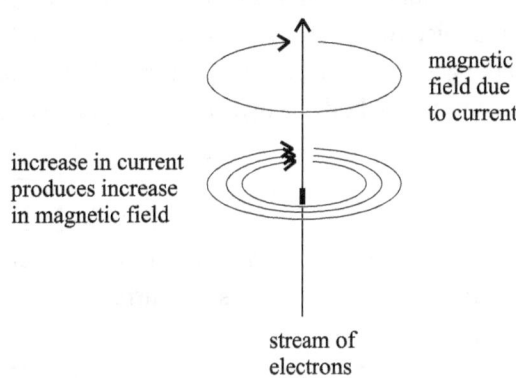

Fig. 2.7 **Magnetic field around electron current**

The 1930s also saw the beginnings of a statistical theory for plasma. Russian physicist Lev Davidovich Landau (1908-68) had been working on damping problems in wave mechanics for several years. In 1936 he produced a kinetic equation for a system of charged particles which made it possible to deduce the velocities and mean free paths of the charged particles in a plasma.[43]

Despite this progress, there were still gaps in the theory. For example: even in 1940, it was still a mystery how the anode fall of potential related to the current between the electrodes.[44] This was partly because it was difficult to obtain pure discharge gases, free of mercury contamination. The best vacuum conditions at that time were achieved using mercury vapour pumps, while pressures were often read using a mercury manometer. Mercury was therefore in the discharge tubes. As vacuum techniques developed in the 1940s, impurities were reduced and the correlation between theory and experiment improved.

New ideas on breakdown began to emerge. During the 1930s, it became clear that

Townsend's 1900 theory, which was based on low values of pd — the product of pressure and electrode separation distance — could not be extrapolated to sparks where pd was more than 200 mmHg cm in air. In this case, suggested John Millar Meek in 1940, the breakdown of a uniform field occurred when an electron avalanche from cathode to anode merged with a self-propagating streamer of positive ions from anode to cathode, to form a conducting filament between the electrodes. The avalanche itself would not constitute breakdown but instead contributed to the sparking mechanism. The anode streamer would develop if photoionisation close to the streamer was sufficient for the space charge at the anode to form a plasma.[45]

In 1900, Townsend had assumed that ionisation by positive ions was responsible for sparking. Although this was now known to be wrong, Leonard Loeb, a colleague of Meek at Berkeley, believed that equations very similar to those devised by Townsend could be used to explain sparking. To sustain the discharge, spark breakdown needed sufficiently intense fields to support both the electron-avalanche process and a secondary process to replace the initial electrons. This secondary process could take several forms, such as photoionisation in the gas, but the most likely source of additional electrons was from the impact of positive ions on the cathode. Townsend's original equation for the discharge current

$$i = i_0 \frac{e^{\alpha d}}{1 - \gamma e^{\alpha d}} \tag{2.4}$$

where α and γ are known as the first and second Townsend coefficients, d is the electrode separation distance and i_0 is the current caused by the initial electrons in the air, had, by the late 1930s, become

$$i = i_0 \frac{e^{\alpha d}}{1 - \gamma (e^{\alpha d} - 1)} \tag{2.5}$$

The quantity γ, often called the *cathode yield* since it represents the number of secondary electrons produced at the cathode, is very much less than 1. By analogy with Townsend's original equation (equation (2.4)), the current i becomes infinitely large when $\gamma e^{\alpha d} = 1$. Townsend and others had interpreted this as being the condition leading to a spark. To Loeb, it marked the threshold for a self-sustaining discharge, independent of i_0, since it ensured that for each electron in the gap, d, sufficient new electrons would be created to maintain the current in its existing state. Thus, said Loeb, one initiating electron can maintain its succession indefinitely. This was not a sparking condition, as had been thought — although, at low pressure, breakdown can occur near to $\gamma e^{\alpha d} = 1$. Instead, it applied to the stable regime occurring after the unstable phase of the spark has passed.[46]

The Second World War (1939-45) diverted both resources and personnel, considerably restricting plasma research. In Stockholm, Swedish physicist Hannes Olof Göst Alfvén (1908-95) found hydromagnetic waves, later called Alfvén waves, in the ionised gas of the Sun while investigating sunspots. As he explained in 1942, if an electrically-conducting fluid is placed in a constant magnetic field, any movement in the fluid induces electric currents which interact with the magnetic field, producing changes in the motion of the liquid. This mutual interaction between electromagnetic and hydrodynamic[f] forces may produce a new type of wave (an Alfvén wave) which travels through the fluid in the direction of the external magnetic field, carrying an induced magnetic field and a velocity field with it. Alfvén devised the set of equations which define the motion of electromagnetic fluids (plasma) and laid the foundations of what became known as *magnetohydrodynamics (MHD)*: how electrically-conducting fluids (ionised gases or liquid) behave in a magnetic field.[47] His equations produced a wave velocity close to that of the sunspot zone as it moves towards the solar equator during the sunspot cycle, indicating that sunspots were associated with a magnetic and mechanical disturbance moving as an electromagnetic-hydrodynamic wave.

Since Maxwell's theory indicated that it was difficult for an electromagnetic wave to penetrate a conductor, Alfvén's discovery was ignored. It was only when he gave a series of lectures in the USA in 1947 that the importance of his work was finally recognised. Two years later, Stig Lundquist, a colleague of Alfvén, demonstrated the existence of MHD waves in mercury.[48] In 1970, Alfvén shared in the Nobel Prize for Physics for his "fundamental work and discoveries in MHD and their fruitful applications" in fields such as controlled nuclear reactors, hypersonic flight and space travel.

By 1950, a century after the first observations, there was still no satisfactory explanation for striations. J.J. Thomson had concluded in 1929 that factors such as impurities, which increase the likelihood of non-ionising collisions would encourage striations.[49] Ten years later, Leonard Loeb suggested they might be associated with plasma oscillations. In 1953, G.D. Morgan, working in Bangor, N. Wales, concluded that striations are a property of the plasma and may be due to a standing wave system associated with plasma oscillations.[50] Donahue and Dieke, working in Baltimore, Maryland, arrived at a similar conclusion. Both oscillations and moving striations occurred in the vast majority of glow discharges in which a positive column was present, irrespective of pressure or current. They must therefore play an essential part in the mechanism of the glow discharge. Moving striations were perhaps "advancing potential waves" or "positive and negative space charge waves".[51] We now know that they are related to fluctuations in electron and ion concentration, and even when a positive column appears uniform, it will probably contain rapidly-moving striations.

Thus, by the mid-1950s, the basic framework of the physics of plasmas was in place and work had already begun on its applications. Considerable progress had been made in the seventy five years since William Crookes first put forward the idea of a fourth state of matter, and there was still more to come.

---

[f]   Hydrodynamics is the science of non-viscous fluids in motion and the forces acting on or exerted by them.

References

1. Thomson, J.J.: *Phil. Mag.* 5th ser., **50**, p.278-83 (Sept. 1900); *Proc. Camb. Phil. Soc.*, **10**, p.74-7 (1898-1900)

2. Townsend, J.S.: *Nature*, **62**, p.340-1 (9 Aug. 1900); *Phil. Mag.* 6th ser., **1**, p.198-227 (Feb. 1901); *The Electrician*, **50**, p.971 (April 1903); *Theory of ionisation of gases by collision*, (1901), Constable & Co., London

3. de la Rue, W., & H.W. Müller: *Phil. Trans.* **171**, p.65-116 (1879)

4. Paschen, F.: *Ann. d. Phys. & Chem.*, **37**, p.69-96 (1889)
   Carr, W.R.: *Proc. Roy. Soc.*, **71**, p.374-6 (1903)

5. Townsend, J.S.: *Phil. Mag.*, 6th ser., **9**, p.289-99 (1905)
   Loeb, L.: *J. Frank. Inst.* **205**, p.305-21 (1928)

6. Thomson, J.J.: *Camb. Phil. Soc. Proc.* **15**, No.4, p.375-80 (1910); *Phil. Mag.* **23**, p.449-57 (1912)

7. Compton, K.T.: *Phys. Rev.*, **7**, pp.489-96, 501-8, 509-17 (1916); *Phys. Rev.*, **8**, p.386-90 (1916)

8. Stark, J.: *Phys. Zeitsch.* **1**, p.439-42 (1900)
   Bispham, J.W.: *Proc. Roy. Soc.*, **81**, p.477-95 (31.12.1908)

9. Thomson, J.J.: *Phil. Mag.* 5th ser., **50**, p.278-83 (1900); *Phil.Mag.* 6th ser., **18**, p.441-51 (1909); *Camb. Phil. Soc.* **15**, p.70 (1909)

10. Crookes, Sir William: *Proc. Roy. Soc.* A, **79**, p.98-117 (1907)

11. Aston, F.W.: *Proc. Roy. Soc.* **79**, p.80-95 (1907); *Proc. Roy. Soc.* **80**, p.45-9 (1907-8)

12. Aston, F.W.: *Proc. Roy. Soc.*, **84**, p.526-35 (1910-11)

13. Ayrton, H.: *Phil. Trans.* A, **199**, p.299-336 (1902)

14. Aston, F.W.: *Proc. Roy. Soc.*, **104**, p.565-71 (1923)

15. Compton, K.T.: *Phys. Rev.* **21**, p.266-91 (1923)

16. Appleton, E.V. & A.G. West: *Phil. Mag.* **45**, p.879-81 (1923)

17. Rayleigh: *Phil. Mag.*, **11**, p.117-23 (1906)

18. Compton, K.T., Louis A. Turner & W.H. McCurdy: *Phys. Rev.* **24**, p.597-615 (1924)

19.   Whiddington, Prof. R.: *Nature*, **116**, p.506-9 (1925); *Engineering*, **120**, p.20-1 (1925)

20.   Langmuir, I.: *J. Frank. Inst.*, **196**, p.751-62 (1923); *Phys. Rev.*, **23**, p.109 (1924) [abstract only: paper presented to Am. Phys. Soc., Dec. 1923]

21.   Langmuir, I.: *Phys. Rev.*, **26**, p.585-613 (1925)
        Stark, J., T. Retschinsky & A. Schaposchnikoff: *Ann. d. Physik*, **18**, p.213-51 (1905)

22.   Langmuir, I.: *Science*, **58**, p.290-1 (1923); Langmuir, I., & H. Mott-Smith: *Gen. El. Rev.*, **27**, pp. 449-55, 538-48, 616-23, 762-71, 810-20 (1924)

23.   Langmuir, I.: *Phys. Rev.*, **2**, p.450-86 (1913); *Gen. Elec. Rev.*, **23**, pp.503-13, 589-96 (1920)

24.   Mott-Smith, H.M. & I. Langmuir: *Phys. Rev.*, **28**, p.727-63 (1926)
        Langmuir, I.: *Phys. Rev.*, **33**, p.954-89 (June 1929)

25.   Childs, E.C.: *Phil. Mag.*, **9**, p.529-46 (1930)

26.   Tonks, L. & I. Langmuir: *Phys. Rev.*, **33**, p.195-210, 990 (Feb., June 1929)

27.   Dittmer, A.F.: *Phys. Rev.*, **28**, p.507-20 (1926)
        Penning, F.M.: *Nature*, **118**, p.301 (1926)

28.   Langmuir, I.: *Nat. Acad. Sci. Proc.*, **14**, p.627-37 (Aug. 1928)

29.   *Nature*, **122**, p.626 (20 Oct. 1928)

30.   Tonks, L. & I. Langmuir: *Phys. Rev.*, **33**, p.195-210 (Feb. 1929)

31.   Tonks, L.: *Am. J. Phys.*, **35**, p.857-8 (1967)

32.   Mott-Smith, H.M.: *Nature*, **233**, p.219 (17.9.1971)

33.   Tonks, L., & I. Langmuir: *Phys. Rev.*, **34**, p.876-922 (Sept 1929)

34.   Eckersley, T.L.: *Phil. Mag.*, **4**, p.147-65 (1927)

35.   Webb, J.S., & L.A. Pardue: *Bull. Amer. Phys. Soc.*, **3**, p.19 (1928)

36.   Debye, P. & E. Hückel: *Phys. Zeitsch.*, **24**, p.185-206, 305-25 (1923)

37.   Langmuir, I.: *J. Frank. Inst.*, **214**, p.275-98 (1932)

38.   Thomson, J.: *Phil. Mag.*, **13**, p.824-34 (1932)

39.   Taylor, J.: *Phil. Mag.* **3**, p.368-82, 753-70 (1927)
      Townsend, J.S.: *Electricity in gases* (1915)

40.   Compton, K.T.: *Phys. Rev.* **21**, p.266-91 (1923)
      Slepian, J.: *J. Frank. Inst.* **201**, p.79-90 (1926)

41.   Langmuir, I. & H.M. Mott-Smith: *Gen. El. Rev.*, **27**, p.449, 538, 616, 762, 810 (1924)
      Compton, K.T.: *A.I.E.E. Journal,* **46**, p.1192-1200 (1927)

42.   Northrup, E.F.: *Phys. Rev.*, **24**, p.474-97 (1907)

43.   Landau, L.: *Phys. Zeits. d. Sowjet Union*, **10**, p.154-64 (1936); *Phys. Rev.*, **77**, p.567-8 (1950) [letter]

44.   von Engel, A.: *Phil. Mag.*, **32**, p.417-26 (1941)

45.   Meek, J.M.: *Phys. Rev.*, **57**, p.722-8 (1940)
      Winstanley Lunt, R., A. von Engel & J.M. Meek: *Rep. Prog. Phys.*, **8**, p.338-67 (1941)
      Zeleny, J.: *J. Appl. Phys.*, **13**, p.444-50 (1942)
      Fisher, L.: *Elec. Engng. (NY)*, **69**, p.613-9 (1950)

46.   Loeb, L.B., & A.F. Kip: *J. Appl. Phys.*, **10**, p.142-60 (1939)
      Loeb, L.B.: *Proc. Phys. Soc. Lond.*, **60**, p.561-73 (1948)

47.   Alfvén, H.: *Nature*, **150**, p.405-6 (1942)

48.   Lundquist, S.: *Nature*, **164**, p.145-6 (1949)

49.   Thomson, J.J.: *Phil. Mag.*, 7th ser., **8**, p.1-29 (1929)

50.   Fowler, R.G.: *Phys. Rev.*, **84**, p.145 (1951)
      Morgan, G.D.: *Nature*, **172**, p.542 (1953)

51.   Donahue, T , & G.H. Dieke: *Phys. Rev.*, **81**, p.248-61 (1951)

Chapter 3

Statistical mechanics and the kinetic theory of gases

3.1   Introduction

Statistical mechanics and the kinetic theory of gases are both based on particle motion. Kinetic theory describes, mathematically, the bulk properties of any gas in terms of the average behaviour of a large number of very small particles in constant and rapid motion. It developed in the mid-1800s to explain the observed response of gases to changes in temperature and volume. A series of papers[1] by James Clerk Maxwell in the 1860s on *The Dynamical Theory of Gases* made an important contribution to kinetic theory and laid the foundations of statistical mechanics.

Before the development of statistical mechanics, classical mechanics and thermodynamics attempted to explain the behaviour of bulk matter. The equations of classical mechanics are based on Newton's laws of motion and describe the motion in time and space of macroscopic bodies acted upon by given forces. They can only be solved accurately when very few bodies are involved. Thermodynamics deals with the general laws governing the effects of heat, such as entropy and conservation of energy, and is concerned only with the very general properties of bulk matter. Statistical mechanics, which predicts the equilibrium properties and behaviour of large collections of microscopic particles from the motion of the constituent particles, provides a bridge between the two. It links the mechanical properties of individual systems — their positions, velocities, energies, etc. — with the thermodynamic properties of the whole assembly: its temperature, pressure, etc. Quantum mechanics is a 20th century extension.

Kinetic theory, in its simplest form, views the molecules of the gas as elastic spheres whose bombardment of the containment vessel walls produces the pressure exerted by the gas. If we assume that the molecules are small in size compared to their average distance apart, and only exert forces on each other during collisions, then the kinetic theory provides a simple explanation of the gas laws.

3.2   The gas laws

A gas expands to fill the space available, exerting pressure on all parts of its container. Robert Boyle, in 1662, defined the relationship between the pressure, p, and volume, V, of a fixed quantity of air at a constant temperature, T, as:

$$pV = constant \text{ (at constant } T)\tag{3.1}$$

Thus, if we double the pressure and maintain the temperature, the gas is compressed to half its original volume. Around 1787, in France, Jacques Alexandre Cézar Charles (1746-1823) used

$$V/T = constant \text{ (at constant pressure)}$$

to relate changes in the volume of a given mass of gas at constant pressure with changes in temperature. As the temperature rises, the volume increases. Some fifteen years later, in 1802, French chemist Joseph Louis Gay-Lussac (1778-1850) independently arrived at a similar result:

$$p/T = constant \text{ (at constant volume)}$$

sometimes called the pressure law. These three laws, known as the gas laws, can be combined to give the general gas law:

$$pV/T = constant\tag{3.2}$$

The constant depends on the sample of gas used and is different for different gases and for different quantities of the same gas. T is the absolute temperature, measured in *kelvin*.

In 1848 Scottish physicist, William Thomson (1824-1907), later Lord Kelvin, devised the absolute temperature scale and formalised the idea of *absolute zero*. Experiments indicated that, if the volume remains constant while gas density, and therefore pressure, is reduced, different gases (oxygen, hydrogen, etc.) lose their individual characteristics and all behave in the same way. When pressure is measured as a function of temperature and the results are plotted on a graph, the resulting straight line, when extrapolated back to zero pressure, gives the same temperature (-273.15°C) for all gases irrespective of the quantity or type of gas used. The SI unit of temperature is the kelvin (symbol: K) and 0 K, or absolute zero, is accepted as being equivalent to -273.15°C. The units of the Kelvin temperature scale are the same size as those of the Celsius scale, so 0°C = 273.15 K. (Notice that K does not take the degree sign (°) — we talk of "273 kelvin" not "degrees kelvin".)

A gas which obeys the gas laws perfectly is described as an *ideal* or *perfect gas*. This is a hypothetical gas in which the molecules have negligible size and exert no intermolecular forces. The behaviour of a real gas approaches that of an ideal gas as the pressure falls. Boyle's law idealises the way gases behave, so an ideal gas would be one for which Boyle's law is exactly true for all pressures and temperatures. Experiments on different types and quantities of gas revealed that it is the number of particles n in the sample that is important, and that the product pV is proportional to n. Thus: $pV \propto nT$. The constant of proportionality used in plasma physics is the molecular gas constant, $k_B = 1.38066 \times 10^{-23}$ J K^{-1}, otherwise known as *Boltzmann's constant*. The resulting equation, known as the *ideal gas law*, is generally written:

$$p = nk_BT\tag{3.3}$$

where n is now the number density per cubic metre of the particles.

3.3 Kinetic theory

Rudolph Julius Emmanuel Clausius (1822-88), working in Germany, made the first serious attempt to formulate a kinetic theory of matter in a paper published in 1857.[2] He regarded a gas as a collection of molecules moving in straight lines in all directions[a] at high velocity and exerting forces on each other only in brief, elastic collisions — what we would now call an ideal gas. Maxwell extended his work, followed later by Ludwig Boltzmann (1844-1906) in Vienna.

The word "molecule" had been used vaguely in the 18th century to refer to particles of matter, but in 1868 Henry Roscoe defined the molecule as a group of atoms which is capable of independent existence. It was thought that molecules behaved like elastic particles, conserving energy in collisions, while the atoms of which they were formed were inelastic, rigid solids. This idea of a molecule as a perfectly elastic sphere played an important role in the development of the kinetic theory.

Four assumptions are made: Firstly, the molecules are "point particles" — tiny spheres whose diameter is small compared to both the average distance between the particles and the size of the container. Secondly, the molecules are in constant motion, obeying Newton's laws of motion and travelling in straight lines between collisions. The effects of gravity are ignored. Thirdly, collisions between molecules, and between molecules and the smooth rigid walls of the container, are perfectly elastic, so kinetic energy is conserved; and finally, the motion of the particles is entirely random, with a range or *distribution* of speeds about a mean.

3.3.1 The link between pressure and kinetic energy

According to kinetic theory, the continual impact of rapidly-moving gas particles on the walls of the containing vessel accounts for the pressure of the gas. Pressure is the magnitude of the force, divided by the surface area perpendicular to the direction of action of the force: $p = F_\perp /A$ (N m^{-2}).

If we take the simple example shown in Fig. 3.1 of a single particle moving in the x-direction with velocity v_x inside a small box, each time the particle collides with the side of the box, its momentum changes. The average rate of change of momentum can be expressed as:

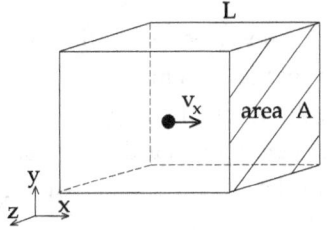

Fig. 3.1 Particle in a box

$$\langle F_x \rangle_{\text{wall}} = - \frac{\text{change in momentum per collision}}{\text{time interval between collisions}}$$

Angular brackets are used because we are considering the average force on the wall, $F_{x\,wall}$. The negative sign indicates that $F_{x\,wall}$ is equal and opposite to the force ($m\,dv/dt$) on the particle. The change in momentum per collision, in the x-

---

[a]   Description of a gas first used in 1738 by Daniel Bernoulli to explain air pressure.

direction is:

$$\text{(final momentum)} - \text{(initial momentum)} = (-mv_x) - (mv_x) = -2mv_x$$

The time interval between collisions with area A, the right-hand wall, is the time for a round-trip from one wall to the other and back: $2L/v_x$. The force exerted on area A by one particle is therefore

$$\langle F_x \rangle_{wall} = -\left(-\frac{2mv_x}{2L/v_x} \right) = \frac{mv_x^2}{L}$$

and by N molecules:

$$\langle F_x \rangle_{wall} = N\frac{m\langle v_x^2 \rangle}{L} \tag{3.4}$$

Angular brackets around v_x indicate that it is now the average speed of N particles. In three dimensions, the speed v of any particle is related to its velocity components v_x, v_y and v_z by $v^2 = v_x^2 + v_y^2 + v_z^2$. Averaged over all particles, this becomes $\langle v^2 \rangle = \langle v_x^2 \rangle + \langle v_y^2 \rangle + \langle v_z^2 \rangle$. Kinetic theory assumes that there is no preference for any particular direction of motion, so $\langle v_x^2 \rangle = \langle v_y^2 \rangle = \langle v_z^2 \rangle$, and therefore $\langle v^2 \rangle = 3\langle v_x^2 \rangle$, or

$$\langle v_x^2 \rangle = \tfrac{1}{3}\langle v^2 \rangle.$$

Substituting for $\langle v_x^2 \rangle$ in equation (3.4), $\langle F_x \rangle_{wall} = Nm\langle v^2 \rangle/3L$, and the pressure on area A in Fig. 3.1 is given by

$$p = \frac{\langle F_x \rangle_{wall}}{A} = \frac{Nm}{3LA}\langle v^2 \rangle = \frac{Nm}{3V}\langle v^2 \rangle$$

since LA is the volume V of the box. The pressure is thus the same in all directions. Written in the form:

$$pV = \frac{1}{3}Nm \langle v^2 \rangle \tag{3.5}$$

we have a means of relating the macroscopic properties of a gas (left-hand side of the equation) to its microscopic properties (the right-hand side), bearing in mind that this is based on the average speed of the particles and there will in fact be a range of velocities.

3.3.2   A definition of temperature

The number of velocity components needed to completely describe the motion of a particle is also the number of *degrees of freedom* possessed by the particle. A monatomic gas (one composed of single atoms) has three degrees of freedom: the thermal energy of each particle is associated with translational motion in three mutually perpendicular directions.

By rearranging equation (3.5), we can link pressure to the kinetic energy of the particles using their average *translational kinetic energy* $\langle \tfrac{1}{2}mv^2 \rangle$:

$$pV = \frac{2}{3}N \langle \frac{1}{2}mv^2 \rangle = \frac{2}{3}N \langle E_{kin} \rangle$$

Comparing this with equation (3.3):

$$pV = nk_BT = \frac{2}{3}n\langle E_{kin}\rangle$$

we obtain

$$\langle E_{kin}\rangle = \frac{3}{2}k_BT \qquad (3.6)$$

Equation (3.6) states that the average translational kinetic energy per molecule of an ideal gas is proportional to the temperature of the gas: if the temperature increases, the average speed of the molecules increases. It thus provides a "molecular" definition of temperature.

According to the *principle of equipartition of energy* (a physical generalisation first stated by Boltzmann) if a system has several degrees of freedom then, on average, its energy is shared equally between them: each component of velocity therefore has an average associated kinetic energy per particle of $\frac{1}{2}k_BT$. The kinetic theory assumptions ignore any rotation or vibration of the particles, so just translational kinetic energy is involved in the definition of temperature, and hence $\langle E_{kin}\rangle = 3(\frac{1}{2}k_BT)$. Equation (3.6) is only strictly true for an ideal gas, but because many real gases behave almost like an ideal gas under most conditions, it can be applied to real gases provided the pressure is not too high or the temperature too low.

Notice that if we write

$$\langle E_{kin}\rangle = \frac{1}{2}m\langle v^2\rangle = \frac{3}{2}k_BT$$

we obtain the result

$$\langle v^2\rangle = \frac{3k_BT}{m} \qquad (3.7)$$

The term on the left is the average or mean-squared speed, with dimensions of [distance/time]^2. To obtain a quantity with the dimensions of speed, we take the square root: $\sqrt{(\langle v^2\rangle)}$. This is the square root of the average of the squared speeds[b] — the *root-mean-squared speed*, v_{rms}. Equation (3.7) shows that it is the average squared speed (*not* the speed) of the molecules that varies with the temperature and mass of the particles. If the gas is in equilibrium, then $\langle v^2\rangle$ can mean either the average of v^2 for one particle taken over a long interval of time or the average of v^2 for all particles at a single instant of time. These two interpretations must be equivalent if the overall state of the gas is not to change with time.

---

[b]   It is <u>not</u> $\langle v\rangle$. There is a difference as the following example shows: Four molecules have velocity components v_x of 200 m s^{-1}, - 400 m s^{-1}, 800 m s^{-1} and 100 m s^{-1}. The average value of v_x is: $\langle v_x\rangle = \frac{1}{4}[200 + (-400) + 800 + 100] = 175$ ms^{-1} while the mean-squared v_x is: $\langle v_x^2\rangle = \frac{1}{4}[(200)^2 + (-400)^2 + (800)^2 +(100)^2] = 2.125$ x 10^5 m^{2}s^{-2} and $\sqrt{(\langle v^2\rangle)} = 460.98$ m s^{-1}.

48

3.4 Maxwell-Boltzmann velocity distribution

If the particles interact with each other many times in the time it takes for particles or energy to be replaced, the effects of individual collisions cancel out and the plasma is in equilibrium. Kinetic theory assumes that the particles have different speeds. In the absence of external forces, therefore, any gas or plasma in equilibrium will contain particles of all velocities: there will be a range, or distribution, of speeds. Maxwell initiated the idea of a velocity distribution in the 1860s, and Boltzmann derived the statistical formula some thirty years later.

If we assume that the spatial distribution of the particles is uniform — there are no density gradients — we can represent the distribution of particle velocities graphically by defining a function $f(v)$, known as a *distribution function*. The function $f(v)$, known as the Maxwell-Boltzmann distribution (often shortened to "Maxwellian distribution") is derived from statistical mechanics and describes the most likely distribution that the particle velocities will have. In any system of particles, the number of particles dn with velocities between v and $v + dv$ is given by $dn = n f(v) \, dv$. The expression $f(v) \, dv$ is the probability that a randomly-selected particle will have a velocity in the interval v to $v + dv$. When $f(v)$ is plotted graphically at various values of v, the area under the curve between v and $v + dv$ represents the proportion of particles in the system which have speeds in that range — their position is irrelevant. In Fig. 3.2, the shaded area between v_1 and v_2 represents the number of particles with speeds between v_1 and v_2.

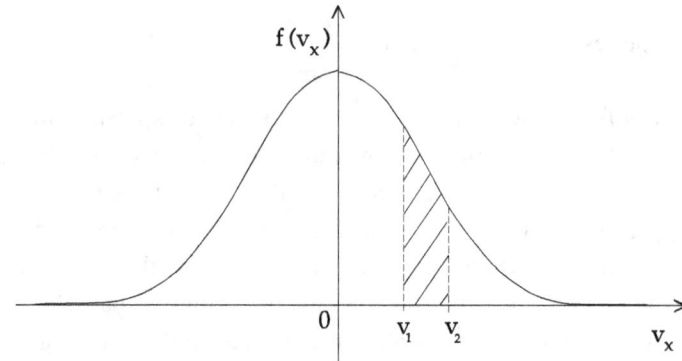

Fig. 3.2   Maxwellian distribution for $f(v_x)$

In one-dimension, for plasma particles which are constrained to move only in (say) the x-direction, the distribution is given by

$$f(v_x) = \frac{dn_x}{dv_x} = n \left(\frac{m}{2\pi k_B T} \right)^{\frac{1}{2}} \exp \left(-\frac{mv_x^2}{2k_B T} \right) \tag{3.8}$$

where k_B is Boltzmann's constant and $f(v_x)$ is the number of particles with velocity v_x. Its form is shown graphically in Fig. 3.2. As a simple example, let us take a plasma with electron density $n_e = 10^{20}\,\text{m}^{-3}$ and a temperature of $1.16 \times 10^7\,\text{K}$. Substituting these values in equation (3.8), the number of electrons with velocity $v_x = 10^7\,\text{m\,s}^{-1}$ is $2.26 \times 10^{12}\,\text{m}^{-3}$, while just one electron would have a velocity of $10^8\,\text{m\,s}^{-1}$. From equation (3.6), the average kinetic energy of the particles is $2.4 \times 10^{-16}\,\text{J}$ and the average velocity is therefore $2.3 \times 10^7\,\text{m\,s}^{-1}$.

In three dimensions, the velocity distribution function becomes a vector function with seven independent variables: $f(r, v, t)$ where r is the position vector and v is the velocity vector. Since the Maxwellian distribution has no preference for any particular direction — it is *isotropic* — this vector function can be simplified by defining a one-dimensional scalar function for the magnitude of v such that

$$\int_0^\infty g(v)\,dv = \int_{-\infty}^\infty f(v)\,d^3 v \qquad (3.9)$$

In effect, we take the three-dimensional form of equation (3.8), with spherical coordinates in velocity space (the volume element is $v^2 \sin\theta\, dv\, d\theta\, d\phi$) and integrate over angles θ and ϕ. So, when direction does not matter, the Maxwell-Boltzmann distribution in three dimensions can be obtained by substituting

$$g(v) = 4\pi v^2 n \left(\frac{m}{2\pi k_B T} \right)^{\frac{3}{2}} \exp\left(-\frac{mv^2}{2k_B T} \right) \qquad (3.10)$$

in equation (3.9). The volume, in velocity space, of each spherical shell containing all the particles with speeds between v and $v + dv$ is $4\pi v^2\, dv$, hence the appearance of $4\pi v^2$ in equation (3.10). The form of this scalar distribution is shown in Fig. 3.3.

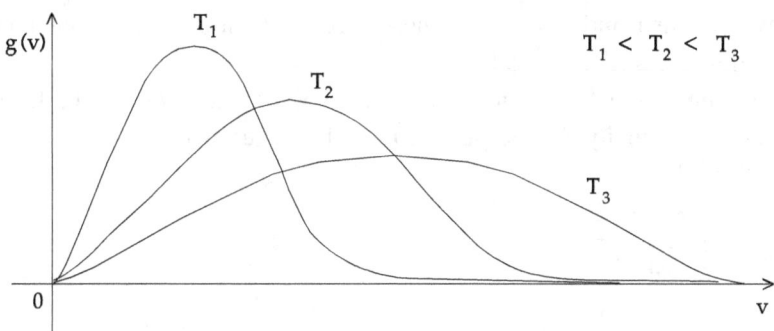

Fig. 3.3   Maxwellian distributions for various temperatures

Using the same values for n, T and v as for equation (3.8), but now considering velocity in three dimensions: the number of particles, $g(v)$, with a velocity of $10^7\,\text{m\,s}^{-1}$ (in our sample plasma) is $2.6 \times 10^{12}\,\text{m}^{-3}$. When $v = 10^8\,\text{m\,s}^{-1}$, $g(v) = 152.5$ particles per cubic metre.

The width of the distribution is determined by the temperature, T. As this increases, so the kinetic energy of the particles, and therefore their velocity, increases. The curve flattens and its maximum moves to higher speeds (Fig. 3.3). The area under the distribution curve remains constant as long as the confining volume stays the same. Taking our sample plasma, with $n_e = 10^{20}$ m^{-3}, and increasing the temperature to 1.16×10^8 K, the number of electrons with a velocity of 10^7 m s^{-1} drops to 10^{11} m^{-3}, while those with a velocity of 10^8 m s^{-1} now number 6.3×10^{11} m^{-3}.

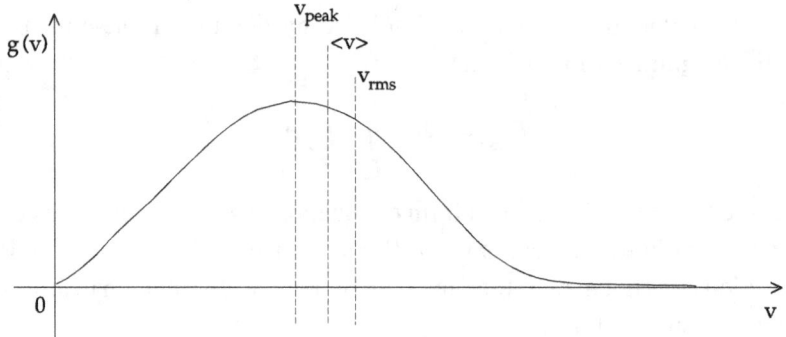

Fig. 3.4   Relationship between velocities for the Maxwellian distribution

As the particle velocity increases, the exponential term in equation (3.10) decreases more rapidly than v^2 increases, producing a maximum in the distribution function. The value of v at which this maximum occurs is the *most probable speed*, known as v_{peak}. Since the distribution curve is rarely symmetrical about this maximum, v_{peak} is less than the average speed, v_{av} or $\langle v \rangle$, and this in turn is less than v_{rms}. Fig. 3.4 shows the relationship between the three on a typical distribution curve. These three velocities can be obtained from the Maxwell-Boltzmann distribution as follows:-

1. v_{peak} is the maximum of the distribution function, the point where $dg/dv = 0$. It represents the most likely thermal velocity for the plasma particles at temperature T. Differentiation of equation (3.10) produces

$$v_{peak} = \left(\frac{2k_B T}{m} \right)^{\frac{1}{2}}$$

(3.11)

2. The average velocity, $\langle v \rangle$, is obtained from

$$\langle v \rangle = \int_0^\infty v\, g(v)\, dv \;=\; \left(\frac{8k_B T}{\pi m} \right)^{\frac{1}{2}}$$

(3.12)

3. The average of v^2 is given by

$$\langle v^2 \rangle = \int_0^\infty v^2\, g(v)\, dv = \left(\frac{3 k_B T}{m} \right) \tag{3.13}$$

This is equation (3.7) obtained by another route.

So:

$$v_{rms} = \left(\frac{3 k_B T}{m} \right)^{\frac{1}{2}} \tag{3.14}$$

This represents the typical speed of a particle since v_{rms} tends to give greater weight to the higher velocities than does $\langle v \rangle$. Comparison of equation (3.12) and (3.14) shows that $\langle v \rangle = 0.92\, v_{rms}$. Most writers use v_{rms} as the thermal velocity; some use v_{peak}. Whilst this may confuse the reader when the algebraic forms are quoted, numerically there is generally little difference: for our plasma with $n_e = 10^{20}\,\mathrm{m^{-3}}$ and $T = 1.16 \times 10^7\,\mathrm{K}$: v_{peak} is $1.875 \times 10^7\,\mathrm{m\,s^{-1}}$; $\langle v \rangle$ is $2.1 \times 10^7\,\mathrm{m\,s^{-1}}$; and v_{rms} is $2.3 \times 10^7\,\mathrm{m\,s^{-1}}$.

3.5   Plasma temperature

The term "temperature" in its strictest sense applies to systems in complete thermodynamic equilibrium, where there are no variations and therefore no temperature gradients. This is the essence of the law of thermal equilibrium: two systems in thermal contact will tend towards a common temperature, where there is no net flow of heat between them.

For a collection of particles, such as a plasma, to be in equilibrium all processes which can exchange energy must be exactly balanced by their reverse process. The physical properties of the plasma will then remain constant over time scales which are significant to the behaviour of the system. In reality this is seldom the case. There are often temperature gradients, preventing a true temperature from being assigned. However, if we define temperature in terms of a Maxwellian distribution and equate it to average energy, temperature becomes a property of the energy distribution. Since this can vary in space, the temperature distribution can also vary in space; so concepts such as temperature gradients are acceptable.

If a system of radiation and particles is totally enclosed within fully reflecting walls, nothing can escape and the system is in *complete thermodynamic equilibrium*. In most plasmas, radiation can escape but the particles are constrained by magnetic or gravitational fields. These radiation energy losses mean that the plasma is not in complete thermodynamic equilibrium. Nevertheless, conditions in parts of the plasma are often close enough to equilibrium to be described by local Maxwellian distributions. It is generally safe, for example, to assume a Maxwellian distribution for free electrons in a high-temperature plasma. If these distributions do not change too rapidly, the plasma is said to be in *local thermal equilibrium* (*LTE*). For *LTE* to exist, all gradients of temperature, density, etc., must be small and collisional processes must be more important than radiative processes, to minimise the effects of energy losses through radiation. In reality, ionisation is not balanced locally by recombination and there are variations in conditions throughout the plasma, but *LTE* can still apply if the time interval between particle collisions is short enough to prevent the plasma from changing too much. In many laboratory plasmas particle densities are too low even for these conditions to be met. The concept of temperature becomes meaningless in very low density plasmas which are not in thermal equilibrium.

3.5.1   Temperature as an energy

It is generally more convenient to express plasma temperature in terms of the mean kinetic energy of the ions and electrons, rather than in degrees Kelvin. The energy unit used is the *electronvolt, (eV)*, defined as the kinetic energy gained by an electron from a potential difference of one volt. It indicates the potential needed to produce a charged particle with a particular energy, for example a 1 keV electron. Although it is convenient , the electronvolt is not an SI unit and temperature in electronvolts must be converted to temperature in joules in any calculations. An electron with charge 1.602×10^{-19} C, accelerated through a potential difference of 1 volt, acquires an energy of 1.602×10^{-19} J. To convert temperature in eV to temperature in joules we therefore multiply by the charge on the electron[c]: a temperature of 1 keV becomes 1.602×10^{-16} J.

In the numerical examples in Section 3.4, temperature was expressed in kelvin. Boltzmann's constant, $k_B = 1.38066 \times 10^{-23}$ J K^{-1}, converted this to units of energy, measured in joules. Our sample plasma, with $T = 1.16 \times 10^7$ K, has a temperature of $k_B T = (1.38066 \times 10^{-23}$ J K$^{-1})(1.16 \times 10^7$ K$) = 1$ keV. To state the obvious: care must be taken in any calculation to ensure that the correct units are used. Some writers omit the Boltzmann constant from their equations, which can cause problems for the unwary.

At room temperature, T is around 300 K and $k_B T$ is approximately $\frac{1}{40}$ eV. The electron temperature in a fluorescent tube is about 10^4 K (approximately 1 eV) but with electron densities in the region of 10^6 m^{-3} (compared to 10^{25} m^{-3} for air) very little heat is transferred to the wall of the tube. A plasma is described as *low-temperature* or *cold* if the temperature is less than about 10 eV.

Ions and electrons in a plasma often have different temperatures and therefore different Maxwellian velocity distributions. This can be caused by infrequent collisions, or by there being more ion-ion or electron-electron collisions than between electrons and ions, so the thermal velocities of the two species cannot equalise. In this case, *electron*

---

[c]   To convert temperature in eV to temperature in kelvin, multiply by e/k_B. To convert to joules for use in equations: if the temperature (written T or $k_B T$) is given in eV, multiply by the charge on the electron, e (C); if temperature is given in kelvin, multiply by Boltzmann's constant, k_B.

temperature, T_e, and *ion temperature, T_i,* are used. A plasma with different electron and ion temperatures is often regarded as a mixture of two gases: an electron gas and an ion gas. Each can exist in thermal equilibrium, with its own Maxwellian velocity distribution, but the two gases are not in equilibrium with each other. If the plasma survives long enough for the two temperatures to equalise, ions and electrons will have the same equilibrium temperature, with their mean (rms) velocities related inversely as the square roots of their masses. For a hydrogen plasma of free positive ions and electrons:

$$\sqrt{\frac{m_p}{m_e}} = \frac{v_e}{v_p} \approx 43 \tag{3.15}$$

In other words, the velocity of the electrons is about 43 times greater than that of the positive ions at the same temperature. The time taken to establish thermal equilibrium between the types of particles is known as the *relaxation time*. It is generally shorter for high-density plasmas.

3.6   Population distributions — The Boltzmann factor

The statistical distribution of large numbers, or populations, of microscopic particles, subjected to thermal agitation and acted on by electric, magnetic or gravitational fields can be determined in a similar way to the velocity distribution. To find the number density of any system of particles in equilibrium, we use an equation of the form

$$n = n_0 \exp\left(-\frac{E}{k_B T}\right) \tag{3.16}$$

where E is the potential energy of a particle in a given region and n_0 is the number of particles per unit volume when $E = 0$. Number density clearly changes exponentially with local changes in particle energy. Equation (3.16) is known as *Boltzmann's distribution* and the exponential term, *$exp(-E/k_B T)$*, is the *Boltzmann factor*. It indicates the relative probability of a particle, *i*, having energy E_i. Substituting $E = \frac{1}{2}mv^2$ gives the factor *$exp(-mv^2/(2k_B T))$* in equations (3.8) and (3.10), which represents the probability of a particle having velocity *v*.

In any system of particles in thermodynamic equilibrium at temperature T, the ratio of the number of atoms of a particular element occupying the two energy levels or states E_2 and E_1 is given by the following form of equation (3.16):

$$\frac{n_2}{n_1} = \frac{g_2}{g_1} \exp\left(\frac{-(E_2 - E_1)}{k_B T}\right)$$

where n is the number density of the respective energy states and g is the statistical weight or *degeneracy*. In quantum theory, the energy of a system is divided into discrete units, or quanta. For any particular energy level of a system with a number of quantum states, g

indicates the degeneracy of the system — the number of quantum states having the same energy.

To relate the population n_r in any given energy level, r, to the population of n atoms of a particular element we use

$$n_r = \frac{n g_r}{P(T)} \exp\left(\frac{-E_{1r}}{k_B T} \right) \tag{3.17}$$

where E_{1r} is the difference in energy between the ground level and the rth level, and g_r is again the degeneracy or weight factor of the rth level. The factor $P(T)$ is a sum-over-states or *partition function* and is a function of temperature. It is based on the idea that the total internal energy of an atom or molecule can be partitioned, or split, into separate components of motion: translational (i.e. movement of electrons within the atom), vibrational and rotational. The need to consider entropy is thus avoided. The physically-possible energy states of a collection of particles form a (generally infinite) series of discrete states for a quantized system — for a classical system the available space must be divided into cells whose dimensions tend to zero. Associated with each state, or cell, there is therefore a definite energy and to each is attached a weight. The partition function $P(T)$ in equation (3.17) is the sum of the weighted Boltzmann distribution functions of all the discrete energy levels within the atom — it tells us how the system is divided up amongst the different energy levels in an equilibrium distribution. Thus:

$$P(T) = g_0 + g_1 \exp\left(\frac{-E_1}{k_B T} \right) + g_2 \exp\left(\frac{-E_2}{k_B T} \right) + \ldots = \sum_{j=0}^{\infty} g_j \exp\left(\frac{-E_j}{k_B T} \right) \tag{3.18}$$

In many atoms the first excited state is several eV above the ground state, so only the first term, g_0, is important. Most simple atoms and ions have $g_0 \approx 1$ and therefore $P(T)$ is approximately equal to 1.

At low temperatures, the number of atoms in the higher excited states is very small. As the temperature increases, more and more atoms appear in the highly excited states and ionisation begins to occur. Equation (3.17) must therefore be extended to take account of ionisation of the atom.

3.6.1   The Saha equation[d]

A simple monatomic gas, in which there is some ionisation due to thermal motion, will consist of three groups of particles: neutral atoms, positive ions and electrons. The number of particles in each group depends on the temperature of the gas; the difficulty lies in calculating the populations in each group. The Indian astrophysicist, Megh Nad Saha

---

[d]   The Saha function has been tabulated by, for example, H.W. Drawin & P. Felenbok in *Data for Plasma in Local Thermal Equilibrium*, Gauthier-Villars, Paris (1965)

(1893-1956) first applied the Boltzmann distribution to ions and neutral atoms in conditions of thermodynamic equilibrium while working at Calcutta University in 1920. The Saha equation, or Saha-Boltzmann distribution, is now used to calculate the population distribution of charged particles and estimate the relative populations of excited and ionised states of the ions in a hot plasma in thermodynamic equilibrium at a temperature T.

Saha developed his equation while studying the spectrum of the Sun's high-level chromosphere. Several of the spectral lines, he suggested, were produced not by radiation from a normal atom of the element, but from an ionised atom — one which had lost an electron. The high-level chromosphere was therefore a region of intense ionisation. He regarded ionisation as a reversible chemical reaction:

atom + ionisation energy ⇌ positive ion + electron

This view ignores excited states of the atom and only applies to a monatomic gas undergoing single ionisation. Nevertheless, it provided a useful starting point. Between 1920 and 1923, Saha developed a formula to calculate the degree of ionisation in terms of absolute temperature, T, total pressure p, and the fraction of atoms ionised, x:

$$\frac{x^2}{1-x^2}p = k_B T \left(\frac{2\pi m_e k_B T}{h^2} \right)^{\frac{3}{2}} \exp\left(-\frac{U}{k_B T} \right)$$

where U is the ionisation energy. Contributions from R.H. Fowler, E.A. Milne and others throughout the 1920s added further refinements.[3] The original formula had been derived on the assumption that the hydrogen atom is either ionised or in its lowest quantum state, the ground state. There were thus equal concentrations of electrons and ions in the ionised gas. As Langmuir and Kingdon pointed out in 1925: in the laboratory, the walls of the enclosing vessel will give off electrons, so there is not necessarily any relation between the ion and electron concentrations. What Saha's equation really gave, as then written, was the equilibrium constant, $n_e n/n_n$. When applied to other elements, with different atomic structures, ionisation potentials, etc., the normal states of the ionised and neutral atom required statistical weighting.

In its generalised form, the Saha equation, as now constructed, relates the densities, n_z and n_{z+1}, of two adjacent stages of ionisation via the formula

$$\frac{n_{z+1}}{n_z} = \frac{2\,P_{z+1}(T)}{n_e\,P_z(T)} \left(\frac{2\pi m_e k_B T}{h^2} \right)^{\frac{3}{2}} \exp\left(-\frac{E_z}{k_B T} \right) \tag{3.19}$$

where n_e is the total number of electrons, not just those resulting from the zth ionisation; h is Planck's constant; E_z is the energy required to ionise the atom from the zth to the $z+1$ stage of ionisation, (the ionisation energy); and $k_B T$ is measured in eV. The factor 2 and Planck's constant both appear because of quantum mechanical effects. The partition function, $P(T)$, as we saw in the previous section, is often approximately equal to 1, so the factor $P_{z+1}(T)/P_z(T)$ can also generally be taken as approximately equal to 1. The next term in brackets is essentially $1/\lambda_{DB}^3$ where λ_{DB} is the de Broglie wavelength[e], the wavelength associated with the momentum of moving particles. For the majority of plasmas, electron density is very much less than this term, and when E_z is approximately equal to k_BT the final term becomes e^{-1} and the degree of ionisation, n_{z+1}/n_z is very high.

Equation (3.19) applies to the ideal case of a homogeneous system of particles in thermodynamic equilibrium and can only provide an approximate distribution for actual ionisation processes. A plasma composed of ions up to charge Z will require Z Saha equations together with the equation of charge neutrality:

$$n_e = \sum_{Z=0}^{Z} Z n_e$$

to describe it. By substituting n_i for n_{z+1} and n_n for n_z we can obtain the number density ratios between ions, electrons and neutral atoms in various excited states for hydrogen or hydrogen-like plasmas containing singly-ionised atoms:

$$\frac{n_i n_e}{n_n} = \frac{2 \, P_i(T)}{P_n(T)} \left(\frac{2\pi m_e k_B T}{h^2} \right)^{\frac{3}{2}} \exp\left(-\frac{E}{k_B T} \right) \tag{3.19a}$$

In this version, excited states of both ions and atoms can be accounted for in P_i and P_n.

Putting some numbers into this equation illustrates the information it can provide. If we start at a low temperature with a total particle density, n_{tot}, of 10^{20} m^{-3} atoms of (say) hydrogen, then, as the temperature rises, some atoms will lose an electron and become ionised. The total number of particles is thus: $n_n + n_i + n_e$. For hydrogen, the ionisation energy, E, is 13.6 eV. We will set $P_i(T)/P_n(T)$ equal to 1.

When $T = 0.5$ eV (so $k_B T = 0.5 \times 1.602 \times 10^{-19}$ J):

$$\frac{n_i n_e}{n_n} = 2 \times 1 \left(\frac{2\pi \times (9.1095 \times 10^{-31}) \times (0.5 \times 1.602 \times 10^{-19})}{(6.626 \times 10^{-34})^2} \right)^{\frac{3}{2}} \exp\left(\frac{-13.6}{0.5} \right)$$

$$= 3.28 \times 10^{15}$$

(Notice that the Boltzmann factor has E in eV so $k_B T$ is left in eV.) If the atoms are singly-ionised, $n_i = n_e$. We can also write n_n, the number of neutral atoms left after ionisation has occurred, as $n_{tot} - n_i = 10^{20} - n_i$.

Therefore:

$$\frac{n_i n_e}{n_n} = \frac{n_i^2}{10^{20} - n_i} = 3.28 \times 10^{15}$$

---

[e] In 1922 Louis de Broglie showed that, if a particle can have wave-like properties, then it must have a wavelength. To satisfy the demands of special relativity, the wavelength is related to particle momentum and is given by $\lambda_{DB} = h/mv = h/(2m_e k_B T)^{1/2}$ by substituting equation (3.11) for v.

From the resulting quadratic equation we find that $n_i = 5.7 \times 10^{17}$ m^{-3} and the percentage ionisation is 0.57%. At $T = 0.75$ eV, $n_i n_e/n_n = 5.22 \times 10^{19}$, $n_i = 5 \times 10^{19}$ m^{-3}, or 50.74% of the total.

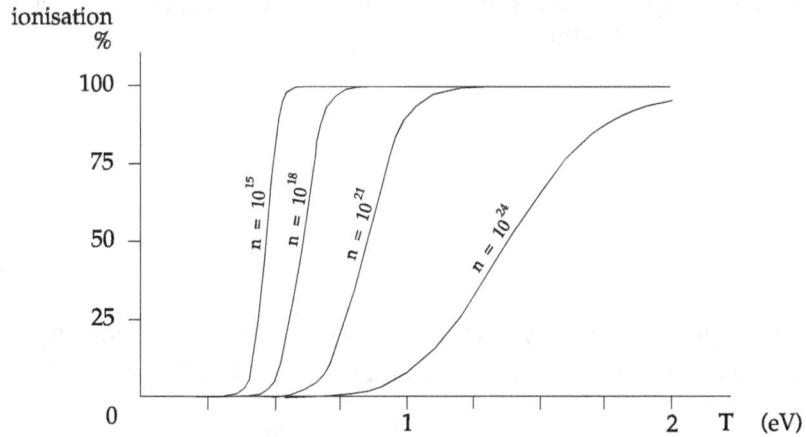

Fig. 3.5   The effect of increasing temperature and density on hydrogen ionisation, as given by the Saha equation

Fig. 3.5 shows how the percentage ionisation of hydrogen increases with temperature and density. The sudden increase follows from the changing shape of the velocity distribution as the temperature increases (see Fig. 3.3). In a cold atomic gas there are very few particles with sufficient thermal energy to produce ionisation by collisions with other particles. As the temperature of the gas rises, the thermal velocity of the particles increases as does the ionisation rate, but as density increases, ionising collisions are reduced.

3.7   Summary

1.   In an ideal gas, molecules are of negligible size and there are no intermolecular forces. Such a gas is theoretical but the behaviour of real gases and plasmas approaches that of an ideal gas as the pressure is reduced. Kinetic theory describes the physical properties of matter in terms of the behaviour or movement of its constituent particles. Particle motion is assumed to be random, with no preference for any particular direction.

2.   The Maxwell-Boltzmann distribution (equation (3.10)) describes the most likely distribution of velocities among the particles of a perfect gas in thermodynamic equilibrium, assuming that the particles have no preference for any particular position in the assembly. Temperature is defined in terms of the Maxwell-Boltzmann distribution and is equal to the average kinetic energy of the particles. Ions and electrons may have different temperatures and therefore different velocity distributions. A temperature of 1 eV = 1.16×10^4 K = 1.602 $\times 10^{-19}$ J.

3.   If LTE exists, the populations of different types of particles can be calculated using

Boltzmann's distribution applied to excitation energies and as many Saha equations as are required for the various ions. Since Boltzmann's distribution and Saha's equation are both statistically derived, as is the Maxwell-Boltzmann velocity distribution on which the idea of LTE is based, they cannot provide information about the detailed processes through which equilibrium is achieved and maintained.

References

1.　　Maxwell, J.C.: *Phil. Mag.* **19**, p.19-32 (1860); *Phil. Mag.* **20**, p.21-37 (1860); *Phil. Trans.* **157**, p.49-88 (1866); *Phil. Mag.* **35**, p.129-47, 185-217 (1868)

2.　　Clausius, R.: *Phil. Mag.* **23**, p.417-35 (1862)

3.　　Saha, M.N.: *Phil. Mag.* **40**, p.472-88 (1920); *Phil. Mag.* **41**, p.267-78 (1921); *Phil. Mag.*, **46**, p.534-43 (1923)
　　　Fowler, R.H.: *Phil. Mag.*, **45**, p.1-33 (1923)
　　　Milne, E.A.: *Phil. Mag.*, **50**, p.547-50 (1925)
　　　Langmuir, I, & K.H. Kingdon: *Roy. Soc. Proc.*, **107**, p.61-79 (1925)
　　　(see also: Cooper, J.: *Rep. Prog. Phys.*, **29**, p.35-130 (1966))

Chapter 4

Motion of charged particles in electric and magnetic fields

4.1 Introduction

Plasma can be described as an electrically-neutral mixture of charged and neutral particles which sometimes behave collectively like a fluid and sometimes as a system of individual particles. Large-scale (macroscopic) electric and magnetic fields frequently extend throughout the plasma. Together with the microscopic fields produced by the individual ions and electrons, these fields influence the movement of each charged particle. Neutral particles are unaffected.

To understand plasma behaviour, we need to know how the charged particles behave, both individually and collectively, in these electric and magnetic fields. Because of the large number of particles involved, a statistical approach is used to predict the behaviour and properties of the system as a whole from the motion of the constituent particles. We will start by considering the forces acting on charged particles in the ideal cases of constant uniform electric and magnetic fields. In reality, there are variations in field strength in both space and time, but these fluctuations are generally large-scale and slow compared to the motions of the individual particles and can, initially, be ignored. For simplicity, the effects of radiation are also ignored.

4.2 The equation of motion: The Lorentz force

A distribution of electric charge at rest creates an electric field, E, in the surrounding space, which exerts a force

$$F_{el} = qE \tag{4.1}$$

on any other charge q in the field. A moving charge, or a current, produces a magnetic field, B, in the surrounding space in addition to its electric field. This magnetic field (the magnetic flux density) exerts a force F_m on any other moving charge present — stationary charged particles experience no magnetic force. The magnitude of F_m is proportional to the component of the particle's velocity v perpendicular to the magnetic field and its direction is always perpendicular to both v and B:-

$$F_m = qv \times B \tag{4.2}$$

Combining these two equations gives the total electromagnetic force acting on a particle of mass m and charge q moving with velocity v through a point (x,y,z) at which there is both an electric field E (x,y,z) and a magnetic field B (x,y,z):

$$F = F_{el} + F_m = q(E + v \times B)$$

or $$m\frac{dv}{dt} = q(E + v \times B) \qquad (4.3)$$

This is the equation of motion for all charged particles. It defines electric and magnetic fields and predicts the forces on any particle moving within an electromagnetic field. It is sometimes referred to as the Lorentz force, after the Dutch physicist Hendrick Antoon Lorentz (1853-1928) who, in 1895, produced the equation of motion for charged particles in a magnetic field, first proposed by Oliver Heaviside in 1889.[1]

4.3   Maxwell's equations

Electric and magnetic fields can be investigated separately, and any interactions between them ignored, if they remain constant in time; for example, electric fields produced by charges at rest, or the magnetic field produced by a steady electric current. However, if either field varies with time, they must be considered together, since a time-varying magnetic or electric field induces a field of the other sort in adjacent regions of space. This interaction of electric and magnetic fields produces an electromagnetic disturbance, a combination of time-varying electric and magnetic fields, which can travel with a definite speed through space, including a vacuum. Unlike mechanical waves, an electromagnetic wave does not need the oscillating particles of a medium, such as air, to enable it to travel — what oscillates is the field itself.

In 1864, Maxwell demonstrated mathematically that electricity and magnetism are aspects of a single force — electromagnetism — and that light is a form of electromagnetic radiation, confirming Faraday's 1845 experiment. Maxwell's equations showed that only transverse disturbances, such as light, will propagate through the electromagnetic field. His ideas proved difficult to understand and acceptance was slow. It was not until 1888, when Hertz discovered electromagnetic (radio) waves and showed that they behave like light waves and have the same velocity, that Maxwell's theory was finally confirmed.

The four equations known as "Maxwell's equations" summarise the classical theory of electric and magnetic fields and describe mathematically the behaviour of these fields and the charges and currents which produce them.[a] Maxwell's genius was to recognise the significance of other people's discoveries and combine their results to predict the existence of electromagnetic waves. The set of equations generally used in plasma physics is that applicable in a vacuum. If particle densities are significant, the permittivity, ε_0, and permeability, μ_0, of free space are replaced by the specific values ε and μ for the plasma.

---

[a]   Maxwell's published work on electromagnetism does not contain the four equations as now written, due to refinements by later researchers.

4.3.1   Gauss' law for electric fields

$$\nabla \cdot \boldsymbol{E} = \frac{\rho}{\varepsilon_0} \tag{4.4}$$

where ρ is the charge per unit volume — the *electric charge density, $e(n_i - n_e)$*. It relates the electric field $\boldsymbol{E}$ to the electric charge density ρ in that the flux of $\boldsymbol{E}$ through any closed surface is proportional to the charge enclosed within the surface. Equation (4.4) basically states that electric fields can be created by distributions of electric charge.

4.3.2   Faraday's law of electromagnetic induction

$$\nabla \times \boldsymbol{E} = -\frac{\partial \boldsymbol{B}}{\partial t} \tag{4.5}$$

This is the general law for the electric field associated with a changing magnetic field. It states that a time-varying magnetic field will induce an electric field unlike the electrostatic field produced by a static charge distribution in that it is non-conservative.[b] The concept of potential therefore has no meaning.

4.3.3   Gauss' law for magnetic fields

$$\nabla \cdot \boldsymbol{B} = 0 \tag{4.6}$$

This is the magnetic equivalent of equation (4.4). It means, in effect, that magnetic field lines are continuous: there are no magnetic monopoles (single isolated magnetic "charges" or poles) to act as sources of magnetic field. However much they wander through space, all magnetic field lines must eventually close on themselves.

4.3.4   Ampère's law with displacement current

$$\nabla \times \boldsymbol{B} = \mu_0 \left(\boldsymbol{J} + \varepsilon_0 \frac{\partial \boldsymbol{E}}{\partial t} \right) \tag{4.7}$$

A changing electric field acts as a source of magnetic field. Equation (4.7) relates the magnetic field to its sources: steady electric currents ($\boldsymbol{J}$) and time-varying electric fields. Both conduction current and displacement current[c], $\varepsilon_0 \, \partial \boldsymbol{E}/\partial t$, act as sources of magnetic field. $\boldsymbol{J}$ is generally much greater than $\partial \boldsymbol{E}/\partial t$, and if the timescale allows collisions between the

---

[b]   In a conservative field, the work done in moving a particle from point A to point B is independent of the path followed.

[c]   A fictitious current invented by Maxwell, which nevertheless exists in dielectrics, such as in a parallel-plate capacitor which is being charged.

62

particles, this will be longer than the time it takes light to travel through the plasma. Speeds of interest will therefore be much less than the speed of light. For both reasons, the displacement current can be neglected, and Ampère's equation, for plasma, becomes:

$$\nabla \times \boldsymbol{B} = \mu_0 \boldsymbol{J} \tag{4.8}$$

Together with equation (4.3), these four equations provide a full description of the behaviour of charged particles in electric and magnetic fields. One limitation of Maxwell's equations is that they are based on classical physics and are not, as they stand, compatible with quantum theory. Only in special cases does quantum mechanics influence the physics of plasmas, so this limitation need not concern us here.

4.4   Motion in a uniform electric field

In the absence of a magnetic field, equation (4.3) reduces to

$$m\frac{dv}{dt} = q\boldsymbol{E} \tag{4.9}$$

which is equation (4.1). The magnitude of the electric force is independent of both the position and the velocity of the charged particle. This constant force gives the particle a constant acceleration

$$\boldsymbol{a} = \frac{dv}{dt} = \frac{q\boldsymbol{E}}{m} \tag{4.10}$$

in the direction of the electric field for a positive charge; and in the opposite direction for an electron, since $q = -e$.

The velocity of the particle can be resolved into two vectors perpendicular ($v_\perp$) and parallel ($v_\parallel$) to the electric field, so $v = v_\perp + v_\parallel$. The electric field has no effect on $v_\perp$, which remains constant in magnitude and direction throughout the motion of the particle. It is $v_\parallel$ which experiences the acceleration. Fig. 4.1 shows a charged particle entering a uniform electric field between parallel plates, at right angles to the field. The combination of its initial velocity perpendicular to the field and the constant acceleration parallel to the field experienced by $v_\parallel$ produces a parabolic trajectory. J.J. Thomson used this idea, together with a magnetic field, to obtain a value for e/m in 1897.

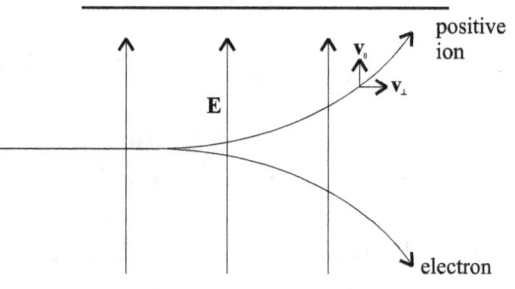

Fig. 4.1   Parabolic path of charged particles moving in a uniform electric field

4.4.1   Drift velocity

If an electric field is applied to a mixture of electrons and positive ions, the more massive ions remain almost stationary and can be treated as if they are at rest. The effect on the electrons is also small since they lose momentum in collisions with other particles. Superimposing the velocity from the field on the random thermal motion of the electrons produces an electron *drift velocity* given by:

$$v_e = \frac{-eE}{m_e \upsilon} \qquad (4.11)$$

where υ is the collision frequency (see Chapter 5). This is not a current in which all particles have velocities in the same direction, but simply a slow drift of charged particles parallel to the field.

4.4.2   The influence of ionisation

When an electric field is applied to a partially-ionised plasma, short-range encounters predominate. Electrons accelerated by the field collide with and ionise slow-moving ions and neutral atoms, releasing more free electrons. These in turn cause further ionisation, producing an *electron avalanche* within the plasma. The process is important in understanding electrical discharges and atmospheric plasmas such as lightning.

In a fully-ionised plasma, fast-moving electrons can acquire energy (and velocity) from an applied electric field quicker than they lose it in collisions. They may gain so much energy that a collision with an ion becomes only a glancing encounter, and the frictional force is too small to influence the electron's velocity. An unstable "*runaway*" situation develops in which the electrons may never make a collision, but form an accelerated beam detached from the main body of the velocity distribution. Such electrons can quickly reach relativistic speeds, causing impact damage to the surface of any containment vessel, as well as producing a large amount of X-rays. Electron runaway becomes significant when the ratio of the mean drift velocity of the plasma to the thermal velocity of the electrons is greater than 0.1.

4.5   Motion in a uniform magnetic field

In the absence of an electric field, $E = 0$ and equation (4.3) becomes:

$$m\frac{dv}{dt} = q(v \times B) \qquad (4.12)$$

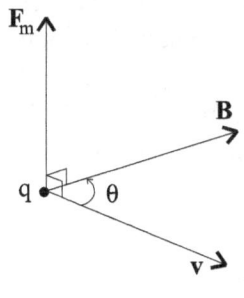

Fig. 4.2 F_m (q positive)

The magnitude of the magnetic force acting on the moving charge q is given by $F_m = |q| \, v B \sin \theta$ (Fig. 4.2). The factor $\sin \theta$ indicates that the magnetic force is proportional to the velocity component perpendicular to B; therefore $v \sin \theta = v_\perp$. The component of v parallel to B makes no contribution to the force on the charge since the angle between them is zero. As a result, $v_\parallel$ remains constant in magnitude and direction throughout the motion of the particle, and $qv_\parallel \times B$ is zero.

The acceleration, $q(v \times B)/m$, acting on the charge q in equation (4.12) is perpendicular to both v and B. Its direction depends on the sign of q. Figure 4.3 shows the situation when the initial velocity is perpendicular to the magnetic field, so $v = v_\perp$ and $v_\parallel = 0$. The acceleration changes the direction of the particle's motion without altering its speed and remains perpendicular to v as v changes direction. This produces a circular motion of the particle, at a constant speed $v_\perp$ in the plane perpendicular to B known

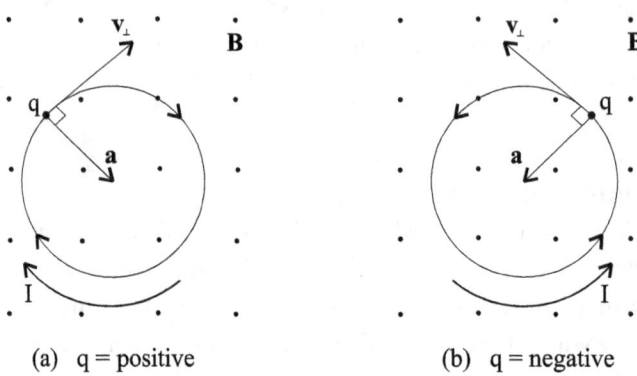

(a) q = positive   (b) q = negative

Fig. 4.3   Circular motion of charged particles in a uniform magnetic field (directed out of page)

as the *Larmor orbit*.[d] It constitutes a circulating electric current, shown as I.

The magnitude of the centripetal acceleration needed to keep a particle with speed $v_\perp$ in a circular orbit of radius R_g is $v_\perp^2/R_g$. The force maintaining this acceleration is the magnetic part of the Lorentz force ($F_m = q\,v_\perp B$), so for a particle of mass m:

$$a = \frac{F}{m} = \frac{|q|\,v_\perp B}{m} = \frac{v_\perp^{\,2}}{R_g} \tag{4.13}$$

Rearranging equation (4.13) gives the radius of the Larmor orbit:

$$R_g = \frac{mv_\perp}{|q|B} \quad (m) \tag{4.14}$$

There is usually a component of particle velocity parallel to the magnetic field, so the path of the charged particle is a combination of circular motion perpendicular to the field and a steady movement parallel to it: a helix (Fig. 4.4). The radius, R_g, of the helix is variously referred to as the *Larmor radius, cyclotron radius, gyro-radius* or *radius of gyration*. The axis of the helix, known as the *guiding centre*, is parallel to the direction of the magnetic field.

The magnitude of the particle's angular velocity of motion, ω_c, is given by

---

[d]   Joseph Larmor (1857-1942) was a major contributor to charged particle theory.

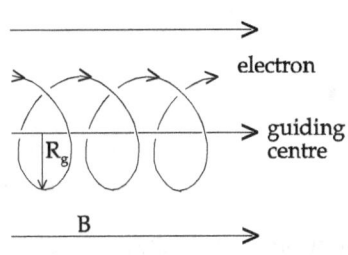

Fig. 4.4 Helical motion of electron in a uniform magnetic field

$$\omega_c = \frac{v_\perp}{R_g} \qquad (4.15)$$

Combining this with equation (4.14) gives

$$\omega_c = \frac{|q|B}{m} \quad (rad\ s^{-1}) \qquad (4.16)$$

This is the frequency at which a charged particle orbits in a uniform magnetic field. Like the orbital radius, ω_c has various names: *Larmor frequency, cyclotron frequency, gyro-frequency* or *angular frequency of rotation.*

In a weak magnetic field the Larmor radius, R_g, is large because the force deflecting the charged particle from its straight-line path is small. As the magnetic field strength increases, ω_c increases and R_g decreases. The smaller the particle mass, the greater is the frequency and the smaller the Larmor radius becomes. While ω_c is independent of particle energy — there is no factor v in equation (4.16) — the radius of the Larmor orbit increases with particle energy (from equation 4.14). So, the more energetic the particle (large $v_\perp$), the less likely it is to be deflected by the magnetic field.

The Larmor orbit determines both a fundamental length: the Larmor radius (equation 4.14), and, via the angular frequency, a fundamental time: the *Larmor period* (also known as the *cyclotron period* or *gyro-period*):

$$\tau = \frac{2\pi}{\omega_c} \quad (s) \qquad (4.17)$$

This is the time taken for the particle to complete one orbit and is independent of the radius of the orbit.

4.5.1 Magnetic moment, M

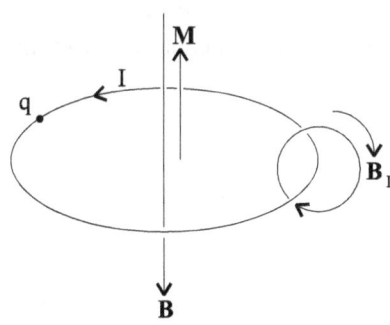

Fig. 4.5 Magnetic moment associated with an orbiting charged particle

The circulating electric current, I, shown in Fig. 4.3 produces a microscopic magnetic field which is in the opposite direction to the externally-applied magnetic field inside the orbit of the particle, but in the same direction outside (Fig. 4.5). The circular motion of the plasma particles therefore slightly reduces the external macroscopic magnetic field: plasmas are very weakly *diamagnetic.*

The magnetic moment, *M*, associated with the circulating current is perpendicular to the area enclosed by the orbiting particle and opposite in direction to the externally-applied magnetic field. Its magnitude is the product of the current produced by the orbiting particle ($I = q/\tau$), and the area enclosed by the orbit (πR_g^2). Thus

$$M = IA = \frac{q}{\tau}\pi R_g^2 \quad (Cm^2 s^{-1}) \tag{4.18}$$

Substituting equations (4.14), (4.16) and (4.17) gives

$$M = \frac{mv_\perp^2}{2B} = \frac{E_{kin\perp}}{|B|} \tag{4.19}$$

where $E_{kin\perp}$ is the kinetic energy associated with the circular motion. The diamagnetic effect thus varies with the perpendicular kinetic energy of the particles, which results from the perpendicular motion, $v_\perp$. The vector form of equation (4.19) is

$$M = -\left(\frac{mv_\perp^2}{2B^2}\right)B$$

the minus sign indicating that the magnetic moment is in the opposite direction to the externally-applied magnetic field.

To get some idea of the magnitudes of these magnetic-field-dependent parameters, imagine that our hypothetical plasma, with a particle density of 10^{20} m^{-3} and a temperature of 1 keV, is immersed in a magnetic field of strength $B = 0.5$ Tesla. If the thermal velocity of the plasma particles (from section 3.4, $v_{rms} = 2.3$ x 10^7 m s^{-1}) is at 45° to the direction of the field, then $v_\perp = v \sin 45° = 1.63$ x 10^7 m s^{-1}. The magnetic force F_m acting on each charge, from equation (4.12), is 1.3 x 10^{-12} N. The electron cyclotron frequency, ω_{ce}, from equation (4.16) is 8.8 x 10^{10} rad s^{-1}; the radius of the electron's orbit, R_g, is 1.85 x 10^{-4} m; while the time taken for the electron to complete one orbit, the Larmor period, is 7.1 x 10^{-11} s. From equation (4.18), the magnetic moment is 2.4 x 10^{-16} C m^2 s^{-1} and the current produced by the electron in one orbit is 2.25 x 10^{-9} A. Using the equation for the magnitude of the magnetic field around a long straight wire ($B = \mu_0 I/2\pi r$) reveals just how weak the diamagnetic effect is: $B_I = 4.5$ x $10^{-16} r^{-1}$ Tesla.

4.6   Motion in combined electric and magnetic fields

The combined effects of electric and magnetic fields which are uniform in space and constant in time produces a drift of particles across the fields. The full equation of motion was given in equation (4.3) as

$$m\frac{dv}{dt} = q(E + v \times B)$$

Resolving the velocity and electric field vectors into their components parallel and perpendicular to the magnetic field, we obtain two equations:

(i) $$m\frac{dv_\parallel}{dt} = qE_\parallel \quad (v_\parallel \times B = 0) \tag{4.20}$$

where $E_\parallel$ is the component of the electric field parallel to the magnetic field. This describes a constant acceleration $q\,E_{\parallel}/m$ along the magnetic field. Integration yields

$$v_{\parallel}(t) \; = \; \frac{q}{m}E_{\parallel}(t) \; + \; v_{\parallel}(0) \tag{4.21}$$

where $v_{\parallel}(0)$ represents the initial velocity parallel to B at $t = 0$.

(ii) $\qquad m\dfrac{dv_{\perp}}{dt} \; = \; q(E_{\perp} + v_{\perp} \times B) \tag{4.22}$

$E_{\perp}$ is the component of the electric field perpendicular to the magnetic field and

$$v_{\perp} = \omega_c \times R_g \tag{4.23}$$

from equation (4.15).

    In the absence of an electric field, $q\,v_{\perp} \times B$ describes circular motion around the magnetic field lines and guiding centre motion parallel to B. Introducing the factor $E_{\perp}$ into the equation affects plasma motion and behaviour, in particular the particle velocity.

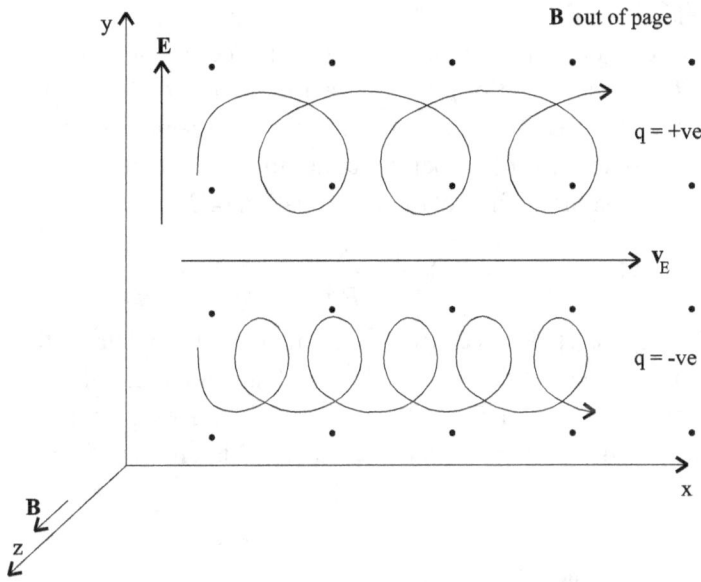

Fig. 4.6 $E \times B$ **drift**

    Fig. 4.6 shows the simple case of mutually perpendicular (crossed) E and B fields (with E in the y-direction and B in the z-direction, so $E = E_{\perp}$). If the initial velocity of the charged particle is parallel to the electric field, then, as the particle is accelerated and gains velocity in the y-direction from the electric field ($qE_{\perp}$), it will feel a magnetic force in the direction $E \times B$ (since v is parallel to E) and will move in the x-direction. The particle's orbital radius therefore increases on the side of the orbit where the electric field accelerates the particle and raises its kinetic energy. On the side where the energy is reduced, the particle moves with a smaller radius of curvature.

The uniform circular motion produced by the magnetic field is thus distorted by a uniform translational velocity in the plane perpendicular to B (the x-y plane in Fig. 4.6). The direction in which the particle gyrates is determined by the sign of its charge; the radius of its orbit is proportional to $v_\perp$. The more massive ions have a larger Larmor radius and smaller Larmor frequency than the electrons. The end result is a drift motion given by:

$$v_E = \frac{E \times B}{|B|^2}$$

(4.24)

This drift motion is known as the $E \times B$ *drift* (pronounced as "e cross b") and can be seen as the drift velocity of the guiding centres of the charged particles. Its magnitude, $v = E/B$, is the same for both ions and electrons — v_E is independent of q, m and $v_\perp$. The direction of drift remains the same irrespective of the sign of the charge, since the direction of both acceleration and gyration change with sign. Crossed E and B fields do not therefore produce charge separation, merely giving a constant drift velocity (in the absence of collisions) to all the charged particles present: the entire mass of ions and electrons drifts across the electric and magnetic fields.

In the special case shown in Fig. 4.6, an electric field E in the y-direction, perpendicular to B (in the z-direction) produces a drift in the x-direction. In the general case, there is a component of E parallel to B, which, as we have seen in equation (4.20), adds an acceleration to the component of velocity parallel to B. By combining equations (4.21), (4.23) and (4.24) we obtain the full solution of equation (4.3):

$$v(t) = (\omega_c \times R_g) + \frac{E \times B}{B^2} + \frac{q}{m}E_\parallel(t) + v_\parallel(0)$$

(4.25)

The velocity of charged particles in constant, uniform electric and magnetic fields is the sum of the circular motion about the magnetic field lines (equation 4.23), the $E \times B$ drift velocity (equation 4.24), and acceleration along the magnetic field lines, plus any initial velocity of the particles (equation 4.21). Fig. 4.7 sketches the resulting motion for an electron.

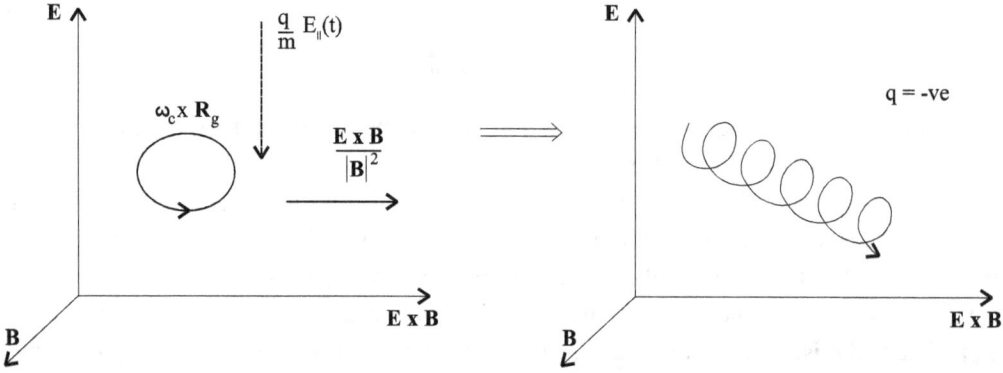

Fig. 4.7 **Motion of electron in perpendicular E and B fields**

4.7 The effect of an external force field

When an external force field F is applied to charged particles in a magnetic field B, the basic equation of motion (4.3) becomes

$$m\frac{dv}{dt} = q(E + \frac{F}{q} + v \times B) \qquad (4.26)$$

If F is constant in time and uniform in space, it has a similar effect to that of the electric field in the previous section, with one important difference. The component of the force perpendicular to B produces a drift motion, v_F, of the charged particles in the direction perpendicular to both F and B of the form

$$v_F \equiv \frac{F \times B}{q|B|^2} \qquad (4.27)$$

We have, in effect, substituted for E in equation (4.24) using equation (4.1). Since v_F now depends on the charge of the particle, the drift is in opposite directions for oppositely-charged particles. An example of such an external force field is a uniform gravitational field: $F = mg$.

4.7.1 Gravitational drift

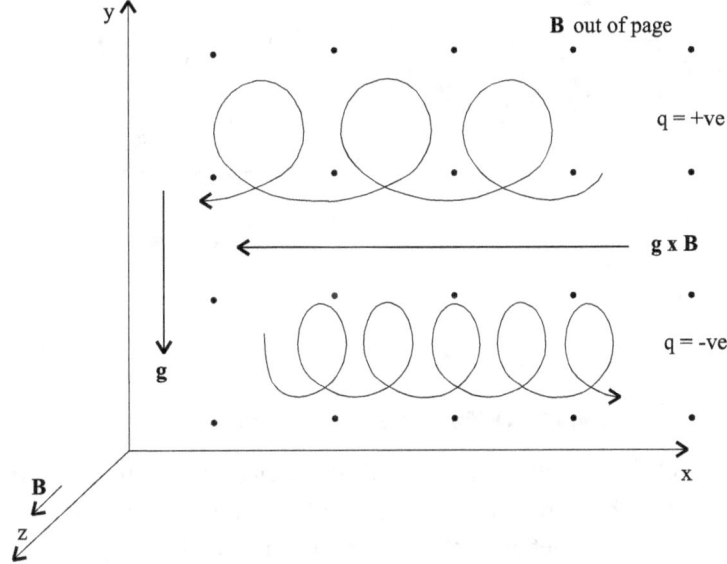

Fig. 4.8 Gravitational drift in uniform magnetic field

In the simplified case of a charged particle moving through a uniform magnetic field which is parallel to the Earth's surface, and in the absence of an electric field, the weight of the particle will displace the centre of its orbit. This makes the particle move with a transverse drift velocity given by

$$v_g = \frac{mg \times B}{q|B|^2} \qquad (4.28)$$

Once again, the drift is perpendicular to the direction of both the magnetic and gravitational fields. The drift velocity of the heavier ions is greater than that of the electrons, and in the opposite direction, as shown in Fig. 4.8.

The magnitude of v_g is usually very small: if $g = 9.8$ m s^{-1} and $B = 5 \times 10^{-5}$ T (an average figure for the Earth's magnetic field strength) then, for an electron, with $m_e = 9.1095 \times 10^{-31}$ kg, $v_g = 1.1 \times 10^{-6}$ m s^{-1}; and for a proton ($m_p = 1.6726 \times 10^{-27}$ kg), $v_g = 2 \times 10^{-3}$ m s^{-1}. Associated with the gravitational drift velocity of the particles is a net electric current density, J_g, in the direction $g \times B$ given by

$$J_g = nqv_g = \rho_m \frac{(g \times B)}{B^2} \qquad (A \ m^{-2}) \qquad (4.29)$$

where $\rho_m = n(m_e + m_i)$ is the total mass density of the charged particles.

4.8 Diamagnetic drift — The effect of temperature or density gradients

Most plasmas contain temperature or density gradients which can produce particle drift or plasma currents. An example is a cylindrical plasma immersed in a uniform magnetic field, with density increasing towards the centre (as found in a discharge tube).

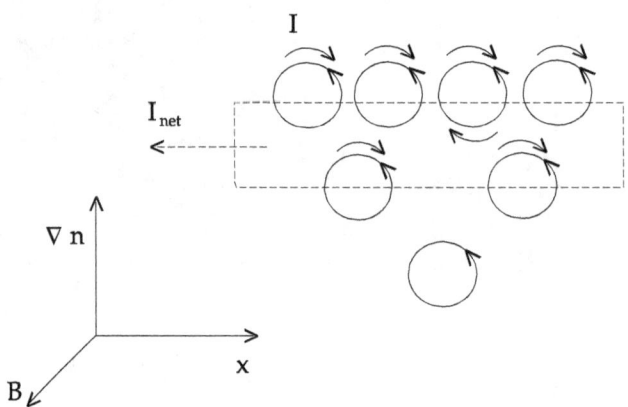

Fig. 4.9   Orbits of electrons in density gradient

If we consider a plasma composed entirely of electrons, with a density gradient, ∇n, as shown in Fig. 4.9, the currents associated with the individual charged particle orbits (Section 4.5) will produce a net current to the left when added together. (In the boxed area in the diagram, there are more left-pointing than right-pointing arrows.) This current occurs even if there is no movement of the guiding centres of the individual particle orbits. The apparent movement of the particles produces a <u>fluid</u> drift, known as the *diamagnetic drift*, perpendicular to both ∇n and B even when the guiding centres are stationary.

If temperature varies across the magnetic field, particle velocities and Larmor radii will also vary, decreasing with decreasing temperature and producing a similar effect. This diamagnetic drift is sometimes described as a fictitious drift because it is a flow in the fluid sense but not in the particle sense, since the guiding centres do not move. The drift changes direction with the charge on the particles because the direction of rotation of the Larmor orbit reverses.

Temperature and density gradients can also be expressed as a pressure gradient, ∇p. The molecular form of the ideal gas law, $p = n k_B T$ (equation 3.3), relates temperature and density, providing a measure of the random thermal energy density of a plasma. By analogy with equation (4.27), the force produced by the pressure gradient results in a drift velocity, v_D, given by

$$v_D = \frac{\boldsymbol{B} \times \nabla p}{nq|\boldsymbol{B}|^2} \qquad (4.30)$$

The influence of the charge, q, means that ions and electrons drift (as a fluid, not as particles) in opposite directions, perpendicular to both the pressure gradient and the magnetic field, as shown in Fig. 4.10. This flow produces a "diamagnetic current" given by

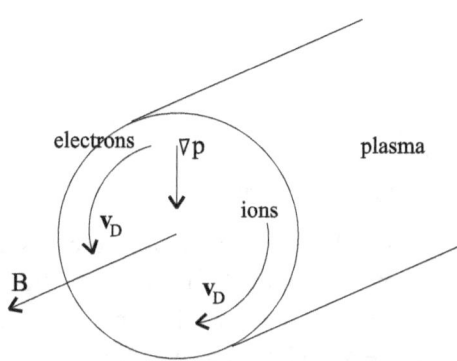

Fig.4.10   Diamagnetic drift in plasma column

$$J_D = nq v_{D\perp} = \frac{\boldsymbol{B} \times \nabla p}{B^2} \qquad (4.31)$$

which induces a magnetic field opposite in direction to the externally-applied magnetic field. The induced magnetic field reduces the total magnetic field within the higher density regions of the plasma — hence "diamagnetic". As the pressure increases, particle density and velocity increase and the diamagnetic current increases, enhancing the diamagnetic effect.

4.9   Non-uniform electric and magnetic fields

Electric and magnetic fields which vary in space or time make solving equation (4.3) very difficult, since particle motion cannot be divided exactly into its components: circular, parallel to E, and $E \times B$ drift. Nevertheless, if the magnetic field varies slowly in time and over distances much greater than the Larmor radius, R_g, and is much stronger than the electric field, the method can still provide an approximation.

4.9.1   Motion in non-uniform magnetic fields

In reality, no magnetic field is perfectly uniform. There will always be perturbations which cause particle drift and limit containment times. We will consider three common effects of non-uniform magnetic fields: a magnetic gradient parallel to B, a magnetic gradient perpendicular to B; and a curved magnetic field.

a) Magnetic gradient parallel to B — The magnetic mirror

Fig. 4.11(a), overleaf, shows a magnetic field increasing in strength towards the right.

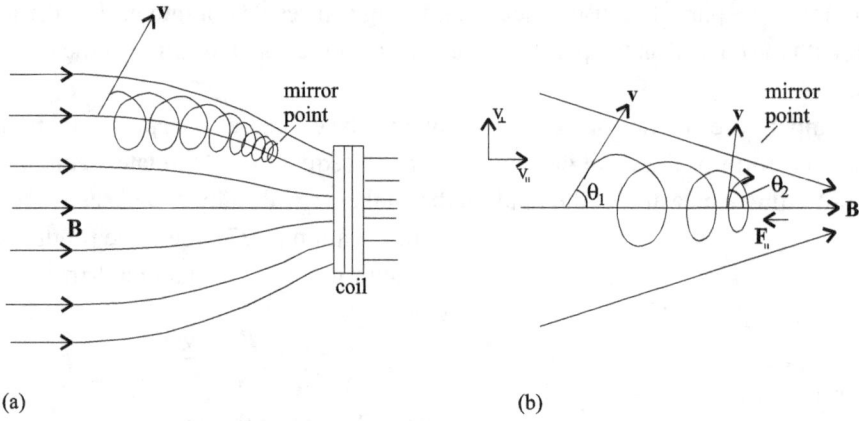

Fig. 4.11   Magnetic mirror — motion of positive charge

A positively-charged particle entering from the left, with velocity v, initially encounters a region of almost-uniform magnetic field. It will follow a helical path. As the magnetic field strength increases, the Larmor radius decreases and the particle turns in smaller circles: from equation (4.14) R_g is inversely dependent on B. The particle's guiding centre motion follows the curved field line. The component of velocity perpendicular to B is $v_\perp = v\sin\theta$, where θ is the pitch angle of the helix — the angle between v and B — defined by $\tan\theta = v_\perp/v_\parallel$ (see Fig. 4.11(b)). From equation (4.19), the magnetic moment is

$$M = \frac{m(v\sin\theta)^2}{2B} \tag{4.32}$$

When the magnetic field varies slowly in space or in time, we can view M as constant.

If there is no electric field and therefore no acceleration of the particle, M is constant, so $(\sin^2\theta)/B$ in equation (4.32) is constant. Thus, in Fig. 4.11, as the magnetic field strength increases, the pitch angle, θ, will also increase to keep M constant. As θ approaches a right angle the helix is compressed. The increase in pitch angle represents a transfer of kinetic energy from parallel to perpendicular motion. Total energy, like M, must remain constant, so as $v_\perp$ increases, $v_\parallel$ decreases to keep v constant in magnitude. This slows movement along the magnetic field, tending to push the particle away from the region in which the field is strongest. The particle loses speed and the orbit becomes flatter, until ultimately, $\theta = 90°$ and $v_\perp = v$. All the motion is perpendicular to the axis of the helix and velocity along the axis is zero: movement to the right ceases as shown in Fig. 4.11.

This is the *mirror point*. Its exact location depends on the strength of the magnetic field and the mass, charge and velocity of the particle when it enters the non-uniform region. The larger the initial value of $v_\parallel/v_\perp$, for a given value of v — and therefore the smaller the initial pitch angle θ — the further the particle will penetrate into the strong-field region before coming to a halt. At the mirror point the converging field continues to exert a retarding force along the magnetic field lines. This force can be defined as

$$\boldsymbol{F}_\parallel = -M\,\nabla_\parallel B \tag{4.33}$$

It results from the diamagnetic properties of plasma (equation 4.19). As the pitch angle increases beyond 90°, the velocity reverses and the particle starts to move towards the left in Fig. 4.11 — it is reflected at the mirror point.

Fig. 4.12 shows a non-uniform magnetic field produced by two coils set a distance apart. The coils create two magnetic mirrors facing one another and the resulting field is called a *magnetic bottle*. Since charged particles can be trapped inside, such a field provides one of the simplest means of confining a plasma, both in the laboratory and in nature, as with the Earth's magnetic

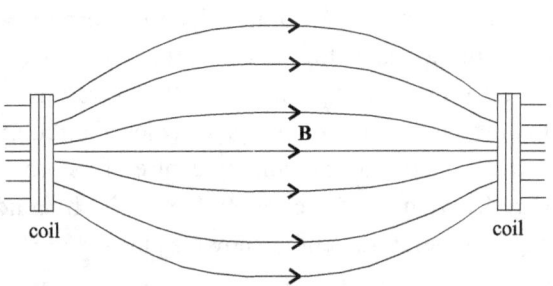

Fig. 4.12 Magnetic "bottle"

field. Charged particles, orbiting the magnetic field lines, are reflected back and forth between the ends of the bottle. The success of the trapping depends on the ratio of $v_\parallel$ to $v_\perp$. If a particle's perpendicular velocity is very small, the angle θ in equation (4.32) is small. The magnetic moment, M, is therefore very small and the retarding force, from equation (4.33), is also small, so the particle can escape from the ends of the bottle. The smaller the value of M, the smaller is the retarding force $\boldsymbol{F}_\parallel$. Particles moving parallel to the magnetic field along the axis of the bottle are most likely to escape.

b) Magnetic gradient perpendicular to *B* — 'Grad B' drift

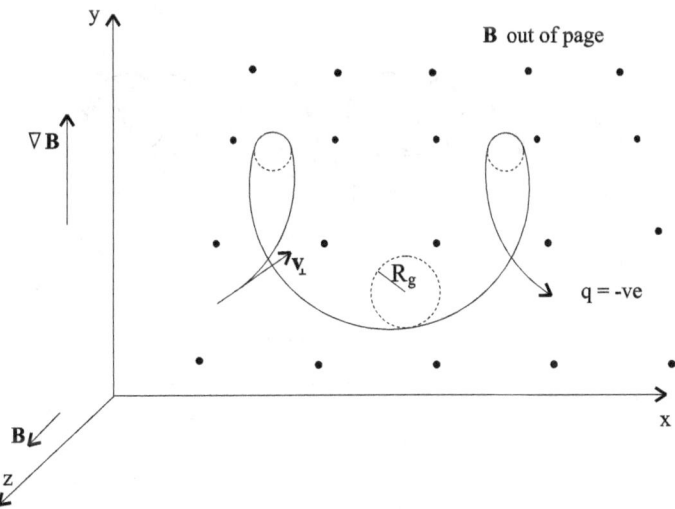

Fig. 4.13 Motion of electron in a non-uniform magnetic field

Any motion which carries a particle across the field is called a transverse drift. Fig. 4.13 shows a non-uniform magnetic field in the z-direction, directed out of the page, and an electron initially moving with velocity $v_\perp$ perpendicular to the magnetic field, in the x-direction. The magnetic field strength increases in the y-direction, so the field strength experienced by the particle varies as it rotates around a particular field line.

The radius, R_g, of the electron's orbit is assumed to be much smaller than the scale length of the magnetic field variation. The electron's path forms part of the circumference of a circle of radius $R_g = mv_\perp/(|q|B)$. The radius is large when the magnetic field is weak, and small when the field is strong, so a closed circular path is impossible. The particle performs a series of loops as the guiding centre of its orbit drifts perpendicular to the magnetic field, (Fig. 4.13), in the direction of $\boldsymbol{B} \times \nabla B$. The guiding-centre drift velocity produced by the magnetic field gradient is known as the *Grad B drift* and is given by

$$v_{\nabla B} = \pm \frac{v_\perp R_g}{2} \frac{\boldsymbol{B} \times \nabla B}{|\boldsymbol{B}|^2} \tag{4.34}$$

or
$$v_{\nabla B} = \frac{1}{2} m v_\perp^2 \frac{\boldsymbol{B} \times \nabla B}{q|\boldsymbol{B}|^3} \tag{4.34a}$$

where the sign in equation (4.34) is determined by the charge on the particle. Positive ions would therefore move from right to left in Fig. 4.13.

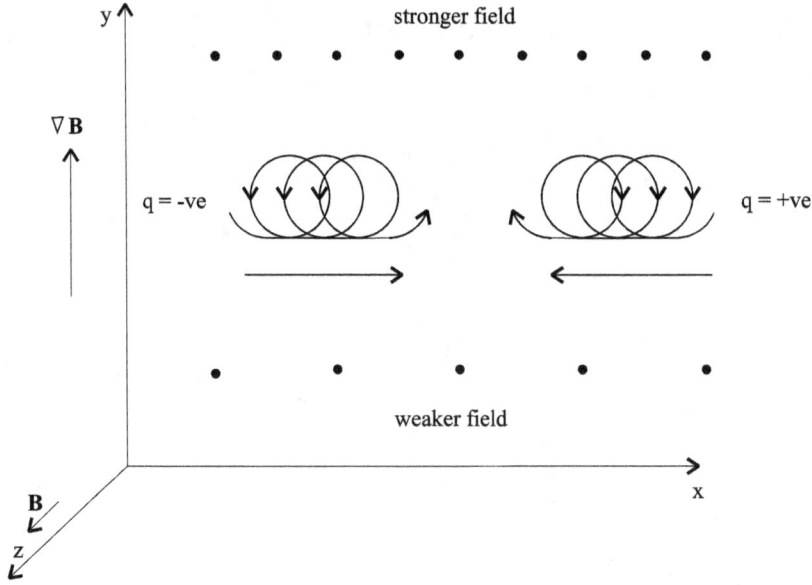

Fig. 4.14   Charged particle motion in non-uniform magnetic field

In a less rapidly changing magnetic field than that of Fig. 4.13, charged particles moving perpendicular to the field will follow almost circular orbits which gradually drift across the field (Fig. 4.14), in a direction perpendicular to both the magnetic field and its gradient. The direction depends on the sense of rotation, so electrons and ions drift in opposite directions producing charge separation and a net electric current.

c) Curvature drift

So far, we have assumed that the field lines are (locally) straight. This is not always the case: the Earth's magnetic field is an obvious example. Movement along a curved path changes the direction of the component of its velocity parallel to the path. This produces acceleration perpendicular to the path and directed towards the centre of curvature even if the particle's speed is constant. A charged particle orbiting a curved magnetic field line thus experiences a centripetal force from the component of its velocity parallel to the magnetic field, $v_\parallel$. The local radius of curvature of the field lines can be defined by a vector $\boldsymbol{R}_c$, directed from the field line to the centre of curvature, as shown in Fig. 4.15. Implicit in this is the assumption that the radius of curvature of the magnetic field is very much greater than the particle's gyroradius.

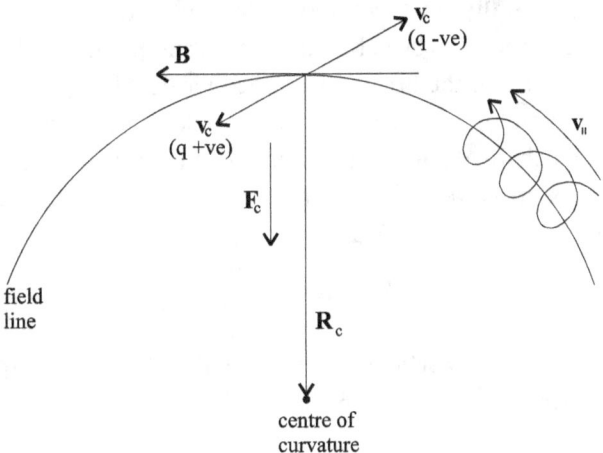

Fig. 4.15   Curvature drift

The centripetal force, $\boldsymbol{F}_c$, acting on the particle is the product of the particle's mass and its centripetal acceleration v^2/R. It is the component of velocity parallel to the magnetic field that changes, so:

$$\boldsymbol{F}_c = \frac{mv_\parallel^2}{|\boldsymbol{R}_c|^2}\,\boldsymbol{R}_c \tag{4.35}$$

The acceleration $v_\parallel^2/R_c$ is directed towards the centre of curvature. $\boldsymbol{R}_c/R_c$ is a unit vector in the direction of $\boldsymbol{R}_c$. By substituting equation (4.35) into equation (4.27), we can obtain the

76

equation for this curvature drift velocity:

$$v_c = \frac{F \times B}{q\,|B|^2} = \frac{mv_{\parallel}^2}{q\,|R_c|^2}\frac{R_c \times B}{|B|^2} \tag{4.36}$$

This shows that v_c is perpendicular to both the magnetic field and its radius of curvature, the direction depending on the particle's charge. Electrons and positive ions will therefore move in opposite directions producing a transverse *curvature current* which flows perpendicular to both the magnetic field and its curvature.

Curvature drift and grad-B drift generally appear together — if the field is curved the magnetic field strength closer to the centre of curvature must be greater than it is further away from the centre. Since both drift velocities are influenced by the charge on the particles, v_c and $v_{\nabla B}$ are in the same direction and can be added together to produce the *gradient-curvature drift*: $v_{gc} = v_c + v_{\nabla B}$. Both v_c and $v_{\nabla B}$ are small compared to the basic velocity along or around **B**. An example of gradient-curvature drift and the current it produces is the ring current encircling the Earth. The decrease of the Earth's magnetic field with altitude and the curvature of the field produces a slow, steady drift of particles around the equatorial plane, westwards for positive particles and eastwards for electrons.

4.9.2   Time-varying spatially-uniform magnetic field

So far, particle motion has been based on the assumption that the fields are constant in time. We will now consider the effect on charged particle behaviour of a magnetic field which changes slowly with time compared to the particle's gyro-orbit.

The energy of a charged particle in a static magnetic field remains constant because the magnetic field does no work on the particle. A magnetic field which varies in time induces an electric field, according to Maxwell's equation:

$$\nabla \times E = -\frac{\partial B}{\partial t} \tag{4.5}$$

The effect of this is to change the particle's energy: it will either gain energy from the induced electric field or lose energy to it. Fig. 4.16 shows what happens to the particle's orbit.

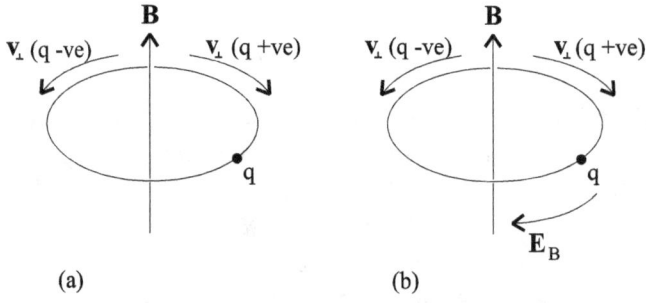

Fig. 4.6   **Particle orbit in (a) constant, and (b) time-varying magnetic fields**

We saw in Section 4.5 that the component of particle velocity parallel to B, $v_\parallel$, is unaffected by B when the magnetic field is constant. This is still the case when B varies slowly with time. The velocity perpendicular to B is affected as follows:

A slow change in the magnetic field creates an electric field which we will call E_B. From equation (4.5), E_B is perpendicular to the direction of B as shown in Fig. 4.16(b) and is therefore parallel to the orbit of the particle. This produces a force qE_B which accelerates the particle around its orbit, so $v_\perp$ changes steadily. As a result, the perpendicular kinetic energy of the particle also changes, since $E_{kin\perp} = \frac{1}{2}mv_\perp^2$. Thus, if B increases, $E_{kin\perp}$ increases. We met this term $E_{kin\perp}$ in equation 4.19, the definition of the magnetic moment, $M = \frac{1}{2}mv_\perp^2 = E_{kin\perp}/B$.

Integrating around the orbit gives the effect of this force qE_B during one particle-orbit:

$$\delta(E_{kin}) = q \int E_B \cdot dR_g$$

where R_g is the gyroradius of the particle. Using Stokes' Theorem[e] the change in E_{kin} in one orbit is:

$$\delta(E_{kin\perp}) = q \int E_B \cdot dR_g = -q \int_s (\nabla \times E) ds$$

$$= -q \int_s \frac{\partial B}{\partial t} \cdot ds$$

where s is the surface of the orbit. The area of the orbit is πR_g^2, and because of diamagnetic effects, $B \cdot ds$ is negative for positive ions, so

$$\delta(E_{kin\perp}) = |q| \frac{\partial B}{\partial t} \pi R_g^2 \tag{4.37}$$

The change in the magnetic field over one gyration period, $\tau = 2\pi/\omega_c$ (equation 4.17), is

$$\delta B = \frac{\partial B}{\partial t} \frac{2\pi}{\omega_c}$$

so $\partial B/\partial t = \delta B \, \omega_c/2\pi$. Substituting equation (4.15): $R_g = v_\perp/\omega_c$, and equation (4.16): $\omega_c = |q| B/m$, we can write equation (4.37) as

$$\delta(E_{kin\perp}) = \frac{mv_\perp^2}{2B} \delta B = M \, \delta B \tag{4.38}$$

Since $E_{kin\perp} = MB$, equation (4.38) becomes $\delta(MB)/\delta B = M$, and therefore $\delta M = 0$. The magnetic moment is thus invariant in a slowly-varying magnetic field. From equation (4.18), $M = \pi R_g^2 q/\tau$, so by simple substitution of equations (4.16) and (4.17) we find that BR_g^2 is also constant.

---

[e]   The line integral of a vector function around a closed curve bounding a surface equals the surface integral of the curl of the function: $\int_L E \cdot dL = \int_s \nabla \times E \, ds$

If the magnetic field increases, the particle's orbit will decrease, as with the magnetic mirror, and its energy will increase, since the magnetic moment is constant. This has the effect of heating the plasma. If the magnetic field decreases, the particle energy decreases, cooling the plasma.

4.9.3   Time-varying electric field

An electric field, varying slowly in time with a frequency very much less than the cyclotron frequency of the plasma (equation (4.16)), will interact with a static magnetic field, producing movement in the charged particles in the plasma. For simplicity, we will assume that the static, externally-applied magnetic field is much stronger than the magnetic field associated with the time-varying electric field.

A charged particle, initially at rest in the static magnetic field, responds to the application of a time-varying electric field, $E(t)$, by moving along the direction of $E(t)$. Having acquired a velocity v, the particle then starts to feel the effect of the Lorentz force, $qv \times B$, due to its motion. If $E(t)$ then reverses, there will be a corresponding momentary drift along the (new) direction of $E(t)$.

The resulting drift velocity, known as the *polarisation drift*, is given by

$$v_{pol} = \frac{m}{q|B|^2} \frac{\partial E_\perp}{\partial t} \tag{4.39}$$

It is additional to the usual $E \times B$ drift in electric and magnetic fields. The drift is along the direction of the applied electric field, perpendicular to the magnetic field and proportional to the frequency of the applied electric field. The magnitude of v_{pol} depends on particle mass, m, so ions and electrons move at different speeds. Ions, with greater mass, have greater effect. The direction of v_{pol} is determined by the charge on the particle, q, so oppositely-charged particles move in opposite directions, producing charge separation and a net current flow in the plasma.

The polarisation effect, analogous to dielectric materials, results from the time-variation of the electric field. In a steady electric field, no polarisation occurs.

4.10   Summary

1.   The movement of charged particles within a plasma both causes and results from the electric and magnetic fields within the plasma.

2.   Charged particles moving in a uniform electric field experience a constant acceleration which produces a slow drift velocity parallel to the direction of the field. There is no effect on $v_\perp$ which remains constant in magnitude and direction. The more massive positive ions remain almost stationary and charge separation occurs.

3.   A charged particle moving in a uniform magnetic field experiences a centripetal

acceleration perpendicular to the magnetic field. The magnetic field has no effect on $v_\parallel$. This produces circular motion around the magnetic field lines (the *Larmor orbit*) and a steady movement of the guiding centre parallel to the field, resulting in a helical motion. Electrons rotate in a clockwise direction, positive ions in an anticlockwise direction. The Larmor orbit defines a fundamental length: the Larmor radius and a fundamental time: the Larmor period.

4.   The *magnetic moment, M,* associated with the orbiting charged particles is opposite in direction to the externally-applied magnetic field. This diamagnetic effect varies with the perpendicular kinetic energy of the particles.

5.   The combined effect of uniform electric and magnetic fields produces an *E × B drift* velocity (with no additional charge separation) in addition to the circular motion around the magnetic field lines and acceleration parallel to *E*.

6.   When an external force, such as gravity, is applied to a plasma in a uniform magnetic field, the component of the force perpendicular to *B* produces a drift motion. Charge separation occurs.

7.   Temperature and density gradients, as well as producing diffusion of the particles, reveal a diamagnetic drift — a fluid flow not a particle flow — producing a "diamagnetic current".

8.   Spatially-varying magnetic fields produce different effects depending on the direction of the gradient: parallel or perpendicular to the magnetic field. A magnetic gradient parallel to *B* produces a *magnetic mirror* and provides a means of containing the plasma. A magnetic gradient perpendicular to *B* results in a (transverse) *Grad B drift* velocity perpendicular to both the field and the gradient; charge separation occurs. A curved magnetic field produces *curvature drift* perpendicular to the magnetic field and its curvature; charge separation occurs.

9.   A time-varying magnetic field induces an electric field which alters the perpendicular component of the kinetic energy of the particles. A time-varying electric field produces a *polarisation drift* velocity along the direction of, and proportional to the frequency of, the applied electric field. In both cases, charge separation occurs.

Reference
1.   *Phil. Mag.*, **27**, p.29-50 (1889)

Chapter 5

Plasma characteristics

5.1 What is plasma?

Any gas contains some charged particles, but while ordinary gases are composed almost entirely of atoms, a true plasma contains very few. When an ordinary gas is heated, the kinetic energy of the neutral atoms increases with temperature and ionising collisions become more frequent. The number of charged particles increases until the gas can be described as an *ionised gas*. This is simply a gas with a larger than normal proportion of charged particles. It is not necessarily a plasma and the transition from one to the other is often not clearly defined.

An ionised gas can be called a plasma only when there are sufficient charged particles for their presence to begin to alter the properties and behaviour of the gas. The true plasma state is reached when charged particles outnumber atoms and there is no overall charge imbalance. This *quasi-neutrality* — approximately equal numbers of positively- and negatively-charged particles — forms the basis of our definition of a plasma as an electrically-neutral mixture of ions and electrons which behave collectively, as a body, in response to electromagnetic forces. It is the combination of individual particle-like and collective fluid-like behaviour which distinguishes plasma from an ordinary gas. Whether a plasma behaves collectively like a fluid or as an assembly of individual particles often depends on the numbers of particles or the frequency of collisions.

For simplicity, plasma is often portrayed as a mixture of electrons and protons (completely-stripped ions), but ionisation and recombination ensure that there are always partially-ionised or neutral atoms present. The predominance of charged particles ensures that plasma, unlike an ordinary gas, is generally an excellent electrical conductor, interacting with electromagnetic fields, radiating and absorbing electromagnetic waves. These electrical properties give rise to many of plasma's distinctive features. Left to itself, a plasma will remain electrically neutral. Charge separation can only be maintained by externally imposed electric fields or the thermal energy of the plasma itself.

An important property of plasmas is that particle motion can be constrained by a magnetic field, with the charged particles being forced to move along the magnetic field in circular orbits around the field lines. In a normal gas, particle motion is limited solely by collisions with other particles or with the containment vessel wall.

5.2   Examples of plasmas

It used to be said that over 99% of the known universe was believed to be in the plasma state, the exceptions being the surfaces of cold planets like the Earth. That is no longer strictly true, since perhaps 80% of the universe is now thought to consist of dark matter and dark energy. Plasma, nevertheless, is still an important constituent of the universe as a whole.

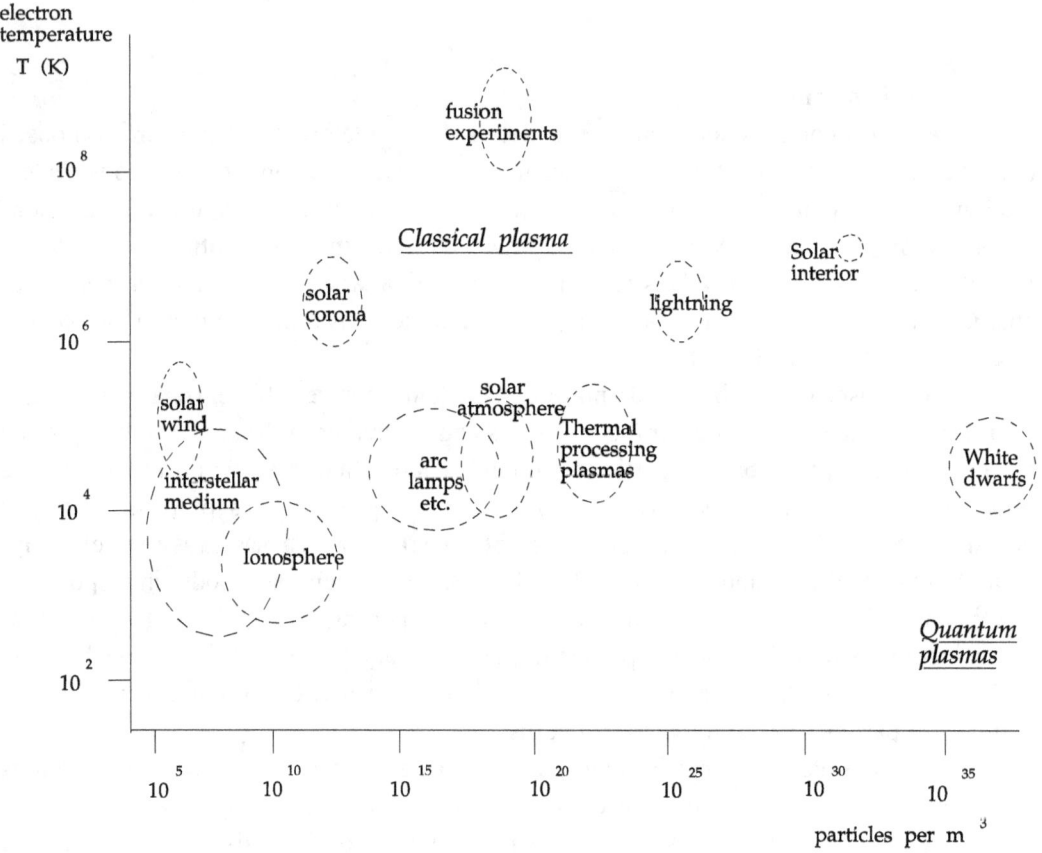

Fig. 5.1   Ranges of temperature and density

Here on Earth, we only experience plasma in its natural form in flashes of lightning, in the aurorae and the ionosphere — a layer of weakly-ionised plasma in the upper levels of the Earth's atmosphere. Naturally-occurring plasmas, such as gaseous nebulae, stellar interiors and atmospheres, and much of the interstellar medium, may contain few neutral particles and are often relatively insensitive to what is going on at their boundaries because of their size. Man-made plasmas, on the other hand, often have quite high neutral particle densities. They occur in arc lamps, gas discharge lasers, fluorescent tubes and "neon" signs, and in laboratory experiments which may be complicated by interactions between the plasma and its containment vessel. The shock-wave-induced plasma surrounding a space vehicle as it re-enters the Earth's atmosphere can be thought of as a man-made plasma. During the re-entry period, the spacecraft decelerates and suffers intense heating from atmospheric friction, becoming surrounded by a sheath of ionised air. Conventional communication through this layer is impossible, producing a "communications black-out" which lasts for several minutes.

Different plasmas are characterised by different combinations of temperature and density, both of which can vary by several orders of magnitude. Their distribution is shown in Fig. 5.1. Almost all are described by the same equations, based on classical physics. Quantum effects are only important at very high densities and very low temperatures.

5.3   Collective behaviour

A feature of plasma, which distinguishes it from ordinary fluids or solids, is the collective nature of much of its behaviour. A neutral particle in an ordinary gas is unaffected by electromagnetic forces. It moves undisturbed through electric or magnetic fields until it collides with another particle.   A charged particle in a plasma, however, interacts simultaneously with many other charged particles over long distances under the influence of electric and magnetic forces, giving the plasma a cohesion which is lacking in an ordinary gas.

Movement of the ions and electrons within the plasma produces charge separation and local concentrations of positive or negative charge can build up, creating localised electric fields. The movement of charged particles also generates electric currents, which produce magnetic fields, exerting a force on other moving charges. These electromagnetic fields can affect not just nearby charged particles but also those at large distances from the initial disturbance through the long-range influence of the Coulomb force — the electrostatic attraction or repulsion between two charged particles. Despite there being no physical contact between the charged particles, they can act together as a body to produce the wide range of physical phenomena peculiar to plasmas.

5.4   Debye length, λ_D

A basic characteristic of plasma is its ability to screen the potential field around a charged particle. A positive ion, placed as a test charge in a cold plasma in which there is no random thermal motion of the particles, will attract a cloud of electrons around it and repel any local positive ions. As a result, it will be completely shielded from the rest of the plasma and no electrostatic field will exist within the body of the plasma, outside the electron cloud.

If the temperature of this cold plasma is raised so that the plasma particles acquire a random thermal motion, the electrons at the edge of the cloud, where the ion's influence is weakest, will have enough energy to escape. The outer limit of the cloud is now the radial distance from the ion at which the potential energy is approximately equal to the thermal energy of the particles. This radial distance is the *Debye length*, or Debye shielding distance, and is represented by λ_D. It is a theoretical length determined by the temperature and density of the plasma and measures the sphere of influence of a charged particle in a plasma. Only particles separated by a distance which is less than λ_D are directly affected by each other's electric field, so individual interactions between particles are only important over distances shorter than λ_D.

The Debye length is therefore the maximum distance over which an electron will be influenced by the field of a given ion, and vice versa. It is also the maximum distance over which spontaneous charge separation occurs. Beyond that distance, in the absence of external forces, charge separation cannot be maintained and collective effects dominate: electrons, in particular, can move freely under the influence of the Coulomb force to neutralise any areas of space charge imbalance — areas where there is an excess of positively- or negatively-charged particles, which create an electric field.

The Debye length is determined by electron temperature since it is the lighter, more mobile, electrons that produce the screening by moving to reduce the effects of local electric fields. When T_e is much greater than T_i:

$$\lambda_D = \left(\frac{k_B T_e \varepsilon_0}{ne^2} \right)^{\frac{1}{2}} \quad (m) \tag{5.1}$$

This is equivalent to $\sqrt{69}(T/n)$ with T in kelvin. The separation distance therefore increases as the temperature increases and decreases as plasma density increases. For our sample plasma, with $T_e = 1$ keV (so $k_B T_e = 10^3\,eJ$) and $n_e = 10^{20}\,m^{-3}$, the Debye length is 2.35 x 10^{-5}m. If T_e increases to 10 keV, $\lambda_D = 7.4$ x 10^{-5} m. If T_e remains constant at 10 keV and n_e increases to $10^{22}\,m^{-3}$, λ_D reduces to 7.4 x 10^{-6} m.

For laboratory plasmas of all types, the Debye length is very much less than 1 mm. In low-density high-temperature naturally-occurring plasmas it can reach several metres in length. At high temperatures, the random thermal motion of the particles in the body of the plasma allows more of the electrons in the screening cloud to escape, so the test charge is less effectively screened. Similarly, if particle density is low, the shielding electrons will have to be drawn from a larger volume in order to adequately shield the test charge. The influence of any test charge is thus greater in a hot diffuse plasma than in a cool dense plasma.

When T_e and T_i are approximately equal the ions can generally move fast enough to take part in the shielding process and the formula for λ_D becomes:

$$\lambda_D = \left(\frac{k_B T_e \varepsilon_0}{2ne^2} \right)^{\frac{1}{2}} \quad (m)$$

For the collection of charged particles to be a plasma, the physical dimensions of the system must be large compared to λ_D. This requirement is normally written:

$$\lambda_D \ll L$$

where L is a characteristic dimension of the plasma. If this condition is met, then whenever local concentrations of charge arise, they are screened out in a distance which is small compared with the dimensions of the plasma, leaving the bulk of the plasma free of large electric fields or potentials. If the condition is not met, there is insufficient space for collective shielding to occur and the collection of charged particles will not behave like a plasma but will remain an ionised gas. In the absence of external forces, this requirement implies charge neutrality within the body of the plasma. At distances larger than λ_D the Coulomb field of a point charge is screened out and the plasma remains unaffected by the charge imbalance. Deviations from charge neutrality can therefore only occur naturally over distances of the order of λ_D. It is this shielding mechanism, through which the plasma tries to maintain internal charge neutrality (its characteristic quasi-neutrality), which sets plasma apart from the other three states of matter.

A useful measure of screening effectiveness is the number of particles, N_D, which lie within a Debye length of the test charge and so contribute to the screening. This constitutes a sphere of radius λ_D, known as the *Debye sphere*. The number of particles contained in such a sphere is obtained from

$$N_D = \frac{4\pi}{3} n \lambda_D{}^3 \tag{5.2}$$

where n represents the number density in the plasma of the electron or ion under consideration. Since the shielding effect comes from the collective behaviour of the particles inside the Debye sphere, the condition for effective screening is that N_D is very large:

$$N_D \gg 1$$

Each charged particle in the plasma interacts collectively only with the charges that lie within its Debye sphere. Its effect on other charges is negligible. So the average distance between the particles must be very much smaller than λ_D. This can be written:

$$n^{-1/3} \ll \lambda_D$$

where $n^{-1/3}$ is the inter-particle distance.

Our sample plasma, with $\lambda_D = 2.35 \times 10^{-5}$ m, has 5.4×10^6 particles in a Debye sphere. With a particle density of 10^{20} m^{-3}, the inter-particle distance is 2.15×10^{-7} m, which is much less than the Debye length.

5.5 Sheaths

A shielding cloud of charged particles forms around anything inserted into the plasma or on any boundary surface such as the wall of the containment vessel. This charge cloud results from the interaction of the plasma with its surroundings and forms a boundary layer, or *plasma sheath*, between the surface and the plasma. All plasmas — natural or man-made — have boundaries at which there are sheaths and anything immersed in the plasma, such as a probe, antenna or spacecraft, will be surrounded by a sheath. The properties of the sheath are important in plasma processing of materials.

86

When charged particles from the plasma reach a surface in contact with it, they are lost from the plasma. Ions tend to recombine and return as neutral atoms; electrons may recombine or may be absorbed into any electrical currents in the surface, if that surface is a metal. Electrons, being smaller, have much higher velocities than ions and so will leave the plasma at a faster rate.

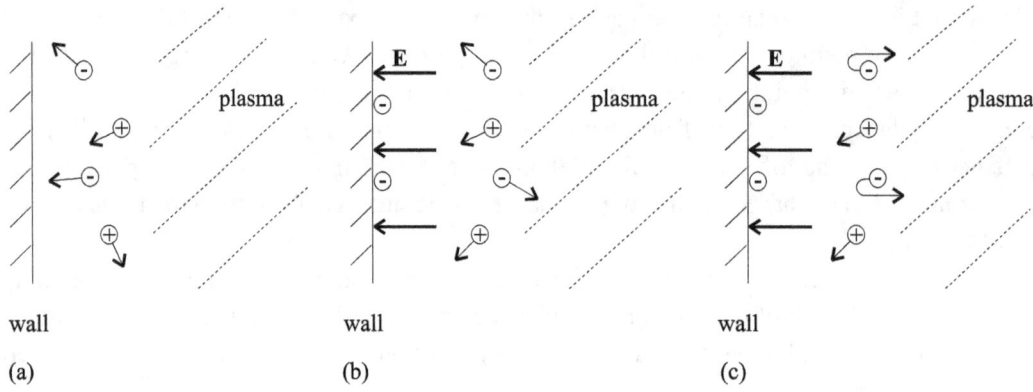

Fig. 5.2 Plasma sheath development

Fig. 5.2(a) shows the simple situation of a plane wall in contact with a plasma. Random thermal motion initially brings electrons to the wall faster than ions and it rapidly acquires a negative charge. This gives the wall a negative potential with respect to that of the plasma. An electric field develops close to the surface, and perpendicular to it (Fig. 5.2(b)) because of charge separation. The field repels further electrons and attracts more of the positive ions (Fig. 5.2(c)), until it is strong enough to ensure that electrons and ions hit the surface at an equal rate. The sheath is thus self-adjusting.

The negative potential at the wall "floats" to maintain an equilibrium between the wall and the plasma, ensuring that the net current at the wall is zero. There must therefore be a steady flow of both ions and electrons towards the wall to replace those lost through recombination at the surface. Only electrons with sufficient thermal energy can penetrate the field and reach the wall. Within the sheath, charge neutrality is not maintained and the electric field is strong. Unlike the body of the plasma, where electron and ion densities are considered equal, the sheath is a region of net positive charge, since most of the electrons are repelled by the negative potential at the wall. The Debye shielding ensures that this region of positive charge is confined to a layer of just a few Debye lengths in thickness. The sheath is therefore too thin to allow time for collisions to take place — it is essentially "collisionless". Particle density within the sheath decreases towards the wall. The rate of flow of ions towards the wall, $n_i v_i$, must remain constant and have the same value as within the body of the plasma, but the closer the ions get to the wall, the greater is the effect on particle velocity — the ions are accelerated towards the wall by the electric field. So, if v_i increases, n_i must decrease to maintain the value of $n_i v_i$.

Within the sheath, the potential increases from a negative value at the boundary, or object, to the value existing in the bulk of the plasma. The sheath potential, V_s, which develops across the sheath, between the positive potential of the body of the plasma and the floating negative potential of the surface in contact with the plasma, is expressed as

$$V_s = \frac{k_B T_e}{2e} \ln\left[2\pi \frac{m_e}{m_i} \left(1 + \frac{T_i}{T_e} \right) \right]$$
(5.3)

For a hydrogen plasma, if $T_i = T_e$, this simplifies to $V_s = -2.5 k_B T_e /e$. This variation in potential is confined to a layer about three or four Debye lengths in thickness, the actual thickness varying with the electron temperature, T_e, of the plasma.

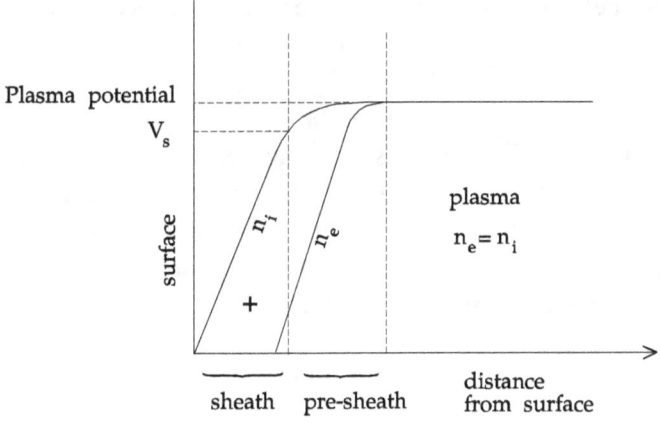

Fig. 5.3 Plasma sheath

The plasma sheath, as shown in Fig. 5.3, is the region where the potential drop, V_s, occurs and electron density is negligible. There is no sharp boundary between the sheath and the body of the plasma. Instead, the number of electrons increases through a transition zone, often called the pre-sheath region, until n_i is approximately equal to n_e in the plasma itself.

5.5.1 Double layers

Double layers are sheaths which are not attached to any physical boundary but are supported by locally non-Maxwellian conditions. They appear as a boundary between two distinct regions of plasma and occur naturally, in such as the Earth's magnetosphere, and in discharge tubes.[1]

A double layer appears in the discharge tube when the current exceeds a critical value, reached when the average drift velocity of the electrons is approximately equal to their thermal velocity. It is seen as a sharp boundary across the positive column of the discharge tube, with the plasma on one side of the layer being of a different colour to that on the other side. In mercury vapour, the plasma has the normal bluish colour on the cathode side of the layer but is distinctly reddish on the anode side. The boundary forming the double layer is several Debye lengths thick, but is less than the average distance travelled by a particle between collisions — the collisional mean free path — so collisions can generally be ignored. Electrostatic potential varies across the layer, changing by more than $k_B T/e$, so quasi-neutrality is not maintained within it. This potential causes localised particle acceleration across the layer, the direction of motion depending on the sign of charge.

5.6   The plasma parameter

The quantity in equation (5.2) which produces the variation in N_D is $n \lambda_D^3$. This quantity, which may be thought of as the number of particles in a cube of side λ_D, a simplified Debye sphere, also defines a dimensionless quantity known as the plasma parameter and given by

$$g = \frac{1}{n \lambda_D^3} \equiv \frac{1}{N_D} \tag{5.4}$$

The condition that N_D is very large implies that g must be very small[a]. Equation (5.4) shows that, as the density decreases or the temperature increases, g becomes smaller since N_D is proportional to $(T^3/n)^{1/2}$. Changes in density or temperature can influence the number of collisions between the particles, so the plasma parameter can indicate collision frequency: as g becomes smaller, the number of collisions decreases.

5.7   Collisions

The exchange of energy between the particles of any gaseous medium occurs mainly through collisions. In plasmas, both short-range and long-range collisions, and their frequency, can significantly affect both the properties of the plasma and its containment. Short-range collisions involve physical contact, usually between electrons and neutral atoms. These collisions are of two types:

(i) *elastic*, in which mass, momentum and energy are conserved. The total kinetic energy of the system and the internal energy of the colliding particles remain the same; they keep their identities and there is no ionisation of the atom. Only translational kinetic energy is exchanged, the energy transfer being determined by the mass ratio of the colliding particles. The results of an elastic collision can be predicted from the laws of conservation of energy and momentum. In collisions with even the lightest atom, an electron, with its much smaller mass, will lose only a fraction of its kinetic energy but can transfer almost all of its momentum to the heavier particle, depending on the angle at which it is scattered.

(ii) *inelastic*, in which translational kinetic energy and internal energy are exchanged. The total kinetic energy of the system is reduced and the particles may change their identity or

---

[a]   Both N_D and g may be described in the literature as the "plasma parameter".

their internal energy state, producing ionisation or excitation. An electron colliding with an atom, for example, may excite the atom and lose kinetic energy while increasing the atom's potential (internal) energy. For this to happen, the incoming electron must have sufficient energy to raise an electron in the atom to a higher energy level. If a less-energetic electron collides with an already-excited atom, further excitation or even ionisation can still take place, because less energy is required to excite the atom. Excitation and ionisation can also result from collisions between ions or neutral particles.

5.7.1   Coulomb collisions

A collision between two charged particles is known as a *Coulomb collision*. These interactions are long-range, weak collisions caused by the electrostatic, or Coulomb, force which each particle exerts on the other, according to Coulomb's law:

$$F_{el} = \frac{q_1 q_2}{4\pi\varepsilon_0 r^2}$$

No physical contact is made. It is the interaction of the electrostatic fields surrounding each charged particle which produces the change in motion. The $1/r^2$ dependence means that the interaction is long-range, so the field around each charged particle can interact with many others, unlike neutral particles which rarely interact with more than one particle at a time.

Most Coulomb collisions are small-angle collisions with the charged particles being only slightly deflected from their original paths. The more-mobile electrons dominate these interactions because they respond more rapidly to electric and magnetic fields. In the weakly-ionised plasmas of, for example, the ionosphere and gas discharge tubes, there are large numbers of atoms, so most electron collisions are non-Coulomb collisions, involving physical contact with neutral atoms. As the number of ions increases, Coulomb interactions become more important, even when the degree of ionisation is still low, since collisions between electrons and ions have a greater effect on the plasma than those involving neutral particles.

When an electron is a distance r from a positive ion it experiences an attractive Coulomb force, given by

$$F_{el} = \frac{e^2}{4\pi\varepsilon_0 r^2} \qquad (5.5)$$

This deflects the electron around the ion on a hyperbolic path, as shown in Fig. 5.4. The ion, with its much greater mass and slower approach velocity than the electron, is usually regarded as stationary.

The presence of the ion, and its Coulomb field, causes the electron to deviate from its original path; by how much depends on the initial velocity of the electron and the distance marked as b. This is the distance of closest approach, or *impact parameter*. It measures what would have been the shortest distance between the two particles if no interaction had occurred. Substituting b for r in equation (5.5) gives the Coulomb force on the electron at this

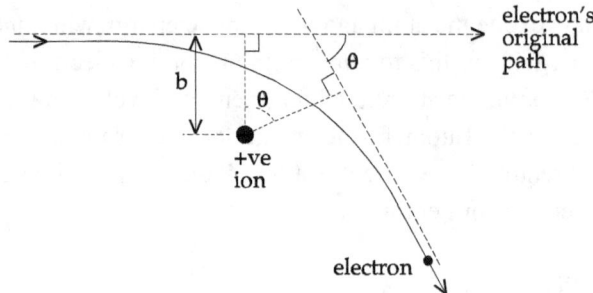

Fig. 5.4 Deflection of electron in Coulomb collision with a positive ion

distance. Its effect lasts for approximately $t = b/v_e$ seconds. Multiplying F_{el} by t provides an estimate of the change in the electron's momentum. The angle of deflection, θ, is given by

$$\tan\frac{\theta}{2} = \frac{Z_i e^2}{4\pi\varepsilon_0 m_e v^2 b} \tag{5.6}$$

where Z_i is the charge on the ion. In large-angle collisions, where θ is close to 90° and therefore $\tan\theta/2 \approx 1$, most of the electron's initial momentum is lost and, by rearranging equation (5.6), we can obtain an estimate for b:

$$b \approx \frac{Ze^2}{4\pi\varepsilon_0 m_e v^2} = \frac{Ze^2}{12\pi\varepsilon_0 k_B T_e} \tag{5.7}$$

(by substituting $\tfrac{1}{2}mv^2 = \tfrac{3}{2}k_B T$ from equation 3.6) where T_e is the electron temperature in eV.

In large-angle collisions in our hypothetical plasma, if $Z_i = 1$ and electron velocity is $2.3 \times 10^7 \text{ m s}^{-1}$, the impact parameter is approximately 4.8×10^{-13} m; the magnitude of the Coulomb force is 10^{-3} N and its effect lasts about 2×10^{-20} s.

5.7.2   Collision cross-section, σ_c

In real life, the outcome of a collision between two particles cannot be determined from their initial trajectories and velocities. Only a prediction based on probabilities and average results is possible. In atomic and nuclear physics, the term "cross-section" means the probability that a particular interaction will take place between particles. The collision cross section, σ_c, therefore indicates the likelihood that one particle will collide with another, and provides a measure of the effective target area. The size of the cross-section will vary with charged particle density and with the relative velocities of the colliding particles because the most significant factor is momentum transfer during the collision.

For collisions between electrons and neutral particles, the collision cross-section expresses the probability of momentum loss or of ionisation through the relation

$$\sigma_{cn} = \frac{1}{n_n \lambda_{mfp}} \quad (m^2) \tag{5.8}$$

where λ_{mfp} is the *mean free path* — the average distance travelled by the particles (in this case the electron) between collisions — and n_n is the density of neutral atoms with which the electrons generally collide. As particle density increases, or the mean free path increases, the collision cross-section decreases. A rough estimate of the collision cross-section for elastic scattering between electrons and neutral atoms can be obtained from $\sigma_{cn} \sim \pi a_0^2$, where $a_0 =$ 5.3 x 10^{-11} m is the Bohr radius of the atom, since it is highly probable that the electron will undergo a large-angle collision with the atom at this distance.

In Coulomb collisions, the effective collision cross-section must take into account not only large-angle collisions ($\theta \approx 90°$) but also small-angle collisions ($\theta < 90°$) where the electron is only slightly deflected by the Coulomb field of the ion. Small deflections are more numerous than large-angle collisions because of the long-range influence of the Coulomb force. By analogy with elastic scattering, we could estimate the Coulomb collision cross-section to be $\sigma_c \sim \pi b^2$. In fact, the effective cross-section is much larger than this, by a factor $4 \ln \Lambda$, because of the cumulative effect of all the small-angle collisions. So the Coulomb collision cross-section becomes:

$$\sigma_c \sim \pi b^2 \approx 4 \ln \Lambda \pi \left(\frac{Z_i e^2}{4 \pi \varepsilon_0 m_e v^2} \right)^2 = \frac{Z_i^2 e^4 \ln \Lambda}{4 \pi \varepsilon_0^2 m_e^2 v^4} \tag{5.9}$$

The quantity $\ln \Lambda$ is known as the *Coulomb logarithm*. The variable Λ is equivalent to the ratio between the maximum impact parameter, b_{max}, and the actual impact parameter, b. Debye shielding, as we saw earlier, suppresses the Coulomb field of the charged particle at distances larger than the Debye length, λ_D. We can therefore take λ_D to be the maximum impact parameter. So:

$$\Lambda \equiv \frac{b_{max}}{b} = \frac{\lambda_D}{b} \approx 12 \pi n \lambda_D^3 = 9 N_D \tag{5.10}$$

where N_D is the number of particles in a Debye sphere. Although Λ depends on density and temperature through λ_D, its logarithm is insensitive to exact values of n and T_e. For most laboratory plasmas and many naturally-occurring plasmas, the value of $\ln \Lambda$ is generally between 10 and 20. Fig. 5.5, overleaf, plots sample values.

For our sample plasma, $\ln \Lambda = 17.7$ ($\lambda_D = 2.35$ x 10^{-5} m, $b = 4.8$ x 10^{-13} m) and σ_c is approximately 5 x 10^{-13} m^2.

5.7.3 Diffusion in plasma

When charged particles collide, the change in their direction of motion is small and random. After many collisions, individual particles may have wandered a considerable distance from their original position. Particles in regions of high density experience more collisions than those in less dense regions and will migrate, or *diffuse*, over a period of time,

92

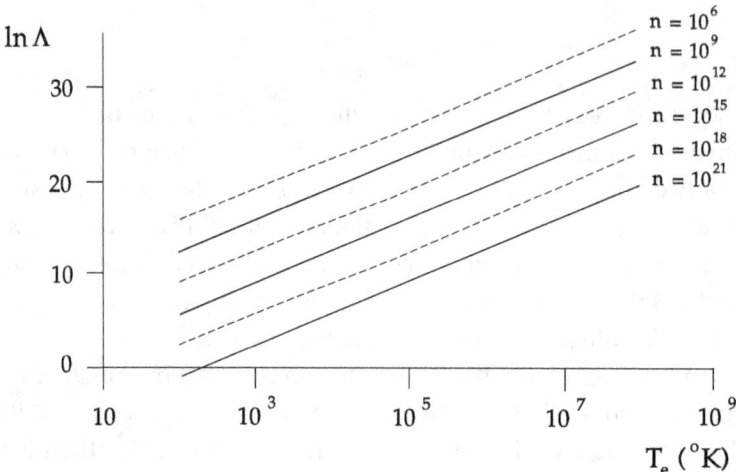

Fig. 5.5   ln Λ for various electron temperature and density values

to the less dense areas.

In the presence of an external magnetic field charged particles move parallel to the magnetic field. Collisions cause the charged particles to move from orbiting one magnetic field line to orbiting another. This movement, or transport, of particles across the magnetic field due to collisions produces diffusion perpendicular to the magnetic field. It is additional to the drift of charged particles across the magnetic field lines, and is usually so slow that charge neutrality is maintained. Ions and electrons tend to diffuse together across the field lines, a behaviour often described as *ambipolar diffusion*.

The direction of movement relative to the magnetic field is important. Parallel to the magnetic field, electrons diffuse faster than ions because of their higher thermal velocity. The rate of diffusion along the field varies inversely with the number of collisions since these interfere with the flow of the particles. Perpendicular to the magnetic field, ions diffuse more rapidly because their Larmor radius is greater than that of the electrons. The rate of diffusion here is directly proportional to the number of collisions, since it is the collisions which cause the migration. As the number of collisions increases, migration along the magnetic field decreases and diffusion across the field increases.

5.7.4   Collision frequency, υ

The number of interactions a particle undergoes in unit time — the collision frequency — depends on the plasma density and the velocity and mean free path of the particles: the higher the density, the more frequent the collisions and the shorter the mean free path, λ_{mfp}, becomes. Similarly, the higher the velocity of the particles, the longer is the distance between collisions. The average time between collisions for an electron moving with velocity v_e is therefore given by

$$\langle t \rangle = \frac{\lambda_{mfp}}{v_e}$$

So the collision frequency can be defined as:

$$\upsilon = \frac{1}{\langle t \rangle} = \frac{v_e}{\lambda_{mfp}} \tag{5.11}$$

Substituting for λ_{mfp} from equation (5.8) provides an estimate for the frequency of collisions between ions and electrons:

$$\upsilon_{ei} = n \langle \sigma_c v_e \rangle \tag{5.11a}$$

where n is the density of the collision particles (atoms or ions) and v_e is the thermal velocity of the electrons. The electron velocity distribution determines the frequency with which electrons collide with ions, so $\langle \sigma_c v_e \rangle$ indicates an average over all the velocities of the Maxwellian velocity distribution. Substituting equation (5.9) for σ_c in (5.11a) gives an average electron-ion collision frequency:

$$\langle \upsilon_{ei} \rangle = \frac{n Z_i^2 e^4 \ln\Lambda}{4 \pi \varepsilon_0^2 m_e^2 v_e^3} \tag{5.12}$$

The $1/v_e^3$ dependence confirms that the faster the electron moves the less frequently it collides with ions. Written in terms of electron temperature, using v_{rms} as given by equation (3.14), the collision frequency becomes:

$$\langle \upsilon_{ei} \rangle = \frac{n Z_i^2 e^4 \ln\Lambda}{4 \pi \varepsilon_0^2 m_e^{\frac{1}{2}} (3 k_B T)^{\frac{3}{2}}} \tag{5.12a}$$

A rough estimate of the total electron collision frequency (including collisions with atoms) can be obtained using the plasma frequency (equation 2.2) and the plasma parameter (equation 5.4):

$$\upsilon_e \sim \omega_{pe} \frac{\ln(n \lambda_D^3)}{n \lambda_D^3} \qquad (s^{-1}) \tag{5.13}$$

With $Z = 1$, our hypothetical plasma ($n = 10^{20}\,\text{m}^{-3}$, $T = 1$ keV) has $\ln\Lambda = 17.7$. Using equation (5.12), the electron collision frequency, υ_{ei}, is 1.17×10^5 s^{-1}, and from equation (5.13), $\upsilon_e = 6.07 \times 10^6$ s^{-1}.

5.7.5 "Collisionless" plasmas

Equation (5.9) shows that the collision cross-section, σ_c, decreases with increasing velocity (σ_c is proportional to $1/v^4$) and therefore with increasing temperature. The faster the particles move, the smaller is the target area for collisions. Equation (5.12) confirms this: the faster an electron moves, the less frequently it collides with ions. At very high temperatures, a plasma becomes effectively collisionless and may be described as "perfectly-conducting" — its conductivity is very high. Much of the plasma in the universe is collisionless because of low density: if there are few particles, the likelihood of a collision is very small. Large deviations from equilibrium can thus persist for a long time.

In a gas, collisions ensure that all the particles have the same temperature, that the velocity distribution is Maxwellian, and changes in temperature or pressure will spread throughout the gas. A collisionless plasma may have a non-Maxwellian velocity distribution and different types of particle may have different temperatures, with little possibility of them equalising. Such plasmas do not mix easily and may form layers separated by definite boundaries or sheets of electric current.

5.8   Plasma conductivity, σ, and resistivity, η

Conductivity is defined as the ratio of the current density, J, in a conductor to the electric field, E, causing the current to flow:

$$\sigma = \frac{J}{E} \quad (\Omega^{-1} m^{-1}) \tag{5.14}$$

J is the number of charged particles per second moving with drift velocity v through a unit area perpendicular to the direction of the electric field, so J is parallel to E and

$$J = nq\mathbf{v} \tag{5.15}$$

A material's ability to resist the flow of electric current — its *resistivity, η* — is the reciprocal of equation (5.14):

$$\eta = \frac{E}{J} \quad (\Omega m) \tag{5.16}$$

The greater the resistivity, the smaller is the current density caused by a given electric field. Equations (5.14) and (5.16) are forms of Ohm's law which apply when conductivity or resistivity is constant and the plasma is isotropic (there is no external magnetic field).

We saw in Chapter 4 that, when an electric field is applied to a fully-ionised unmagnetised plasma, electrons and ions will drift in opposite directions along the electric field. Most of the collisions that occur involve an electron and transfer a momentum, $m_e v_e$, to the other particle. The collision frequency, υ, determines the frictional force, $m_e v_e \upsilon$, which slows the electron and provides the resistivity. Without collisions, the particles respond unhindered to any applied electric field: "perfectly-conducting" plasmas have no resistivity.

In equilibrium the forces acting on the electron are equal and opposite:

$$m_e v_e \upsilon = -eE \tag{5.17}$$

giving the electron drift velocity in an electric field, equation (4.11):

$$v_e = \frac{-eE}{m_e \upsilon}$$

From equation (5.15), the current density associated with the electron motion is:

$$J = -n_e e v_e = \frac{ne^2 E}{m_e \upsilon} \tag{5.18}$$

Substituting this expression for J in (5.16) gives the plasma resistivity

$$\eta = \frac{1}{\sigma} = \frac{m_e \upsilon}{n e^2} \tag{5.19}$$

In the presence of a magnetic field the standard Ohm's law, $J = \sigma E$, must include the electric field induced by particle motion across the magnetic field. From equation (5.3), the frictional force of equation (5.17) becomes:

$$m_e v_e \upsilon = -e(E + v \times B)$$

and Ohm's law is written

$$J = \sigma(E + v \times B) \tag{5.20}$$

Conductivity in a weakly-ionised plasma is low simply because there are few free electrons to act as charge carriers. Collisions occur mainly between electrons and atoms, so resistivity results from collisions with neutral particles. The more neutrals there are, the more collisions the electrons will suffer and the lower the electron velocity will be. Electron velocity is therefore inversely proportional to neutral particle density, so current density, J, is proportional to n_e/n_n. In a fully-ionised plasma, with $n_e = n_i$, current density increases with n_e since there are more charge carriers, and the frictional drag due to the ions increases with n_i. From equation (5.12), the collision frequency, υ, varies with density so the $1/n$ dependence of η in equation (5.19) disappears — the two n's cancel out. This means that, in a fully-ionised plasma, with $n_e = n_i$, plasma resistivity is independent of electron density.

An approximate value for η, which can also be used in the presence of a magnetic field, is given by:

$$\eta = \frac{Z_i e^2 m^{\frac{1}{2}} \ln\Lambda}{4\pi\varepsilon_0^2 (3k_B T_e)^{3/2}} \tag{5.21}$$

where υ_{ei} from equation (5.12a) is substituted into equation (5.19) and Z_i is the ionic charge ($Z = 1$ for hydrogen). Equation (5.21) shows that, for a given electron temperature, plasma resistivity is proportional to the mean charge, Z, per ion. Impurities in a hydrogen plasma will therefore increase its resistivity.

A simplified formula, the *Spitzer resistivity*[2] uses electron temperature, T_e, in eV:

$$\eta_\| = 5.2 \times 10^{-5} Z_i \left(T_e\right)^{-\frac{3}{2}} \ln\Lambda \quad (\Omega m) \tag{5.21a}$$

The subscript $\|$ refers to particle motion parallel to a uniform magnetic field. In motion perpendicular to the magnetic field, slow electrons with a small Larmor radius contribute more to the resistivity than when motion is parallel to the magnetic field; so $\eta_\perp = 2\eta_\|$. Collisions between charged particles produce random scattering, causing the plasma to diffuse relative to the magnetic field lines and increasing resistivity. Frequent collisions between particles rotating about the magnetic field lines means that very little current will be produced perpendicular to the magnetic field and resistivity across the field is greater. Diffusion along the magnetic field lines also contributes.

Our sample plasma with $T_e = 1$ keV has $\eta = 4.18 \times 10^{-8}$ Ωm using equation (5.21) with $k_B T = 10^3 e$; and $\eta_{\parallel} = 2.9 \times 10^{-8}$ Ωm using the Spitzer formula (5.21a) with $T_e = 10^3$ eV. Using the formula: $\eta = (\eta_{\parallel}^2 + \eta_{\perp}^2)^{\frac{1}{2}}$, η becomes 6.4×10^{-8} Ωm.

The fact that plasma resistivity can be very small implies that a small voltage difference will produce very large currents. Naturally-occurring plasmas often have greater resistivity but their large size compensates and even here, small electric fields can produce large currents.

The ideal state of a perfectly-conducting ("collisionless") plasma is approached in very high temperature plasmas, where resistivity is negligible because of the lack of collisions. A fully-ionised "collisionless" plasma can have a resistivity of almost 6×10^{-11} Ωm — about 1/300th that of copper at room temperature. Equation (5.21) indicates that plasma conductivity increases with increasing temperature. The conductivity of a fully-ionised plasma is proportional to $T^{3/2}$ (since $\sigma = 1/\eta$) and will therefore become very great at high temperatures. The greater the conductivity, the greater is the response to electromagnetic fields.

Collisional frequency, and therefore resistivity, depends on the collision cross-section and thermal agitation of the electrons. Since the collision frequency in a plasma decreases with increasing temperature, resistivity will similarly decrease. This limits the use of a simple method of heating plasma in the laboratory: that of passing an electric current through it to dissipate some energy as heat — so-called *ohmic heating*. Just as $I^2 R$ losses appear as heating of electrical wires, so a plasma is heated by the transfer of energy from the electric field to the electrons, at a rate of $\mathbf{J} \cdot \mathbf{E} = J^2 \eta$ per unit volume. The temperature dependence of η means that above $T_e = 1$ keV, when the plasma becomes a good conductor, ohmic heating is much reduced.

Equation (5.21) is valid provided the electrons' mean drift velocity, $\mathbf{J}/n_e e$, is small compared to their mean thermal velocity. In reality, there is a Maxwellian distribution of electron velocities. If an electric field is applied to the plasma, electrons in the high energy tail of the distribution will be more easily accelerated by the electric field. As electron velocities increase, the collision cross-section shrinks rapidly (υ_{ei} varies inversely with v^3), so the high-speed electrons are unaffected by collisions. This distorts the velocity distribution and a "runaway" situation may develop in which the fast electrons form an electron beam detached from the main distribution. The current is then carried mainly by these electrons, rather than by the slower electrons in the body of the distribution.

5.9   Summary

1.   Plasma is an electrically neutral mixture of charged particles which act collectively. Each charged particle interacts simultaneously with several others so their movements are interlinked. What distinguishes any plasma from a simple collection of charged particles is the law of charge neutrality — the approximate equality of positively and negatively charged

particles (quasi-neutrality) — and the formation of a sheath at all plasma boundaries.

2.　　　The Debye length, λ_D, measures the sphere of influence of a charged particle in a plasma — the distance over which the electric field of the particle extends before it is shielded by oppositely-charged particles. The action of charged particles to reduce the effect of a local electric field, caused by an excess of positive or negative particles, is called Debye shielding and restores quasi-neutrality. The number of particles, N_D, in a Debye sphere is a measure of the screening capabilities of a plasma.

3.　　　Criteria for an ionised gas to be a plasma:
　　　(i)　　　$\lambda_D \ll L$ — required for quasi-neutrality.
　　　(ii)　　$N_D \gg 1$ — required for collective behaviour.
　　　(iii)　　$n_e = \Sigma n_i$ — macroscopic charge neutrality.

4.　　　Collisions between charged particles are Coulomb collisions. The cross-section for these long-range collisions is always much larger than that for elastic scattering since the charged particles interact at a distance and the cumulative effect of many small-angle deflections is more important than the relatively few large-angle deflections.

5.　　　Collision frequency depends on plasma density and particle velocities: the higher the density the more frequent the collisions; the higher the velocity the smaller the collision cross-section and the less frequent the collisions.

6.　　　In the presence of a magnetic field, diffusion along the field varies inversely with collision frequency because the collisions impede particle flow. Diffusion across the field is directly proportional to collision frequency since the collisions produce the diffusion.

7.　　　Plasma resistivity results from collisions between the particles. Collision frequency in plasma decreases with increasing temperature and resistivity levels fall as a result. At low temperatures, ohmic heating provides a means of raising the temperature of the plasma using its resistivity. Above $T_e \approx 1$ keV the plasma becomes a good conductor and ohmic heating is ineffective.

References
　　1.　　　Langmuir, I.: *Phys. Rev.,* **33**, p.954-89 (1929)

　　2.　　　Spitzer, L.: *Physics of Fully Ionised Gases* (1962)

Chapter 6

Plasma waves and oscillations

6.1   Waves in plasma

Physical systems often respond to disturbances by emitting waves. In a plasma, any movement can produce oscillations which may propagate as waves. The resulting fluctuations and turbulence affect the transport of particles and energy within the plasma and, especially in low-temperature plasmas, often cause the loss of both particles and energy.

An electromagnetic wave travelling through a plasma exerts an accelerating force on the ions and electrons, while the collective behaviour of the charged particles in turn affects the propagation of the wave. The properties of both wave and plasma can thus be considerably altered. Individual collisions between particles have little effect and can be ignored. Electromagnetic waves can be used to investigate a plasma by comparing the known characteristics of a transmitted wave with its received (distorted) form to obtain information about the plasma through which the wave has passed.

When the velocity of the charged particles is very much less than the speed of light, c, the electric component of the electromagnetic wave has a much greater effect than the magnetic component on the plasma particles. As a result, the wave's magnetic field can be neglected when considering electromagnetic wave propagation.

6.2   Simple harmonic motion and dispersion

Waves occur when a medium, such as a plasma, is disturbed from its equilibrium position and the pattern of the disturbance can travel from one region of the medium to another. To set a system in motion requires an input of energy and it is this energy that is transported. The velocity of propagation is determined by the mechanical properties of the medium through which the wave travels.

Simple harmonic motion — periodic movement about an equilibrium position — is described by amplitude, period and frequency. Position is a sinusoidal function of time and using angular frequency, ω, rather than frequency, f, simplifies many of the equations.[a]

---

[a]   Frequency, f, is the number of cycles per unit of time, measured in hertz (Hz).
Angular frequency, $\omega = 2\pi f$, is the rate of change of an angular quantity, measured in radians per second, and is not necessarily linked to a rotational motion.

Individual particles either move back and forth (a *longitudinal* wave, e.g. sound waves) or up and down (a *transverse* wave, e.g. electromagnetic waves) around their equilibrium position. In longitudinal waves the particle motion is along the direction of propagation, producing regions of increased pressure (compression) and reduced pressure (expansion or rarefaction). With transverse waves, particles are displaced perpendicular (transverse) to the wave's direction of travel.

A periodic wave has a definite frequency, f, and wavelength λ. The number of waves in a unit of distance is represented by the *wave number* or *propagation constant*, $k = 2\pi/\lambda$. Expressing k as a vector incorporates the direction of propagation of the wave. The speed of propagation, v, is related to wave number, frequency and length via:

$$v = \lambda f = \frac{\omega}{2\pi}\lambda = \frac{\omega}{k} \tag{6.1}$$

In vacuum, v equals c, the speed of light, but is usually less than c in the presence of matter.

Waves in plasma can be classified according to the orientation of $\boldsymbol{k}$ with respect to the wave's electric field, $\boldsymbol{E}_{wave}$, since this is always present due to the effect of the charged particles. *Electromagnetic* waves are transverse waves with $\boldsymbol{k}$ perpendicular to $\boldsymbol{E}_{wave}$, so particle movement is perpendicular to the wave's direction of travel. Longitudinal waves, such as sound waves, may be termed *electrostatic*, since $\boldsymbol{k}$ is parallel to $\boldsymbol{E}_{wave}$.

We can describe the position and motion of individual particles in the medium using a wave function of the form $y = A \sin(\omega t - kz)$, where y is the transverse displacement of the particle at time t and z is the direction of propagation. The quantity $(\omega t - kz)$ is the *phase* of the motion: an angular quantity, measured in radians. It indicates which part of the sinusoidal cycle is occurring at time t at a given point z. *Phase velocity*, v_p, is the velocity of a point of constant phase in the wave. This, in effect, is the velocity at which we must travel to keep up with a particular point on the wave. The phase velocity of a wave travelling in the z-direction propagates according to

$$(\omega t - kz) = \text{constant} \tag{6.2}$$

Differentiating with respect to t:

$$\frac{dz}{dt} = \frac{\omega}{k} \tag{6.3}$$

So, from equation (6.1):

$$v_p = \frac{\omega}{k} \tag{6.4}$$

In a vacuum, plane electromagnetic waves all travel with the same velocity, c. In matter, their velocity depends on their frequency. This dependency, or *dispersion*, is a property of the medium through which the wave passes. A group of waves propagating within a narrow range of frequencies forms a *wave packet*: a superposition of waves with different wave numbers k and angular frequencies ω, travelling at a *group velocity*, v_g, given by

$$v_g = \frac{d\omega}{dk} \tag{6.5}$$

This represents the rate at which energy or information can be transmitted by a wave in a dispersive medium. It measures the velocity of propagation of a signal and so can never exceed the velocity of light, c, without violating relativity theory.

Phase velocity (equation 6.4) is the velocity at which a point of constant phase moves, which in the case of a wave packet could be the individual crests. These points can move faster or slower than the bundle of energy and information making up the wave packet. In short: the envelope of the wave packet carries the signal and travels at the group velocity, v_g. The individual frequency components making up the signal contained within the wave packet travel at different velocities represented by v_p. So, although c is the maximum velocity at which energy can move in electromagnetic radiation, the phase velocity can and generally does exceed this value in ionised gases and plasmas.

If the ratio of the change in phase velocity to the change in wavelength, $\partial v_p / \partial \lambda$, is greater than zero, the wave packet spreads out progressively — it disperses. The dispersive properties of any medium are given by the dispersion relation

$$\omega^2 = c^2 k^2 + constant \tag{6.6}$$

This relates the angular frequency, ω, of the wave to the wave number k and indicates the ranges of frequencies over which wave propagation can occur: it describes the response of the medium to a given wave. The constant depends on the medium through which the electromagnetic wave travels.

Most plasmas are confined by magnetic fields, either externally-applied or self-generated by internal currents, but some oscillations behave as if there is no magnetic field present. A plasma with no external magnetic field is *isotropic* since its properties do not vary with direction. We will consider this situation first. For such a plasma, the constant in equation (6.6) is represented by ω_p^2:

$$\omega^2 = c^2 k^2 + \omega_p^2 \tag{6.7}$$

Rearranging this produces:

$$k = \frac{\omega}{c} \left(1 - \frac{\omega_p^2}{\omega^2} \right)^{\frac{1}{2}} \quad (rad\ m^{-1}) \tag{6.8}$$

Differentiating equation (6.7) with respect to k yields:

$$2\omega \frac{d\omega}{dk} = 2c^2 k$$

or

$$\frac{\omega}{k} \frac{d\omega}{dk} = c^2 \tag{6.9}$$

which is equivalent to the vacuum situation, $v_p v_g = c^2$. Rearranging equation (6.9) gives an alternative definition of the group velocity:

$$v_g = \frac{d\omega}{dk} = c^2 \frac{k}{\omega}$$

Substituting the expression for k given in equation (6.8) produces a definition of the plasma group velocity:

$$v_g = c \left(1 - \frac{\omega_p^2}{\omega^2} \right)^{\frac{1}{2}} \tag{6.10}$$

while, from equation (6.4), the phase velocity, v_p, can be written

$$v_p = \frac{c}{\left(1 - \frac{\omega_p^2}{\omega^2} \right)^{\frac{1}{2}}} \tag{6.11}$$

The *refractive index*, N, for any transparent medium is defined as the ratio of the phase velocity of electromagnetic waves in free space to that in the medium, $N = c/v_p = ck/\omega$. Substituting for v_p from equation (6.11) gives the refractive index for an isotropic plasma:

$$N = \left(1 - \frac{\omega_p^2}{\omega^2} \right)^{\frac{1}{2}} \tag{6.12}$$

A refractive index of N less than one means that radiation entering a plasma from free space is deflected away from the normal. As with conventional optics, electromagnetic waves can only propagate if the plasma is transparent.

The factor $(1 - \omega_p^2/\omega^2)$ is often called the *dielectric constant* of the plasma. It measures the amount by which the plasma alters the values of the various parameters from their value in vacuum, just as a dielectric affects capacitance and potential difference when inserted between the plates of a capacitor. The dielectric constant is used to determine the effective permittivity of the plasma in the absence of collisions and with no magnetic field:

$$\varepsilon_p = \varepsilon_0 \left(1 - \frac{\omega_p^2}{\omega^2} \right) \tag{6.13}$$

6.3   Plasma frequency, ω_p

A basic property of plasma is its ability to maintain charge neutrality on a macroscopic scale. To demonstrate what happens when the equilibrium is perturbed, we will take the theoretical case of a cold plasma in which the electrons and ions are uniformly distributed and stationary: the plasma is electrically neutral everywhere and there is no random thermal motion of the particles. If this stable system is disturbed by, for example, moving a group of electrons from one part of the plasma to a neighbouring region, there will be a net positive charge in the area from which we have removed the electrons, and a net negative charge in the region to which they have been transferred. This charge imbalance produces an electric field. The electrons, being much lighter than the ions, respond more rapidly to the electric field; so we will assume, in the following discussion, that the ions remain stationary.

As the electrons are accelerated by the electric field they gain kinetic energy. When they return to their original position, the electric field disappears and the plasma is again electrically neutral. However, all the electrostatic potential energy associated with the initial disturbance has been converted into kinetic energy of the displaced electrons, causing them to overshoot and creating another disturbed density distribution. The resulting charge separation again creates an electric field, this time in the opposite direction, and the electron motion is retarded as the kinetic energy is converted into potential energy. Eventually, electron velocity is zero and the situation is identical to that produced by the initial disturbance, except that the direction of the electric field is reversed. The process continues as the electrons oscillate back and forth.

Thus, when charge neutrality is disturbed, space charge fields within the plasma produce bulk oscillations of electrons, and to a much lesser extent ions, which tend to restore the balance. These are the high-frequency electrical oscillations which led to Langmuir's development of the framework of plasma physics in 1928-9 (Chapter 2). Often called *Langmuir oscillations*, these fluctuations are driven by the electric field which is created through the electron displacement. No externally-applied electric field is needed. The angular frequency with which the electrons oscillate about their equilibrium position is a natural frequency of oscillation known as the *electron plasma frequency*, ω_{pe}. It depends only on the electron density and is given by

$$\omega_{pe} = \left(\frac{ne^2}{m_e\varepsilon_0} \right)^{\frac{1}{2}} \quad (rad\ s^{-1}) \tag{6.14}$$

where n is the electron density. (For a mathematical derivation of ω_p see Appendix A.) The frequency in hertz is:

$$f_p = \frac{\omega_p}{2\pi} = 8.98n^{\frac{1}{2}}\ Hz \tag{6.15}$$

This is an oscillation in which there is an electric field but no magnetic field. The period, $1/f$, of the oscillation is very short, making it difficult, but not impossible, for the ions to respond. Since m_i is very much greater than m_e, the amplitude of the ion oscillations is much smaller. This results in a slight correction to the overall plasma frequency of

$$\omega_p = \omega_{pe} + \omega_{pi} = \omega_{pe} \left(1 + \frac{m_e}{m_i} \right)^{\frac{1}{2}} \tag{6.16}$$

i.e. $\omega_p = 1.00027\ \omega_{pe}$. The oscillation occurs if the plasma is homogeneous and isotropic and provides a basis for describing the collective rather than individual behaviour of the plasma particles. It is a fundamental parameter of plasmas. Since m_e is small, ω_p is usually very high. The plasma frequency is sometimes referred to as *plasma resonance*, or plasma-electron resonance, because the electrons behave like a resonant system.

For our sample plasma, with an electron density of $10^{20}\,\mathrm{m}^{-3}$, $\omega_{pe} = 5.6 \times 10^{11}\,\mathrm{rad\,s}^{-1}$; giving a frequency of $8.9 \times 10^{10}\,\mathrm{Hz}$ and a wavelength of just over $3.3 \times 10^{-3}\,\mathrm{m}$. Thus ω_p lies in the microwave range. The period of the oscillation is $1.1 \times 10^{-11}\,\mathrm{sec}$. From equation (6.8), if the propagating electromagnetic wave has an angular frequency, ω, of 10^{12} rad s^{-1}, the wave number, k, is 2.76×10^3 rad m^{-1}, group velocity, v_g (equation 6.10), is 2.5×10^8 m s^{-1}, and phase velocity, v_p (equation 6.11), is 3.6×10^8 m s^{-1}.

An approximate value for ω_p may be obtained from the ratio of the mean thermal speed of the electrons, $\langle v_e \rangle$, to the Debye length, since:

$$\lambda_D \omega_p = \left(\frac{\varepsilon_0 k_B T}{n e^2} \frac{n e^2}{\varepsilon_0 m_e} \right)^{\frac{1}{2}} = \left(\frac{k_B T}{m_e} \right)^{\frac{1}{2}} \approx \langle v_e \rangle \tag{6.17}$$

where $\langle v_e \rangle$ is taken as v_{rms} (equation 3.14). Using equation (6.17), our sample plasma has v_{rms} = 2.3×10^7 m s^{-1} and $\lambda_D = 2.35 \times 10^{-5}$ m, which gives a value for ω_p of 9.8×10^{11} rad s^{-1}, compared with the previously calculated 5.6×10^{11} rad s^{-1}.

6.3.1 Langmuir waves

Since ω_p is independent of k, the group velocity given by equation (6.5) is zero and the disturbance, in theory, does not propagate. In practice, the electric field produced by the local charge separation affects neighbouring regions of the plasma and the oscillation tends to spread through the plasma. In a warm plasma, the thermal motion of the electrons can cause these long wavelength Langmuir waves to propagate with a frequency given by

$$\omega^2 = \omega_p^2 + \left(\frac{3 k_B T_e}{m_e} \right) k^2 \tag{6.18}$$

Here, the c^2 in equation (6.7) is replaced with $(v_{rms})^2$ from equation (3.14). Langmuir waves can be described as longitudinal electrostatic waves since the direction of propagation is parallel to the wave electric field ($k \| E_{wave}$).

6.3.2 Significance of ω_p

The dispersion relation of equation (6.7) expresses the angular frequency, ω, of an electromagnetic wave propagating in a plasma, in terms of the wave number, k, and the plasma frequency, ω_p. Fig. 6.1 represents this relation diagrammatically. The curve $\omega(k)$ is the dispersion relation for real values of k; ω_p marks the minimum point of the curve: the point where $k = 0$. At any given point, P, the group velocity, v_g, is the slope of the tangent to the curve $\omega(k)$ at that point, and the phase velocity, v_p, is the slope of the line from the origin to P. The dotted line marked $\omega = ck$ is the non-dispersive, vacuum situation.

If the frequency, ω, of the wave is greater than the plasma frequency, then k is a real number and the electromagnetic wave will propagate in the plasma without any reduction (*attenuation*). As the wave frequency increases, the plasma electrons become unable to respond to the oscillating electric field of the wave. The value of k approaches its free space value, ω/c, and the wave behaves as if the plasma were not present: $v_p = v_g = c$. Thus, visible light can pass freely through a plasma whereas radio waves may be strongly attenuated or reflected.

From equation (6.14), the plasma frequency is proportional to electron density, so as density increases, ω_p will increase and k will decrease (from equation 6.8). At the point where $\omega = \omega_p$, the wave number, k, is zero and the wave is reflected. Propagation of the wave through the plasma ceases, often quite suddenly: the plasma frequency acts as a *cut-off frequency* for electromagnetic waves in an unmagnetised plasma. The cut-off point can be used to measure plasma density if the frequency, f, of the electromagnetic wave is known. Using equations (6.14) and (6.15), the critical electron density, n_{ec}, (and therefore the electron density of the plasma) when $\omega = \omega_{pe}$ is:

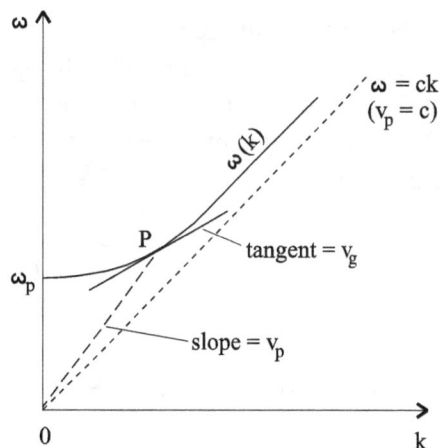

Fig. 6.1 **Dispersion relation for a wave propagating in cold isotropic plasma for real values of k**

$$n_{ec} = \omega_{pe}^2 \frac{m_e \varepsilon_0}{e^2} = \frac{(2\pi f)^2 m_e \varepsilon_0}{e^2} = 1.24 \times 10^{-2} f^2 \ \ electrons \ m^{-3} \qquad (6.19)$$

If electron density is less than n_{ec}, the incident electromagnetic wave will pass through the plasma with little effect. If electron density is greater than n_{ec}, the incident wave will interact strongly with the plasma electrons at the surface and the wave will be reflected or absorbed. Equation (6.12) shows that the refractive index decreases with increasing density, so electromagnetic waves tend to bend away from more highly-ionised regions.

Another way of looking at the cut-off effect follows from equation (6.17). If ω_p is approximately equal to $\langle v_e \rangle / \lambda_D$, then the mean time for an electron to move a distance equal to a Debye length is $1/\omega_p$. The Debye length, as we saw in Chapter 5, is the distance beyond which the plasma is unable to maintain charge separation. If the frequency of the electromagnetic wave is lower than the plasma frequency, the electrons can move fast enough to shield the plasma from the wave. This is sometimes described as the "particle interaction regime" since individual electrons can respond to the electric field of the electromagnetic wave. As the wave frequency increases, the electrons in the plasma find it increasingly difficult to screen out the disturbance.

The group velocity, v_g (equation (6.10)), is always less than c and goes to zero when $\omega = \omega_p$ and k is zero. It becomes imaginary as ω drops below ω_p, as does the index of refraction (equation 6.(12)). From equation (6.4), the phase velocity, v_p, approaches infinity as k tends to zero. Phase velocity also increases with increasing electron density because of the influence of ω_p on v_p (equations (6.11) and (6.14)).

If the frequency of the wave is less than ω_p, the ratio ω_p^2/ω^2 in equation (6.8) is greater than one and k becomes an imaginary number; phase velocity is infinite and group velocity is zero. The electromagnetic wave will be unable to propagate and will rapidly decay

— it becomes *evanescent*. The decay is exponential and the amplitude of the wave decreases rapidly but the phase does not change. The evanescence is not caused by wave energy dissipation (or *damping*); the wave and its energy are quite simply reflected by the plasma. The rate of attenuation below the cut-off is such that the field strength of the wave is reduced to e^{-1} (= 0.368) of its initial value when it has penetrated a skin depth d where

$$d = \frac{c}{\omega_p} \times \frac{1}{\left(1 - \frac{\omega^2}{\omega_p^2}\right)^{\frac{1}{2}}} = \frac{c}{(\omega_p^2 - \omega^2)^{\frac{1}{2}}}$$ (6.20)

Thus, if $\omega_p = 10^{11}$ rad s^{-1} and $\omega = 10^{10}$ rad s^{-1}, $d = 3$mm.

6.4   Damping mechanisms

In reality, wave energy can be lost or dissipated due to viscosity of the medium or collisions between particles. These energy loss mechanisms reduce the amplitude of the wave oscillations, causing them to die away. The reduction in amplitude is accompanied by a slight increase in the period of the oscillation (and therefore a change in phase) and is known as *damping*. The effect is shown in Fig. 6.2. There are two main damping mechanisms in plasma: one is collisional; the other involves a transfer of energy.

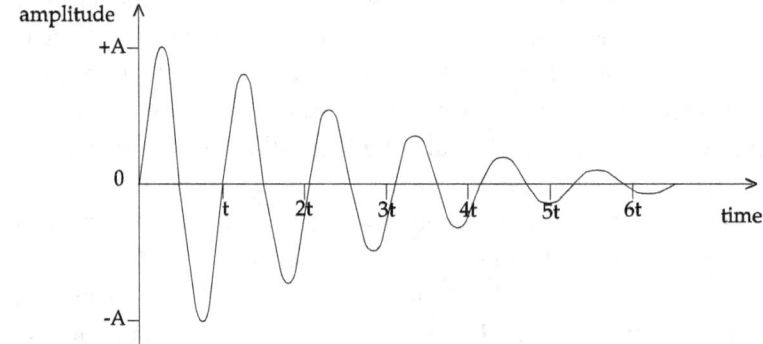

Fig. 6.2   Damped oscillation

6.4.1   Collision damping

With plasma oscillations, the electrons move collectively, this movement being superimposed on the random thermal motion of the individual particles. If an oscillating electron collides with another particle it will be randomly scattered, gaining energy but playing no further part in the oscillations. Collisions between electrons and neutral particles thus tend to damp the oscillations, gradually reducing their amplitude as energy is transferred from the wave field to the particles. In many low-density plasmas the damping effect of collisions is small. At higher densities, the characteristic plasma properties may be completely destroyed. To keep collision damping to a minimum, the collision frequency must be very much less than ω_p, otherwise the electrons will be unable to act independently and will be forced, by collisions, into equilibrium with the neutral particles. The plasma will become just an ionised gas.

This requirement that the collision frequency is much less than the plasma frequency provides a fourth criterion, to add to the three given in Chapter 5, for an ionised gas to be a plasma. The level of ionisation is important: few neutral particles compared to electrons means there is less chance of an electron and neutral particle colliding than when neutral particle density is greater than electron density.

6.4.2 Landau damping

A second damping mechanism exists which does not involve collisions between particles, so no energy is dissipated. It was predicted by Lev Landau in 1946 as part of his theory of plasma wave-particle interactions.[1] At long wavelengths, waves propagate with high phase velocity but low group velocity and very little damping. As the wavelength decreases, the phase velocity approaches the electron thermal velocity and damping increases.

The effect responsible is known as *Landau damping* or "collisionless" damping and involves particles with a velocity almost equal to the phase velocity of the wave, ω/k. These particles, because they are travelling at almost the same speed as the wave, experience a relatively static rather than a fluctuating electric field: they resonate with the oscillations of the wave and can therefore easily exchange energy with the wave. An electron travelling with the wave will be accelerated or slowed down depending on where it is on the waveform. Particles moving slower than the wave take energy from it and are accelerated; those moving faster than the wave give up energy to it and slow down. If the velocity distribution is Maxwellian, there are generally more slower than faster particles so more particles are being accelerated than decelerated. This results in a net transfer of energy from the wave to the particles. The wave loses energy and is damped without any collisions having occurred.

Landau damping is fundamental to a wide range of plasma-wave interactions. It occurs whenever there are particles whose velocity is almost equal to the phase velocity of a plasma wave. When this happens the particles are said to be in resonance with the wave. Electron plasma waves (Langmuir waves) are strongly damped even when there is no resistivity and therefore no conventional mechanism for dissipation.

6.5 Ion oscillations

An oscillation similar to the electron oscillation, but involving positive ions, can develop at the much lower *ion plasma frequency*:-

$$\omega_{pi} = \left(\frac{Z_i^2 \, |e|^2 \, n_i}{m_i \varepsilon_0} \right)^{\frac{1}{2}} \qquad (6.21)$$

where Z_i is the charge on the ions. Each type of ion in the plasma will have its own plasma frequency, determined by its atomic number, Z. Taking our hypothetical plasma as an example, with $n_e = n_i = 10^{20}$ m^{-3}, the electron plasma frequency was calculated to be 5.6 x 10^{11} rad s^{-1}. By comparison, the hydrogen ion plasma frequency, ω_{pi}, is 1.3 x 10^{10} rad s^{-1}. This gives a frequency, f, of 2.1 x 10^9 Hz and a wavelength of approximately 0.14 m.

In most cases, the term "plasma frequency" (ω_p) is taken to mean the electron plasma frequency, since, if electron and ion densities are equal, the plasma electron and ion frequencies vary as

$$\frac{\omega_{pi}}{\omega_{pe}} = \left(\frac{m_e}{m_i} \right)^{\frac{1}{2}} \approx \frac{1}{43} \tag{6.22}$$

and so ω_{pe} is very much greater than ω_{pi}.

6.5.1   Ion acoustic waves

Low-frequency longitudinal ion oscillations, first observed in the late 1920s,[2] can propagate as *ion acoustic waves* if the angular frequency of the wave, ω, is less than or equal to the ion plasma frequency, and if the electron temperature, T_e, is much greater than the ion temperature, T_i. The frequency of these waves is given by

$$\omega_s = k \left[\frac{k_B T_e}{m_i(1 + k^2 \lambda_D^2)} + \frac{3 k_B T_i}{m_i} \right]^{\frac{1}{2}} \tag{6.23}$$

Ion acoustic waves occur when displaced ions form regions of compression and rarefaction within the plasma. The ions in the compressed regions tend to expand into the areas of rarefaction and oscillate back and forth, forming a wave similar to an ordinary sound wave. Sound waves in ordinary fluids need collisions to transmit energy. In plasma, particles interact and transmit energy via Coulomb collisions. The restoring force is therefore electrostatic rather than collisional. The wave travels at the *ion sound speed*, $v_s = \omega_s/k$, given approximately by

$$v_s = \langle v_i \rangle \left(\frac{T_e}{T_i} \right)^{\frac{1}{2}} \equiv \left(\frac{T_e}{m_i} \right)^{\frac{1}{2}} \tag{6.24}$$

where $\langle v_i \rangle$ is the average thermal speed of the ions, and T_e is measured in joules. For our sample plasma, $v_s \approx \sqrt{(10^3 \, e/m_i)} - 3$ x 10^5 m s^{-1}.

These low-frequency waves involve both electron and ion oscillations. The electrons, with a much smaller mass, experience a greater acceleration in a force field and therefore play an important role in the ion oscillations. If an ion is displaced from its equilibrium position, the change in the charge distribution will produce electric fields within the plasma. Electrons can move in these fields, causing the ion oscillations to propagate. The electrons move with the ions as a single fluid, tending to shield out electric fields produced by the concentrations of ions.

Electron oscillations produce constant frequency (ω_{pe}) waves. At long wavelengths (with small wave-number k) ion acoustic waves are constant velocity waves, with phase velocity v_p similar to the ion sound speed, v_s, and $v_g = v_p$. At wavelengths shorter than a Debye length (large k), the sound wave becomes a constant frequency wave at the ion plasma frequency, ω_{pi}. Unlike electron plasma waves, ion acoustic waves are always slightly damped, even at long wavelengths. Although the electrons make a small contribution, most of this damping comes from the ions through *ion Landau damping*. If the phase velocity of the wave is similar to the thermal velocity of the ions, and the ion temperature, T_i, is greater than or equal to the electron temperature, T_e, then strong ion Landau damping can occur. As the frequency of the wave approaches the ion plasma frequency, the phase velocity decreases and the wave resonates with more ions, becoming more heavily damped. Ion Landau damping of these acoustic waves can be severe when the difference in electron and ion temperatures is very small.

Artificial propagation of ion acoustic waves can be used to measure electron and ion temperatures. A qualitative estimate of Landau damping can indicate whether T_e is equal to or greater than T_i. In the latter case, increasing the wave frequency until the damping rises sharply provides a means of estimating the ion plasma frequency and from this, the ion density. Reducing the wave frequency until additional non-Landau damping occurs, as a result of ion-ion and ion-neutral collisions, gives an indication of the collisional frequency.

6.6   The influence of magnetic fields

The important feature in an unmagnetised plasma is the cut-off for electromagnetic waves at the plasma frequency: an electromagnetic wave with a frequency less than the plasma frequency will not propagate. A plasma immersed in a static, uniform magnetic field is anisotropic and other types of waves are possible. Electromagnetic waves can propagate at frequencies much less than the plasma frequency and their behaviour is more complicated.

Electromagnetic waves are transverse waves with an oscillating electric field component, labelled earlier as E_{wave}. Perpendicular to E_{wave} is an oscillating magnetic field component which we will call B_{wave}. The simplest form of electromagnetic wave is the ordinary light wave, in which the wave vector, k, is perpendicular to both E_{wave} and B_{wave}. The dispersion relation for electromagnetic waves in vacuum is given by $\omega^2 = c^2 k^2$. In a plasma with no background magnetic field, B_{ext}, this becomes: $\omega^2 = c^2 k^2 + \omega^2_p$. An external magnetic field adds further modifications. The direction and magnitude of the wave propagation vector, k, now becomes important. The wavelength is short in directions in which the component of k is large, since k varies with $1/\lambda$, so quantities vary rapidly in space. In directions in which k is small, and the wavelength is therefore long, quantities will vary slowly in space.

Ignoring the effects of collisions, we will consider some of the simple cases of electromagnetic wave propagation parallel and perpendicular to the magnetic field.

6.6.1 High-frequency wave propagation parallel to B — the effect of polarisation

In ordinary (unpolarised) electromagnetic radiation, the electric and magnetic fields vibrate in all directions perpendicular to the direction of propagation. Waves in which the wave fields remain fixed in direction during propagation are said to be plane-polarised. When a plane-polarised electromagnetic wave travels through a plasma in the direction of an external magnetic field, the wave's electric and magnetic fields interact with charged particles in the plasma which are gyrating about the magnetic field at their respective cyclotron frequencies (equation 4.16). This affects propagation of the wave at all frequencies. The directions of the wave fields and the plane of polarisation both rotate in a plane perpendicular to the direction of propagation. If the wave electric field rotates about the direction of propagation with constant magnitude as the wave travels through the medium (Fig. 6.3), the wave is *circularly-polarised* and the plane of electromagnetic vibration is, in effect, helical. If the wave frequency equals the particle cyclotron frequency, the wave electric field can rotate with the particles, producing resonance and allowing efficient exchange of energy between the two systems.

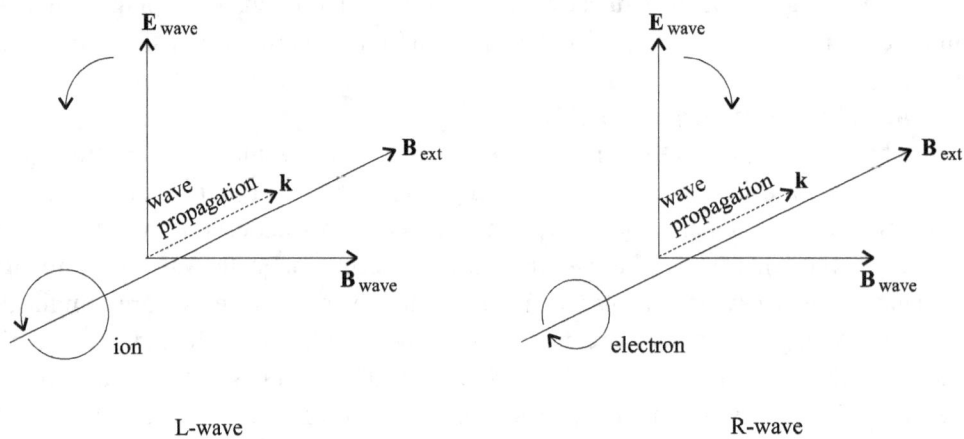

Fig. 6.3 Circularly-polarised electromagnetic wave propagation parallel to an external magnetic field

If the electromagnetic wave rotates in an anti-clockwise direction as it travels through the plasma (left-hand diagram of Fig. 6.3) it is said to be left-circularly-polarised. If it rotates in a clockwise direction it is right-circularly-polarised. Since ions and electrons gyrate about the magnetic field lines in opposite directions, we can discuss the behaviour of the left and right components of the circularly-polarised wave separately. The left-circularly-polarised component, the *L-wave* (or *L*-mode), rotates in the same direction as the positive ions, and can couple to the motion of the ions; the right-circularly-polarised component, the *R-wave* (or *R*-mode), rotates with the electrons and can couple to the gyration of the electrons around the magnetic field lines. (Note that the "handedness" of the wave is defined relative to the direction of the magnetic field, since this determines the orbital motion of the charged particles.)

Since electrons have smaller mass than ions, the left and right components have different angular frequencies, (from equation (4.16), $\omega_c = qB/m$), and therefore different propagation constants, k. For the R-wave, the dispersion relation given by equation (6.8) becomes:

$$k_R = \frac{\omega}{c}\left(1 - \frac{\omega^2_{pe}}{\omega(\omega - \omega_{ce})} - \frac{\omega^2_{pi}}{\omega(\omega - \omega_{ci})}\right)^{\frac{1}{2}} \tag{6.25}$$

Since ω_{pe} is much greater than ω_{pi}, the electrons have a greater effect than the ions on k and the final term in equation (6.25) can usually be ignored. We saw in Section 6.3.2 that, when the wave frequency equals the plasma frequency, $k = 0$ and the wave is reflected: ω_p is a cut-off frequency for the electromagnetic wave. A similar situation arises with L- and R-waves. The cut-off frequencies, ω_L and ω_R, are found by setting $k = 0$ in the relevant dispersion relation (the solution is given in Appendix B). With $k_R = 0$ in equation (6.25), the cut-off frequency, ω_R, is:

$$\omega_R = \frac{1}{2}\left[\omega_{ce} + (\omega_{ce}^2 + 4\omega_{pe}^2)^{\frac{1}{2}}\right] \tag{6.26}$$

The dispersion relation for the L-wave is given by:

$$k_L = \frac{\omega}{c}\left(1 - \frac{\omega^2_{pe}}{\omega(\omega + \omega_{ce})} - \frac{\omega^2_{pi}}{\omega(\omega + \omega_{ci})}\right)^{\frac{1}{2}} \tag{6.27}$$

Ignoring the effect of the ions, and hence the final term, the cut-off occurs at:

$$\omega_L = \frac{1}{2}\left[-\omega_{ce} + (\omega_{ce}^2 + 4\omega_{pe}^2)^{\frac{1}{2}}\right] \tag{6.28}$$

When the wave frequency approaches the ion cyclotron frequency, ω_{ci}, the ions have greater effect on k than do the electrons and so the final term in the brackets in equation (6.27) is usually much greater than the middle term and cannot be ignored.

The L-wave has a simple dispersion curve, as shown in Fig. 6.4 (overleaf). It only exists when the wave frequency, ω, is greater than ω_L. The L-wave cut-off ω_L is always less than ω_R, the R-wave cut-off. In a dense plasma, the plasma frequency is generally much greater than the electron cyclotron frequency[b] and the cut-off frequency, ω_L, is also higher than ω_{ce}. In low-density plasma, when the plasma frequency is less than ω_{ce}, (right-hand diagram) the cut-off frequency is lower than ω_{ce}.

There are now discontinuities with frequency in the propagation constant k, with "pass" and "stop" bands for the propagating wave. The R-wave has two pass bands: one below ω_{ce} — often called the electron-cyclotron wave — and one above ω_R, with a stop band

---

[b] From equation (6.14) $\omega_{pe} = 56.4\sqrt{n}$ and from equation (4.16) $\omega_{ce} = 1.76 \times 10^{11} B$

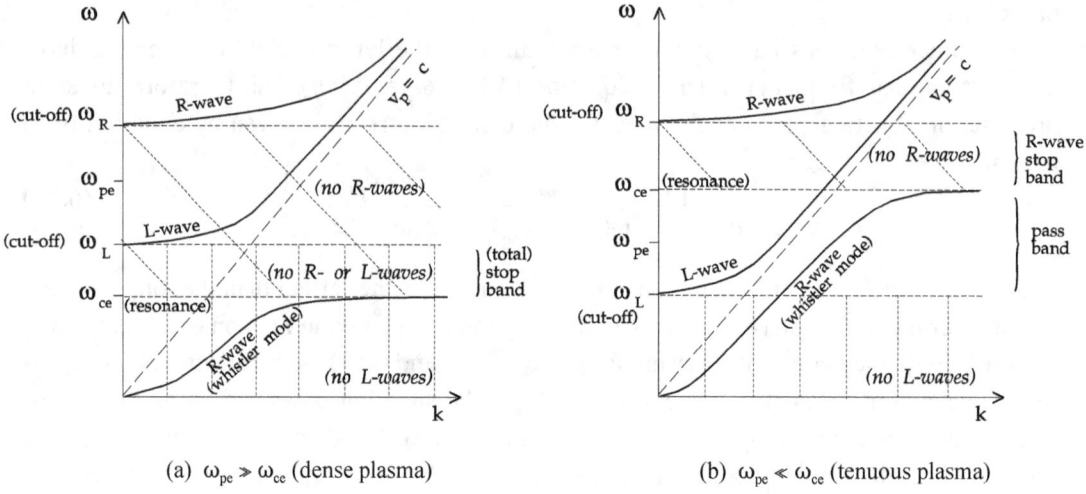

(a) $\omega_{pe} \gg \omega_{ce}$ (dense plasma) (b) $\omega_{pe} \ll \omega_{ce}$ (tenuous plasma)

Fig. 6.4 Left- and right-circularly-polarised electromagnetic waves propagating parallel to B_{ext}

between them. Stop bands occur when the square of the index of refraction (equation 6.12) is imaginary, i.e. $N^2 < 0$. The charged particles in the plasma try to screen out the oscillating electric field of the electromagnetic wave. These screening effects, which occur when k goes to zero, become increasingly evident as ω_{ci} and ω_{ce} become small compared with the frequency, ω, of the electromagnetic wave.

Cut-offs occur when $k = 0$. At the other extreme, as k becomes very large, the wavelength of the wave becomes very small, but its frequency remains constant since, from equation (6.1), $\omega = k\lambda f$. This produces a strong interaction, a *resonance*, between the wave and the plasma particles. A resonance occurs when a periodic driving force is applied to an oscillating system at the same frequency as the system's natural vibrating frequency.

Thus, as k approaches infinity, resonances are produced by the interaction of electrons and ions with the external magnetic field and the electromagnetic wave. These resonances occur when

$$\left(\frac{\omega}{ck}\right)^2 = 0 \qquad (6.29)$$

or when the frequency, ω, of the wave equals ω_{ce} or ω_{ci}. The electromagnetic wave "pumps" the oscillating system at the resonant frequency, rather like pushing someone on a swing in time to the natural rhythm of the swing. Close to a resonance the waves travel with very small phase velocity, dispersion becomes significant and the electromagnetic wave is strongly absorbed by the plasma.

When the frequency, ω, of the wave is equal to the electron cyclotron frequency, ω_{ce}, the R-wave resonates with the electrons because the wave's electric field rotates in phase with the gyrating electrons. The electrons can gain energy from the wave's electric field and the R-wave no longer propagates. This is the lower branch shown in Fig. 6.5, also known as the

Whistler mode. Whistlers propagate at frequencies below the cyclotron frequency and are important in atmospheric plasmas.

The *L*-wave does not resonate with the electrons because the wave field is rotating in the opposite direction. If ω_{pe} is much less than ω_{ce}, the *L*-wave cut-off, ω_L, may be low enough to allow the *L*-wave to resonate with the ions when it rotates at the same frequency as the ion cyclotron frequency ω_{ci} (when $\omega = \omega_{ci}$). The ions will then acquire energy from the wave. This principle is used to heat the ions in a plasma by applying an electric field at the ion cyclotron resonance frequency.

Faraday rotation

Since the left and right circularly-polarised components of the electromagnetic wave have different propagation constants they also have different wave velocities, although both have the same frequency. For large values of ω, the L-wave travels slower than the R-wave because it is coupled to the ions. This causes the plane of linear polarisation to rotate slowly along the ray, an effect known as *Faraday rotation*, which occurs in the range of frequencies in which both *L*- and *R*-waves propagate.

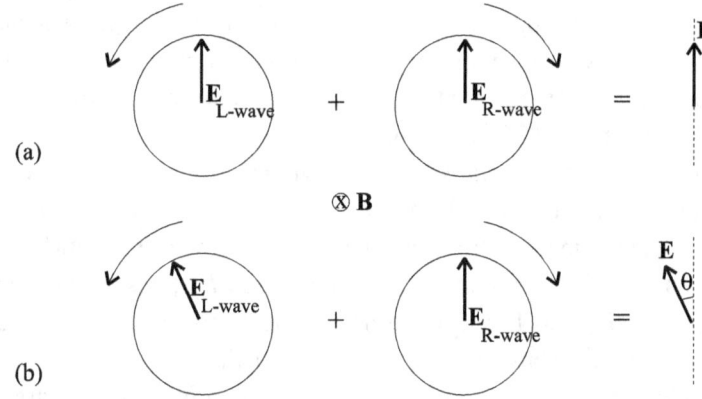

Fig. 6.5   Faraday rotation (magnetic field directed into page)

Fig. 6.5 represents a plane-polarised wave as a superposition of the L- and R-waves shown in Fig. 6.3. The position at the start is shown in (a), with the electric field components both pointing in the same (upward) direction. After a certain number of cycles, the wave electric field vectors of the *L*- and *R*-waves will have returned to their initial positions but the *L*-wave will not have travelled as far as the *R*-wave — the waves are out of phase. When the two components have travelled the same distance (as shown in Fig. 6.5(b)), the *L*-wave will have undergone more cycles than the *R*-wave, causing the plane of polarisation to rotate. The angle of rotation is given by

$$\theta = (k_R - k_L)\frac{x}{2} \quad rad \tag{6.30}$$

where x is the distance travelled. The effect is small unless density is very high.

Faraday rotation becomes important in the low-density plasmas of interstellar space because the path lengths are very long. The rotation of radio emissions from pulsars, for example, has been used to measure the magnetic field of the interstellar medium. A measurement of the angle of rotation provides information about electron density along the wave path: the larger the angle of rotation, the higher is the density and the magnetic field strength. Faraday rotation has also been used to investigate the ionosphere by transmitting polarised radiation from orbiting satellites.

6.6.2 Very low frequency wave propagation parallel to B — Alfvén waves

For very low frequency electromagnetic waves, cyclotron resonance and plasma oscillations can be ignored. We can think of the plasma as a collection of charged particles acting together, as a unit — an "electrically conducting fluid", rather than as a system of individual charged particles. Various types of waves occur in a conducting fluid immersed in a magnetic field. Waves with frequencies very much less than the cyclotron and plasma frequencies are called *hydromagnetic waves*. These are the very low frequency limit of the electromagnetic waves discussed in the previous section and are produced by the bulk motion of plasma across the external magnetic field. The effect of the ions is now included in finding cut-off solutions to equations (6.25) and (6.27).

The simplest hydromagnetic waves, known as *Alfvén* waves, have relatively long wavelengths, often between one and five metres. They are important in both fusion plasmas and space plasmas (solar wind, Earth's ionosphere, etc.). A hydromagnetic wave propagating along the direction of a uniform, constant magnetic field B_{ext}, has its wave magnetic field, B_{wave}, perpendicular to B_{ext} (as Fig. 6.3). Ions and electrons move transversely, producing sinusoidal ripples along the static magnetic field rather like waves on a string. In vacuum, these ripples would propagate at the velocity of light; in a plasma they are slowed by the inertia of the plasma ions and propagate at the phase velocity ω/k. Plasma and magnetic field lines appear to move together as if the lines of force are frozen into the plasma. Alfvén waves propagate as transverse waves along the external magnetic field at a constant speed, known as the *Alfvén speed*, given by

$$v_A = \frac{B}{\sqrt{\mu_0 \rho_m}} \tag{6.31}$$

where ρ_m is the plasma mass density ($m_e n_e + m_i n_i$). An important property of these waves is the absence of any fluctuations in either mass density or pressure, because the guiding centre ($E \times B$) drift is perpendicular to the direction of propagation — the plasma in this case is incompressible. The phase velocity of these waves is approximately equal to v_A.

Shear Alfvén waves occur when the magnetic field lines twist in the x-y plane, like the thread on a screw, with different phases of rotation in the z-direction. The group velocity, and therefore the wave energy, is parallel to the magnetic field; their phase velocity is given by $v_p = v_A \cos\theta$ where θ is the angle between $\boldsymbol{k}$ and $\boldsymbol{B}_{ext}$. Shear Alfvén waves set the plasma in motion in the direction perpendicular to the plane containing both $\boldsymbol{k}$ and $\boldsymbol{B}_{ext}$.

In the high-frequency waves discussed in the previous section, and shown in Fig. 6.4, there was only one pass band for the L-wave. At low frequencies, when ion motion is included, there is a second pass band for the L-wave below the ion cyclotron frequency, ω_{ci}, sometimes called the Shear Alfvén L-wave.

The Alfvén speed is a characteristic of MHD plasmas (Chapter 8), being the typical speed to which a plasma can be accelerated by magnetic forces. If our sample plasma, with electron and ion densities of 10^{20} m^{-3}, is immersed in a magnetic field where $B = 0.5$ T, the plasma mass density, ρ_m, is 1.67×10^{-7} kg m^{-3} and the Alfvén speed, v_A, is 1.1×10^6 m s^{-1}. In the solar wind, where $B \approx 10^{-9}$ Tesla and $n_e = n_i \approx 10^6$ m^{-3}, ρ_m is 1.67×10^{-21} kg m^{-3} and the Alfvén speed is 2.2×10^4 m s^{-1}.

6.6.3 High-frequency wave propagation perpendicular to B
a) ordinary waves $(\boldsymbol{E}_{wave} \parallel \boldsymbol{B}_{ext})$

With electromagnetic wave propagation perpendicular to an external magnetic field, $\boldsymbol{B}_{ext}$, where the wave electric field is parallel to $\boldsymbol{B}_{ext}$, there is little effect on the charged particles. Particle motion due to the wave electric field is parallel to $\boldsymbol{B}_{ext}$ and particle motion due to the wave magnetic field is negligible. The wave propagates as if there were no magnetic field. These waves are known as *ordinary waves* or *O-waves*. Their dispersion relation is given by equation (6.7) and the effects discussed in sections 6.2 and 6.3 still apply.

b) extraordinary waves $(\boldsymbol{E}_{wave} \perp \boldsymbol{B}_{ext})$

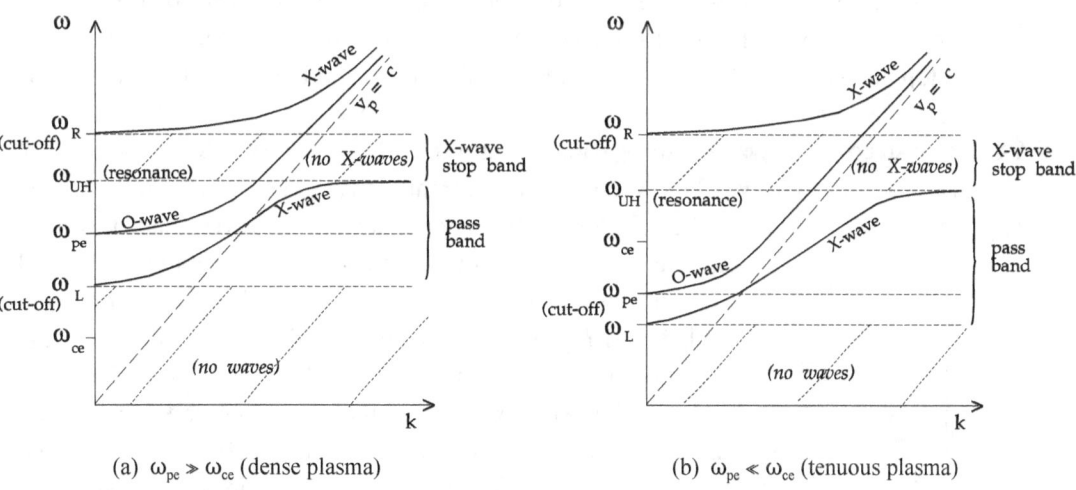

(a) $\omega_{pe} \gg \omega_{ce}$ (dense plasma)   (b) $\omega_{pe} \ll \omega_{ce}$ (tenuous plasma)

Fig. 6.6 O- and X-waves propagating perpendicular to B_{ext}

116

When the electric field of the electromagnetic wave oscillates perpendicular to the external magnetic field, charged particles, particularly electrons, are driven across the magnetic field. The wave vector, k, is perpendicular to B_{ext} and is both perpendicular and parallel to E_{wave}, depending on the frequency, ω, so there are both transverse and longitudinal aspects to the wave. These waves are known as *extraordinary* or *X-waves*. They are high-frequency waves so the effects of the ions can be ignored because they are too massive to respond. The dispersion relation is given by

$$\frac{c^2 k^2}{\omega^2} = 1 - \frac{\omega^2_{pe}(\omega^2 - \omega^2_{pe})}{\omega^2(\omega^2 - \omega^2_{UH})} \tag{6.32}$$

where ω_{UH} is known as the *upper hybrid frequency* and is defined as

$$\omega_{UH} = (\omega^2_{ce} + \omega^2_{pe})^{\frac{1}{2}} \tag{6.33}$$

This is clearly a mixture of the plasma and cyclotron properties of the electrons. The resonant frequency occurs when k goes to infinity and from equation (6.32) this occurs when ω approaches ω_{UH}. The upper hybrid frequency is therefore the resonant frequency for X-waves. As the X-wave approaches ω_{UH} its phase and group velocities go to zero and the electromagnetic energy in the wave is converted to electrostatic oscillations — the wavefronts "pile up". At frequencies close to ω_{UH}, the wave is referred to as an *upper hybrid wave*. These are high-frequency, longitudinal (electrostatic) waves, which travel across the magnetic field, with $k \parallel E_{wave}$. Frequency is independent of wave number. As the external magnetic field goes to zero, the cyclotron frequency disappears and $\omega = \omega_p$: plasma oscillations appear.

The cut-off occurs when k is zero in equation (6.32). In Appendix B, this is shown to be the cut-off frequencies given in equations (6.26) and (6.28). There are therefore two branches to the X-wave: one propagating above the ω_L cut-off and being absorbed at ω_{UH}, the second propagating above the ω_R cut-off frequency (shown in Fig. 6.6). Once again, there are pass and stop bands for the X-waves as there were for the L- and R-waves of Section 6.6.1.

6.6.4 Low-frequency wave propagation: Magnetosonic waves *(k ⊥ B_{ext})*

Low frequency waves, below the ion plasma frequency, ω_{pi}, can propagate across the external magnetic field, B_{ext}. They are low-frequency versions of X-waves, since O-waves are cut off at the plasma frequency, ω_p. Like a conventional electromagnetic wave, the wave electric field is perpendicular to both k and B_{ext} and its magnetic field is perpendicular to the direction of propagation, but is parallel to B_{ext} (Fig. 6.7).

Magnetic pressure in the plane perpendicular to B_{ext} produces compressions and rarefactions (like a longitudinal wave) in the oscillation, causing variations in density. The lines of magnetic force are "frozen into" the plasma at low

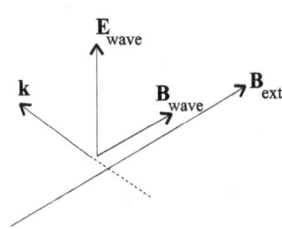

Fig. 6.7   Magnetosonic wave

frequencies, alternately compressing and expanding the external magnetic field as the wave propagates across it, rather like a sound wave. These waves are therefore called magnetosonic or magneto-acoustic waves, and sometimes "compressional Alfvén waves". If the fluid pressure is much greater than the magnetic pressure, the magnetosonic wave is essentially a sound wave.

The magnetosonic wave speed can be estimated from the sound speed v_s and the Alfvén speed:

$$v_M \approx \sqrt{v_s^2 + v_A^2} \tag{6.34}$$

Its resonant frequency, known as the *lower hybrid frequency*, is given by:

$$\omega_{LH} = (\omega_{pi}^2 + \omega_{ci}\omega_{ce})^{\frac{1}{2}} \tag{6.35}$$

This forms the low-frequency limit of the stop band between these waves and the lower X-wave branch.

6.7   Summary

1.   Plasma waves are high frequency oscillations in which electrons move back and forth relative to stationary ions. An electromagnetic wave propagating through a plasma has a different phase shift from one propagating through a vacuum. Phase shift is a measure of density; attenuation is a measure of the collision frequency. In the absence of collisions, the electromagnetic wave cannot transfer any of its energy to the plasma.

2.   With no external magnetic field, the dispersion relation, $\omega^2 = c^2 k^2 + \omega_p^2$, provides all the information about propagation of a wave or oscillation in an isotropic plasma. Electromagnetic waves propagate when $\omega > \omega_p$. Oscillations at the plasma frequency can occur along an external magnetic field, with the magnetic field playing no part in the oscillations.

3.   A cut-off occurs when $\omega = \omega_p$ and both the index of refraction, N, and the wave number, k, go to zero. Phase velocity v_p becomes infinite. A resonance occurs when the N and k become infinite and $v_p = 0$. A wave is reflected at a cut-off and absorbed at a resonance. When $\omega < \omega_p$, k is imaginary, v_p is infinite and $v_g = 0$. The wave decays exponentially. Longitudinal waves (ion acoustic waves) are resonant at ω_p; transverse waves (Langmuir waves) are absorbed below ω_p.

4.   Collision damping and Landau damping cause oscillations to die away as energy is transferred from the wave field to the particles.

5.   An external magnetic field enables electromagnetic waves with $\omega < \omega_p$ to propagate at certain frequencies. Faraday rotation occurs in the range of frequencies where both left-circularly-polarised and right-circularly-polarised waves propagate.

6.   Alfvén waves are oscillations of a magnetised plasma which transport information about magnetic field disturbances at frequencies less than the ion cyclotron frequency.

7.

Electrostatic waves	**Electromagnetic waves**
(longitudinal) $k \parallel E_{wave}$	(transverse) $k \perp E_{wave}$
associated with plasma frequencies	associated with cyclotron frequencies
$B_{ext} = 0$ or $k \parallel B_{ext}$	$k \parallel B_{ext}$, $E_{wave} \perp B_{ext}$
Langmuir (plasma) waves ($\omega \geq \omega_{pe}$)	L- and R-circularly polarised waves
ion acoustic waves ($\omega \leq \omega_{pe}$)	Alfvén waves (low frequency)
$k \perp B_{ext}$	$k \perp B_{ext}$, $E_{wave} \parallel B_{ext}$
upper hybrid waves	ordinary waves

part electrostatic, part electromagnetic

$k \perp B_{ext}$, $E_{wave} \perp B_{ext}$

extraordinary waves

magnetosonic waves (low frequency)

References

1.   *Collected papers of L.D. Landau* (ed. D. Ter Haar). Pergamon Press (1965) p.445-60

Dawson, J.: *Phys. of Fluids*, **4**, p.869-74 (1961)

2.   Compton, K.T., & I. Langmuir: *Rev. Mod. Phys.*, **2**, p.123-242 (1930)

Chapter 7

Plasma Radiation

7.1   Introduction

In the region just above the cut-off frequency, ω_p (Section 6.3.2), the collective behaviour of the plasma particles can have a major impact on the spectrum of electromagnetic radiation. Elsewhere, spectroscopy can provide information on the temperature, density and life history of both laboratory and naturally-occurring plasmas. Absorbed or emitted radiation reveals a plasma's composition and estimates of radiated power losses, plasma resistivity and impurity concentrations can be made.

To use radiation for these so-called "diagnostic" purposes we need to know how it is produced within a plasma and what effect the plasma has as it travels to an external detector. The form the radiation takes is determined not just by the structure of the individual particles, for example the energy levels in the atom, but also by the properties of the plasma itself: its density, temperature, and whether there are any magnetic fields present. Much of the radiation emitted from plasma comes from ionisation and recombination processes. Some is produced when a charged particle is accelerated, the amount increasing rapidly with the acceleration. Electrons, with their smaller mass, may reach very high speeds at fusion temperatures and will radiate strongly, causing serious loss of energy.

7.2   Ionisation

Ionisation — removal of electrons from atoms — can be achieved in high-density plasmas by collisions; and in low-density plasmas by the action of ultraviolet and X-ray radiation. Here, electromagnetic radiation exchanges energy with the plasma particles through the absorption and emission of photons. In the simple case of a hydrogen atom, the energy of an absorbed photon excites the electron in the atom, moving it from the ground state to a higher energy state. If the electron returns to its original position the same amount of energy is released as another photon. Radiation released by excited atoms returning to ground state is emitted at particular frequencies, usually at ultra-violet or visible wavelengths. If the energy of the absorbed photon is greater than the threshold ionisation energy of the atom, the electron is ejected from the atom — ionisation occurs. Any excess energy becomes kinetic energy of the electron. To ionise an electron from the ground state of the hydrogen atom requires an input of energy equal to 13.6 eV and wavelength[a] 91nm. The reverse process, that of recombination of an electron and a hydrogen ion to the ground state of the atom, depends on electron density and is accompanied by the emission of a photon of similar energy to the one absorbed, detectable in the ultra-violet region of the spectrum.

The Coulomb fields of neighbouring charged particles reduce the ionisation potential of any particle immersed in a plasma from that of an isolated particle. A pure hydrogen plasma with an electron temperature of 13.6 eV will be almost fully-ionised. Even at temperatures as low as 2 eV more than half of the atoms in the plasma will have been ionised because of the distribution of particle velocities (equation 3.10).

The rate of ionisation can be estimated by multiplying the product of electron and neutral particle densities, $n_e n_n$, by $\langle \sigma_{ion} | v_e | \rangle$, where σ_{ion} is the ionisation cross-section.[1] The cross-section (the probability that ionisation will take place) is strongly affected by electron velocity and so is averaged over a Maxwellian distribution of electron velocities. An estimate of this averaged cross-section is given by

$$\langle \sigma_{ion} | v_e | \rangle \approx \frac{2 \times 10^{-13}}{6 + \chi} \chi^{\frac{1}{2}} \exp(-\chi^{-1}) \quad (m^3 s^{-1}) \tag{7.1}$$

where $\chi = T_e / 13.6$ (T_e in eV). Similarly, estimates of recombination rates can be obtained using $n_e n_i \langle \sigma_{rc} | v_e | \rangle$, where σ_{rc} is the recombination cross-section, and:

$$\langle \sigma_{rc} | v_e | \rangle \approx 7 \times 10^{-20} \chi^{-\frac{1}{2}} \quad (m^3 s^{-1}) \tag{7.2}$$

In plasmas containing impurities, ionisation and recombination can involve ions which still have some bound electrons. Free electrons may recombine with such ions and still produce an ionised particle, so estimating recombination rates is more difficult.

7.3   The effect of impurities

Laboratory plasmas are polluted by impurities which come mainly from the walls of the containment vessel. These impurities may be classified according to the atomic number, Z, of the element. Low-Z impurities include carbon, nitrogen and oxygen. Elements with an atomic number greater than and including that of iron ($Z = 26$) are generally considered to be high-Z.

Pollution increases both the plasma resistivity, which from equation (5.21) is proportional to the charge on the ion, and the radiative energy losses. These losses cool the plasma and result from inelastic collisions (ionisation, recombination, etc.) with free electrons. Impurities will be ionised to varying degrees, depending on density, temperature, position within the plasma and the time the impurity spends in the plasma. The average charge carried by the plasma ions — the effective ionic charge, Z_{eff} — gives an indication of the impurity

---

[a] $\lambda = c/f$ and 1 eV = 2.418 x 10 ^{14} Hz.

content of a plasma:

$$Z_{eff} = \frac{\sum_i n_i Z_i^2}{n_e}$$ (7.3)

where, for quasi-neutrality, $n_e = \Sigma Z_i n_i$. A pure hydrogen plasma has $Z_{eff} = 1$. Any impurities which have been ionised more than once will cause this value to increase. So for a hydrogen plasma with impurities, Z_{eff} gives a measure of the degree of contamination. In a commercial fusion reactor, Z_{eff} will need to be in the region of two or less.

7.3.1 Impurity production

Impurities can enter a laboratory plasma in several ways. Arcing across the sheath which forms on surfaces exposed to plasma will vaporise the surface, releasing impurities into the plasma. The electric field which develops in the sheath also accelerates positive ions from the plasma towards the surface. Incoming ions may release surface atoms into the plasma if they arrive with sufficient energy. This process, known as *sputtering*, is used deliberately in thin-film deposition and surface-cleaning but when it occurs spontaneously it can be an unwelcome source of impurities.

Thermal desorption, or outgassing, is another source of impurities. Here, the spontaneous release of gases such as oxygen, nitrogen and carbon from a metal surface in contact with the plasma is caused by the thermal energy of particles which may have been absorbed by the surface during assembly of the vessel. As impurities are removed from the surface and taken into the plasma, more impurities diffuse to the surface and the outgassing continues.

Depending on the plasma-surface chemistry, a chemical reaction between a surface particle and an incident particle may produce a new particle to contaminate the plasma: for example, oxygen reacting with a carbon surface can produce carbon monoxide.

There is often a rapid recycling of impurities in the edge regions, close to the wall of the containment vessel. The impurities enter the plasma from the containment vessel, mainly as neutral atoms. They radiate as they become ionised, then return to the wall where they are neutralised. How far the neutral particles can penetrate the plasma before being ionised depends on their velocity and the density of the plasma. Since most of the collisions will be with electrons, the neutral particle mean free path is given by:

$$\lambda_{mfp\,n} = \frac{v_n}{n_e \langle \sigma_{ion} v_e \rangle} \quad (m)$$ (7.4)

As an example, let us take our hypothetical plasma with electron density of $10^{20}\,m^{-3}$ and assume an electron temperature of 15 eV in the edge-region of the plasma. If the containment vessel wall is at room temperature (approximately 300 K), a carbon atom (mass: $2 \times 10^{-26}\,kg$), entering the plasma from the vessel wall would have a thermal velocity, v_n, of $788\,m\,s^{-1}$, from equation (3.14). The averaged ionisation cross-section at 15 eV, $\langle \sigma_{ion} v_e \rangle$, is $1.2 \times 10^{-14}\,m^3\,s^{-1}$ from equation (7.1), and so $\lambda_{mfp} = 6.6 \times 10^{-4}\,m$. Under these conditions, the

122

carbon atom would travel, on average, just over half a millimetre into the plasma before being ionised. Surface-released impurity atoms are generally ionised far beyond the three or four Debye lengths of the sheath region which forms between the surface and the body of the plasma. From equation (5.1), the Debye length for our sample plasma, at a temperature of 15 eV, is 2.9×10^{-6} m.

7.4   Radiation processes

Radiation emitted by a self-luminous plasma results from interactions between electrons and atoms or ions. A collision between two non-relativistic electrons will not usually produce radiation. In a weakly-ionised unmagnetised plasma the radiation comes mainly from collisions between electrons and neutral particles. As ionisation levels increase, collisions between ions and electrons become more significant.

The radiation can be classified under three main headings, depending on the type of electronic transition responsible. Examples are shown in Fig. 7.1. In *free-free transitions*, which produce bremsstrahlung (see below) and cyclotron radiation, the electron remains free after the collision. *Free-bound transitions*, where the free electron becomes bound to the atom, result in recombination radiation. In *bound-bound transitions*, in which the excited electron in the atom returns to a lower energy level, producing line radiation at specific frequencies.

Free-free and free-bound transitions give rise to a continuum of radiation which, for medium and low density plasmas, is relatively weak when compared with line radiation. How much line, recombination and continuum radiation is emitted depends on the populations of the various energy states of the particles present in the plasma. A fully-ionised plasma, for example, will produce little or no line radiation.

The intensity of recombination radiation decreases with increasing electron energy, and therefore temperature. Up to temperatures of the order of 1 keV the intensity of recombination radiation exceeds that of bremsstrahlung; beyond that,

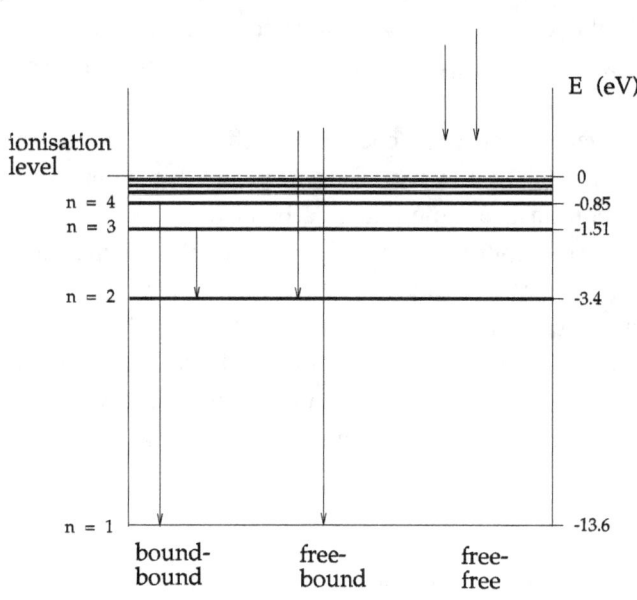

Fig. 7.1   Electron transitions producing radiation in hydrogen

bremsstrahlung predominates. Once the ions have become fully stripped, only bremsstrahlung is important, since all transitions are from one free state to another. For high-Z ions, recombination radiation may be of greater significance.

7.4.1 Bremsstrahlung

Bremsstrahlung literally means "braking radiation". It is the name given to the electromagnetic radiation emitted from a plasma at all frequencies resulting from the random thermal motion of the electrons in the plasma. When a free electron, with kinetic energy E_1, is scattered or decelerated in the field of a positive ion — a Coulomb collision (Section 5.7.1) — the change in velocity moves the electron from one free state to another with lower energy E_2. The excess energy (E_1 - E_2) is released as a photon, generally when the electron is closest to the ion as this is where the change in velocity is greatest (Fig. 7.2). The wavelength is usually short — in the ultraviolet or X-ray region of the spectrum — but if the plasma frequency lies in

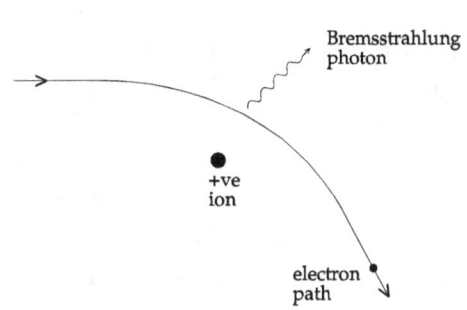

Fig. 7.2 Bremsstrahlung

the microwave region, as often happens, bremsstrahlung can also be detected in the infra-red and visible regions. Using $E_{kin} = hf$, a rough estimate of the frequency of the radiation can be obtained from

$$f = \frac{\frac{3}{2} k_B T_e}{h} = 3.6 \times 10^{14} T_e \quad (Hz) \qquad (7.5)$$

with T_e in eV. Thus, when T_e is 10 keV, the radiation is in the form of X-rays. The spectrum is continuous because the energy of each emitted photon is determined by the specific change in energy of an individual electron, which can take any value. Since the electrons are free both before and after the Coulomb collision, bremsstrahlung is often described as "free-free" or "electron retardation" radiation.

Significant quantities of bremsstrahlung are only produced when temperatures exceed 10 eV, the amount emitted increasing with Z_{eff}. At lower temperatures other forms of radiation, such as line radiation, are more important. A rough estimate of the radiated bremsstrahlung power density — the energy released per second per unit volume — from a fully-ionised hydrogen plasma is given by

$$P_B \sim 10^{-38} Z_i^2 n_i n_e \sqrt{T_e} \quad (W\ m^{-3}) \qquad (7.6)$$

where T_e is measured in eV. If the plasma contains several types of ion with different charge Z_i, then $n_i Z_i^2$ is the sum of the $n_i Z_i^2$ for each ion type. An analysis of the continuum spectrum is useful for estimating plasma density since electron and ion densities are usually considered

124

to be equal and bremsstrahlung varies with the product $n_e n_i$. Bremsstrahlung can also be used to investigate electron temperature, containment time and energy losses.

7.4.2 Cyclotron radiation

A plasma immersed in a steady magnetic field emits radiation from the centripetal acceleration of particles, mainly free electrons, orbiting the magnetic field lines (cyclotron motion). This cyclotron radiation, sometimes called "magnetic bremsstrahlung", results from changes in particle velocity; no collisions are involved (see Fig. 7.3). Accelerated electrons can cause both cyclotron radiation and bremsstrahlung, and, at high temperatures, significant amounts of cyclotron radiation can be detected.

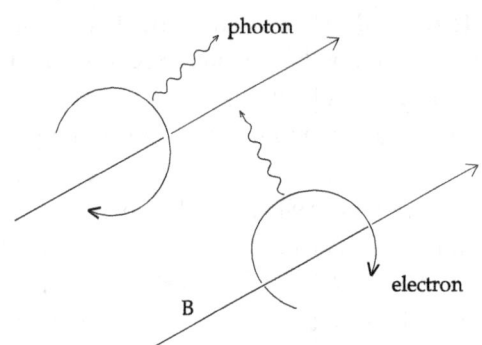

Fig. 7.3 Cyclotron radiation

Like bremsstrahlung, cyclotron radiation is continuous because the particles are free and can radiate at all frequencies. The spectrum can, nevertheless, have the appearance of line radiation because it is composed of frequencies which are harmonics of the cyclotron frequency ($\omega_c = qB/m$). At low velocities, in weak magnetic fields, the radiation is concentrated in a frequency band, occasionally a single spectral line, at the cyclotron frequency. The width of the frequency band depends on Doppler broadening (see below) from motion along the magnetic field and will therefore be affected by the viewing angle. As the magnetic field strength increases peaks may appear at the electron cyclotron frequency, ω_{ce}, and the first few harmonics of that frequency, especially if the particles have relativistic velocities.

Although some radiation is re-absorbed by the plasma, power losses can be significant when the magnetic field strength makes the cyclotron frequency comparable with the plasma frequency. In laboratory plasmas the wavelength for maximum power emission is usually in the infra-red or microwave regions of the spectrum. An estimate of the power radiated per unit volume of plasma can be obtained from

$$P_c \approx 6.2 \times 10^{-20} \, n_e \, T_e \, |\boldsymbol{B}|^2 \quad (W \, m^{-3}) \tag{7.7}$$

where T_e is measured in eV. Since radiated power varies with temperature and B^2, significant amounts of energy can be radiated from high-temperature plasmas immersed in strong magnetic fields. Energy losses from cyclotron radiation increase more rapidly with increasing electron energy and can often be greater than from bremsstrahlung.

Cyclotron radiation can provide information about the magnetic field in which the plasma is immersed and which is accelerating the particles. If the form of the magnetic field is known, an estimate can be made of the electron density and average energy.

The intensity of the radiation depends on the direction of observation with respect to the magnetic field direction and on the uniformity of the magnetic field throughout the plasma. The frequency and angular distribution of the radiation changes as electron energy increases. For relativistic particles, the radiation becomes truly continuous and is known as *synchrotron radiation*, because it was first observed in electron synchrotron machines. The radiation is concentrated into a narrow polarised beam, with the maximum intensity being in the direction of the electron motion (Fig. 7.4). An electron

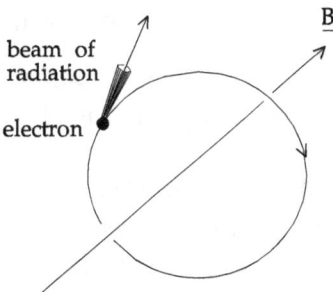

Fig. 7.4   Synchrotron radiation

circling a field line sweeps out a narrow cone of radiation in front of it, rather like a headlamp. This cone concentrates the radiation in the direction perpendicular to the magnetic field and has a half-angle given by $\theta = \cos^{-1}(v/c)$, where v is the velocity of the electrons.

7.4.3   Recombination radiation

Recombination of ions and electrons varies with pressure and velocity. At low pressures, the mean free path is long and the particles, particularly electrons, may gain too much energy to recombine. As pressure increases, recombination becomes more likely, but declines at high pressures as collisions reduce ion velocities. There are three principal sources of recombination radiation: radiative, collisional and dielectronic.

(i)   **Radiative recombination** is a two-particle process involving the capture, into a bound state, of free electrons by ions. The recombining particles have various velocities and therefore different kinetic energies when they collide. The accompanying emission of photon energy is the sum of the free electron kinetic energy and its binding or potential energy. The binding energy is small (13.6 eV for hydrogen) and discrete compared to the continuum form of the kinetic energy. So the recombination spectrum, although continuous, has discontinuities which correspond to the ionisation energies of the relevant bound levels. The radiation is always present, regardless of density, chiefly in the visible and ultra-violet region of the spectrum.

If we assume a Maxwellian distribution, the electron temperature can be estimated from the radiative recombination continuum using the ratio of the intensities at two wavelengths. The intensity of the continuum at a given wavelength depends on the number of electrons with the appropriate kinetic energy, the positive ion concentration, and the recombination coefficient, which, for hydrogen-like ions, is[2]:

$$\alpha_r = 5.2 \times 10^{-20} \, Z \, \lambda^{\frac{1}{2}} \left(0.43 + 0.5 \ln \lambda + 0.47 \lambda^{-\frac{1}{3}} \right) \quad (m^3 \, s^{-1}) \qquad (7.8)$$

where $\lambda = 13.6/T \, Z^2$ (T in eV).

The reverse of radiative recombination is the process of *photo-ionisation*, in which an atom absorbs a photon with sufficient energy to ionise the atom.

(ii)   **Collisional recombination**, or three-body recombination, becomes important at high electron densities and involves three particles: two electrons and an ion, Z^+. The process

$$e + e + Z^+ \rightarrow Z + e$$

produces a neutral atom and a free electron which carries away the excess energy. An estimate of the collisional recombination rate for a hydrogen-like plasma is

$$\alpha_3 = 8.75 \times 10^{-33} T_e^{-\frac{9}{2}} \quad (m^3 s^{-1}) \tag{7.9}$$

(T_e in eV). At particle densities less than 10^{22} m^{-3} radiative recombination is much greater than collisional recombination.

(iii)   **Dielectronic recombination** occurs when a free electron encounters a partially-ionised atom but has insufficient energy to escape from the ion's Coulomb field, and is captured in an outer orbital of the ion. The excess energy excites one of the ion's original electrons to a higher energy level. Both electrons then decay to lower levels, emitting photons of different energies and reducing the total internal energy of the recombined ion. The process can be written:

$$e + e + Z^+ \rightarrow Z^{**} \rightarrow Z + hf_1 + hf_2$$

where Z^{**} indicates an intermediate state. The emitted photons tend to be of low energy. Dielectronic radiation is a very short-lived process which does not occur for hydrogen-like ions. Increasing density reduces its effect as the captured electron suffers collisional ionisation before it can decay. If present, it may appear as faint fuzzy features in the spectrum.

The reverse process, *auto-ionisation*, involves a doubly-excited atom in which two electrons have been raised to higher energy levels in successive collisions. Instead of both electrons returning to their pre-excited states one may be ejected as the other returns to the ground state. If their total excitation energy is greater than that required for ionisation, the extra energy is transferred to the escaping electron. No radiation is emitted.

7.5   Line radiation

Radiation at discrete frequencies is produced when an excited electron moves from one energy level to a lower level within a partially-ionised or neutral atom. Both collisional and spontaneous decay processes are therefore involved. The energy of the emitted photon is the difference between the energies of the two levels. The occurrence rate of a particular radiative decay depends on the populations of the different energy levels and determines the intensity of that particular spectral line.

In magnetically-confined laboratory plasmas, high-Z impurities from the containment vessel walls will not be fully ionised even in the centre of the plasma. The higher the atomic number of the element the higher is the energy (i.e. temperature) that the plasma must have to remove all the electrons from the ions. Line radiation can therefore be an important source of energy loss in laboratory plasmas. All normal gas discharges emit a line spectrum, produced by electron transitions, together with a (generally) weak continuum due to recombination and sometimes to bremsstrahlung. The stronger the line spectrum, the further from equilibrium the discharge is likely to be.

7.5.1   Effect of external fields

External electric and magnetic fields can noticeably affect line radiation, broadening the lines or causing them to shift or split. Investigating line intensity, shape and width can provide information on field strengths and particle density and temperature.

An external magnetic field will split a single spectral line into groups of closely-spaced lines, as shown in Fig. 7.5. This *Zeeman effect* varies with the strength of the magnetic field. In weak fields, the splitting is small and there are usually just three lines: a Zeeman triplet. The centre line of the triplet marks the site of the original line and is plane polarised with its *E* vector parallel to the magnetic field lines when viewed perpendicular to the magnetic field. The outer lines are symmetrically spaced either side of the original line and are plane polarised with their *E* vectors perpendicular to the magnetic field. The spacing of the lines varies with the magnetic field strength, *B*, and is given by

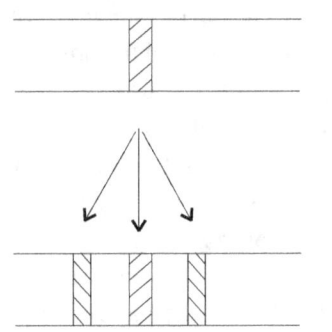

Fig. 7.5   Spectral lines showing Zeeman effect (exaggerated)

no magnetic field

with magnetic field

$$\Delta\lambda \approx \frac{e}{4\pi m_e c^2}\lambda^2 B \qquad (7.10)$$

Only the outer two lines of the triplet are visible when viewed along the magnetic field. From this direction, the lines are circularly polarised in opposite directions. In strong fields, extra lines may appear on either side of the triplet. The amount of splitting therefore indicates the magnitude of the magnetic field — the greater the shift of the outer lines, the stronger the field. The Zeeman effect can be used to measure the magnetic fields of stellar plasmas.

A similar effect, produced by an external electric field and known as the *Stark effect*, was discovered in 1913 by Johannes Stark. Spectral lines split into groups of closely-spaced lines in a strong electric field. Once again, the splitting is proportional to the strength of the applied field, but unlike the Zeeman effect, the Stark effect is not symmetrical about the original line. Its use in analysing a spectrum is therefore limited.

7.5.2   Spectral line broadening

Line radiation is defined as radiation at discrete frequencies, implying that the radiation is observed at an exact frequency, f_0. In reality, and for a number of reasons, the line broadens into a band of frequencies around f_0, and measurement of the spreading can provide information about the plasma. The two main line-spreading processes in plasmas are Doppler broadening and Stark broadening. Collisions between particles may also broaden spectral lines by disturbing the phase of the emission. Measurements can therefore provide an estimate of collision frequencies. Spectral line broadening is a function of local density and temperature and can be used to determine local conditions within the plasma. Unfortunately, all the effects which cause the broadening tend to occur together, so to obtain useful information about the plasma, a situation must be found in which one effect dominates.

Doppler broadening

Doppler broadening is directly related to the average energy of the ions or atoms producing the radiation and is present in most emission spectra. The net effect is a broadened spectral line whose width is temperature-dependent. The Doppler effect, in which the frequency of radiation appears to change with the relative motion of the source and the observer, is responsible. Each particle has a different velocity with respect to the observer and so is accompanied by a different Doppler shift: a particle moving away from the observer appears to radiate at a slightly lower frequency than one moving towards the observer. This random thermal motion results in a range of emission frequencies on either side of a discrete emission line, the degree of broadening increasing with temperature. The resulting profile is similar to that shown in exaggerated form in Fig. 7.6. For plasma, the bandwidth of the line joining the points where the line intensity, I, is half its maximum value (FWHM = Full Width at Half the Maximum) is given by

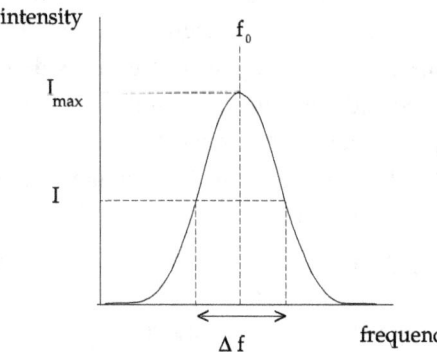

Fig. 7.6   FWHM and spectral line broadening

$$\frac{\Delta f}{f} = \frac{1.67}{c}\left(\frac{2k_B T}{M}\right)^{\frac{1}{2}} \qquad (7.11)$$

where M is the mass of the emitting atom or ion. For a hydrogen atom this reduces to $\Delta f/f = 7.16 \times 10^{-7}\, T^{1/2}$, (T in kelvin). For small widths, $\Delta f/f = \Delta\lambda/\lambda$. Thus, in terms of wavelength, for hydrogen at room temperature ($H_\alpha = 656.3$nm, $T = 300$ K) $\Delta\lambda$ is less than 0.01nm. Doppler broadening becomes significant at high temperatures for light elements and is therefore useful for estimating ion temperature.

Stark or pressure broadening

Spectral lines emitted by a plasma become broader as pressure and current density increase. This broadening is due to the Stark effect and is generally insensitive to temperature; it is affected solely by particle density. Although a plasma is electrically neutral, with approximately equal numbers of positive and negative charges, individual charged particles will produce their own microscopic electric fields which vary in magnitude and direction, according to the motion of the particles, and broaden the spectral lines. An emitting atom close to a charged particle will be perturbed by the field $E = e/(4\pi\varepsilon_0 r^2)$ of the charged particle, where r is the separation distance. The Stark effect therefore produces a statistical broadening of the spectral lines — the more charged particles there are, the greater the effect. In low-temperature, high-density plasmas, electron density can be estimated, even when electron temperature is not known, by measuring the width of the broadened lines.

7.6 The effect of plasma on radiation

Electromagnetic radiation produced in a plasma by any of the above mechanisms can be absorbed and re-emitted several times before it reaches the edge of the plasma and escapes as a continuum of radiation. This can happen if the mean free path for absorption (the *optical mean free path*) is very much less than the dimensions of the plasma. The radiation may then reach a state of equilibrium with itself and the plasma known as "black-body" radiation. It is identical in form to the radiation emitted from a solid-state black-body whose temperature is equal to that of the plasma.

A black-body is one which completely absorbs any radiation falling on it. Any real body approximates a black-body only if its reflectivity is low and its absorption is strong at all frequencies. A gas usually meets the first condition but not the second at visible or near visible wavelengths, unless it is at very high pressure or large in size. In other words, it is usually transparent. In magnetically-confined laboratory plasmas, the optical mean free path is very much greater than the dimensions of the plasma and it is generally safe to assume that the radiation escapes without being re-absorbed in the plasma. Such plasmas are described as *optically thin* or transparent.

A plasma is normally *opaque* — any radiation emitted within the plasma is quickly absorbed — for frequencies below the plasma frequency, and transparent above this frequency. In the opaque state, a plasma can only radiate from its surface. The immense physical size of some astrophysical plasmas can allow a radiation equilibrium to develop. The plasma then approaches the black-body state with its characteristic temperature, and black-body radiation becomes significant over large frequency ranges.

7.7 Information obtainable from radiation

The spectrum of the radiation, usually a series of discrete lines superimposed on a weaker continuous spectrum, can reveal the chemical composition of the plasma. A laboratory plasma, in contact with its containment vessel walls, will have lines in its spectrum characteristic of the wall material. By identifying and measuring the intensity of line radiation emitted by impurities, we can estimate plasma resistivity, since this depends on the effective ionic charge: the higher the charge on the ions, the greater is the resistivity, from equation (5.21).

The hottest parts of a laboratory plasma are usually the most dense and much of the

130

radiation will come from these regions. The wavelength of the radiation — whether it appears as X-ray, ultra-violet, visible radiation, etc. — is determined by plasma temperature. Intensity depends on density and temperature. Average electron temperature can be assessed from the ratio of intensities of selected spectral lines. Atom and ion number density can be estimated from the absolute intensities of the spectral lines. The width of the spectral lines can also be a useful "diagnostic tool": in hot tenuous plasmas, Doppler broadening provides an estimate of temperature; in cold dense plasmas, Stark broadening is useful for density measurements.

In any plasma, only a few ions will dominate over a given temperature range, but because electron temperature generally decreases towards the plasma boundary, the number of electrons still retained by the ions of a particular element will vary throughout the plasma, with the highest ionisation states being found at the centre. These various ionisation stages give a rough indication of electron temperature across the plasma, and can be used, in the laboratory, to reconstruct three-dimensional images of the plasma using data collected from line-of-sight detectors around the plasma. Fractional abundance graphs which plot ionisation state against electron temperature are also useful in this context. An example of such a graph, for carbon, is shown in Fig. 7.7.

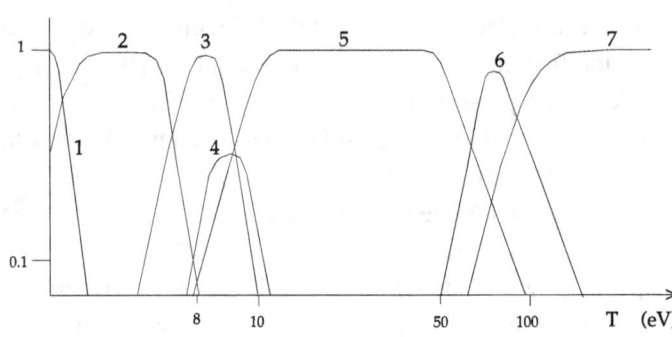

Fig. 7.7   Carbon abundances

The situation may be complicated by the presence of a large temperature range in which there is basically only one ionisation state of a particular element. This occurs when the remaining electrons in the ion form a closed shell[b], making ionisation difficult. Carbon, with six electrons in the neutral atom, is an example: at electron temperatures around 9 eV most carbon atoms have lost all four outer electrons, giving an ion with charge $Z = 4$ (known as C V — "carbon five" — CI being the neutral state). Some C IV ions remain until the temperature is about 11 eV; C VI starts to appear around T_e = 50 eV, but C V is the dominant ion until the temperature approaches 80 eV, when fully-stripped carbon nuclei (C VII) begin to appear. So, from 10 eV to 80 eV there is one dominant ionisation state for carbon: C V. In situations such as this, other means of estimating the plasma temperature must be found.

The total radiated power loss from the plasma can be calculated by summing all the

---

[b]   Grouping of electrons around the nucleus. A closed shell has its full complement of electrons.

spectral intensities, plus any continua from recombination radiation and bremsstrahlung. To estimate the radiated power from a particular type of impurity ion requires knowledge of the excitation rate for the relevant transitions and the fractional abundance of the particular ion. Excitation rates are a function of electron temperature and can be found in the literature. The fractional abundance of the various ionisation states depends on electron temperature, density and impurity confinement.

Radiation from an external source, which is scattered from plasma particles, can be used to measure both electron and ion energy distributions and temperatures, and electron density. The scattering is mainly affected by electron density fluctuations. If the incident radiation has a wavelength much less than a Debye length for the plasma, the scattered line radiation has a spectral distribution which is characteristic of the random thermal motion of the electrons. For wavelengths greater than a Debye length, the scattered line radiation may show a spectral distribution characteristic of the ion velocity distribution function.

7.7.1 Limitations to the accuracy of measurement

A steady-state of ionisation is frequently assumed for astrophysical plasmas but is rarely found in laboratory plasmas. At low temperatures, the time needed for ionisation and recombination processes to take effect is often a significant proportion of the plasma lifetime or particle confinement time. With very short-lived plasmas it is possible that, given the plasma temperature, the final ionisation state for a particular element is not reached, especially if the element has a high atomic number. Any estimate of the ionisation state of a plasma which is not in local thermal equilibrium must take account of the processes affecting ionisation and recombination, such as migration of ions, plasma flow and compression. In such plasmas, only an estimate of the temperature is possible, and different methods will produce different results.

Impurity evolution models, such as the various *coronal models*, can be used to calculate impurity ionisation states, and thus radiation intensity, in low density plasmas which are not in thermodynamic equilibrium. In the very low density plasma of the solar corona, most atoms and ions are in their ground state. Collisional ionisation is balanced by radiative recombination, both processes being dependent on electron density. When a similar situation is observed in other plasmas it may be described as the "coronal domain" of the plasma. Here, the plasma is considered to be optically thin, so that radiation absorption and photo-ionisation processes are unimportant; three-body collisions are rare because of the low density; and the ionisation equilibrium is maintained solely by the balance between collisional ionisation and radiative recombination. In the coronal model the ionisation balance is between processes which are not the true reverse of each other (the reverse of radiative recombination being photo-ionisation), so local thermodynamic equilibrium does not prevail.

Measurement of spectral line intensities requires a light collector (such as a telescope), a monochromator and a detector such as a photomultiplier. High resolution spectroscopic devices such as interferometers are used for measuring the very small line widths ($\sim 10^{-3}$ nm) of spectral lines in non-equilibrium plasmas. The measured strength of any signal is affected by the sensitivity and resolving power of the equipment, the acceptance cone of the system (the cone within which all incident radiation reaches the detector) and the volume of plasma viewed by the detector: how much radiation is being collected and where from. A basic problem in measuring the intensity of any line or continuum is the statistical fluctuation in the number of photons detected in the resolving time of the measurement. There may also be background interference from scattered light, continuum or weak neighbouring lines, often from other elements.

7.8   Summary

1.   Radiation is produced within the plasma either from the centripetal acceleration of charged particles orbiting magnetic field lines or from collisional and recombination processes involving ions and electrons.

2.   Radiation, either emitted from ionisation and recombination processes within the plasma, or as an external source passing through the plasma, is a useful tool for investigating the composition of the plasma. Electron and ion temperatures and density, resistivity and impurity concentrations can all be studied via the output of radiation and the effects the plasma has on the signals received.

3.   At high temperatures, impurities can cause serious loss of energy, reducing the temperature of the plasma.

4.   Plasmas in general will emit a line spectrum due to the various transitions, plus a relatively weak continuum from recombination and bremsstrahlung. In cool plasmas (T_e below about 10 eV) line and recombination radiation are the dominant sources of radiation. At higher temperatures, bremsstrahlung and cyclotron radiation become significant.

5.   Spectral lines are not single discrete lines. There will always be very small variations either side of the dominant frequency quoted for each element. These lines can be spread into a band of frequencies by the Doppler effect, which increases with temperature, and by the Stark effect, due to electric fields and electron density. A strong line spectrum indicates that the radiation is far from the equilibrium black-body form.

References

1.   McWhirter, R.W.P.: *Spectral Intensities*; in *Plasma Diagnostic Techniques*, ed. R.H. Huddlestone & S.L. Leonard, (1965)

2.   Seaton, M.J.: *Mon. Not. R. Astron. Soc.*, **119**, p.81 (1959)

Chapter 8

Mathematical Models

8.1   Introduction

Plasma properties and behaviour are modelled in various ways. In Chapter 4 we used the Lorentz force to investigate charged-particle interactions and determine their individual equations of motion. This is the basis of particle-orbit theory, but the equations of motion cannot be solved for each particle in the plasma because essential information about its initial position and velocity is unavailable. Two alternative options are: the single-fluid model; and the kinetic or statistical model. The first exploits the fact that many observable plasma properties result from the collective, fluid-like behaviour of a large number of particles; the second adopts a statistical approach and investigates what behaviour is most likely for the particles, since we cannot know what actually happens to each individual particle as it interacts with others in the plasma. The two models use different approximations to deal with different circumstances but produce similar results. Both require Maxwell's equations.

8.2   The single-fluid model — Magnetohydrodynamics (MHD)

Magnetohydrodynamics, usually abbreviated to MHD, looks at the motion of an electrically-conducting fluid in the presence of a magnetic field, adapting many of the techniques of ordinary hydrodynamics to take account of the electromagnetic properties of plasmas.[1] An external magnetic field induces currents in the moving fluid which modify the magnetic field. Interaction between these currents and the field produces mechanical forces which modify the motion of the fluid. Disturbances in MHD systems are slow and large-scale, while wave speeds are determined by the sound speed (equation 6.24) and the Alfvén speed (equation 6.31).

The single-fluid model therefore combines electromagnetic theory with fluid mechanics and treats the whole plasma as one conducting hydrodynamic fluid acted on by electric and magnetic forces. The separate identities of the ions and electrons do not feature in the model. A basic property of any fluid is that each particle is influenced by other neighbouring particles. Energy transfer is mainly via collisions. For this to be the case in a plasma the Debye length, λ_D, must remain small compared to the dimensions of the plasma, and particle density must be such that the average distance travelled between collisions is not too large. So, for the single-fluid model to be valid, the plasma must be collision-dominated and the distribution functions of the various particles must be locally Maxwellian. It is

134

inappropriate for low-density plasmas where collisions are infrequent.

There are several versions of the MHD equations, based on a variety of assumptions. Principal amongst the latter is that conditions of local thermal equilibrium prevail. To keep our model simple we will consider a small volume of plasma, with averaged values of velocity, density, temperature and magnetic field, but will ignore the effects of viscosity. Within a real plasma there may be many types of charged and neutral particles, but for simplicity let us imagine a pure hydrogen plasma. We can then think of the plasma as a mixture of three non-viscous "fluids": electrons, one type of positive ion and one type of neutral particle. Some models require electron and ion temperatures to be identical and the number densities of both particles to be approximately equal, but this is not always the case. By separating the plasma into its component fluids we can take account of differences in particle densities, temperatures, etc.

The single-fluid model replaces the Lorentz force (equation 4.3) with a set of hydro-dynamic equations of state. These describe plasma density, and conservation of mass, energy and momentum, without reference to individual particles. Taking these equations in turn:-

8.2.1 Momentum transfer equation

The momentum transfer equation, or fluid equation of motion, describes momentum conservation in fluid terms. It determines the time rate of change of momentum per unit volume using $F = m\,dv/dt$, where F is the sum of the fluid pressure, the electric and magnetic forces acting on the fluid, and the friction due to particle interactions. Gravitational forces are usually much weaker than electromagnetic forces and so can be ignored, as can inter-particle nuclear forces which are only important if the plasma is highly compressed (in which case a quantum-mechanical analysis is required).

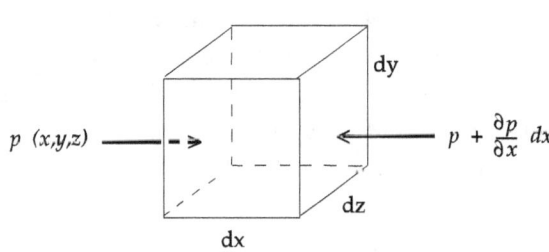

Fig. 8.1   Fluid pressure in a small volume

To obtain the momentum transfer equation for electrons, we will use a small cube, $V = dx\,dy\,dz$, of "electron fluid" (Fig. 8.1). The equations for ions and neutral particles are derived in a similar way.

The fluid pressure, or "momentum flux", is proportional to the number of electrons, n_e, in the volume, V. From Fig. 8.1, the net pressure in the x-direction is the difference between the force p on the left and the force $(p + \partial p/\partial x)$ on the right of the volume of fluid:

$$dF_x = -\frac{\partial p}{\partial x}\,dx\,dy\,dz$$

and similarly for the y- and z-directions. In three dimensions this becomes a total fluid pressure force:

$$dF_p = -(\nabla p)\ dV \qquad (8.1)$$

The electric and magnetic forces acting on the small volume are accounted for by the Lorentz force, F_L on each electron (equation 4.3), so the total electric and magnetic force, in three dimensions, is given by

$$dF_L = n_e\ q\big(E + v_e \times B\big)\ dV \qquad (8.2)$$

where v_e is the average electron velocity — the velocity of the electron fluid.

The frictional or damping force is produced by interactions between the electron fluid and the ion and neutral fluids, equivalent to collisions between the particles. From equation (5.17), this force is the product of the collision frequency, υ_{ei}, and the momentum, $m_e v_e$, which is transferred from the electron to the other particle in the collision:

$$F_f = m_e v_e \upsilon_{ei}$$

From equation (5.11a), collisional frequency is proportional to the density of the collision particles. So, if there are n_e electrons in our small volume, V, the frictional force becomes

$$dF_f = m_e v_e n_e \upsilon_{ei}\ dV$$

Using equations (5.19) and (5.20) this frictional force can be expressed in terms of plasma resistivity, η, and current density, J:

$$dF_f = \eta\, n e J\ dV$$

So we can describe the transfer of momentum between ions and electrons and between neutrals and electrons as a frictional or collisional term:

$$f_{ce} = \eta\, n e J \qquad (Nm^{-3}) \qquad (8.3)$$

This is generally similar in magnitude to the actual collision frequency for electrons since they will lose most of their momentum in each collision with another particle.

The total force acting on the volume of fluid is the sum of these three forces:

$$F_{tot} = F_p + F_L + f_{ce} = m_e n_e\ \frac{dv}{dt}$$

By combining equations (8.1 - 8.3) we obtain the momentum transfer equation for the individual particle fluids:-

electrons: $\qquad m_e n_e\ \dfrac{dv_e}{dt} = -\ \nabla p_e - n_e e\ (E + v_e \times B) + f_{ce} \qquad (8.4)$

ions: $\qquad m_i n_i\ \dfrac{dv_i}{dt} = -\ \nabla p_i + n_i Z_i e\ (E + v_i \times B) + f_{ci} \qquad (8.5)$

neutrals: $\qquad m_n n_n\ \dfrac{dv_n}{dt} = -\ \nabla p_n + f_{cn} \qquad (8.6)$

Neutral particles are unaffected by electric and magnetic forces, so equation (8.6) has no Lorentz force term. In a partially-ionised plasma, the three equations (8.4 -8.6) are linked via their collisional terms f_{ce}, f_{ci}, f_{cn}. In a fully-ionised plasma, equation (8.6) does not apply and equations (8.4) and (8.5) can be combined to produce the single-fluid equation of motion:

$$\rho_m \frac{dv}{dt} = \rho(E + v \times B) - \nabla p \qquad (8.7)$$

where $\rho_m = n_i m_i + n_e m_e$ is the mass density of the plasma and $\rho = e(n_i - n_e)$ is the charge density. Other forces, such as gravity, can be included on the right-hand side if necessary. There is no collision term in equation (8.7) because total momentum must be conserved in the collisions between the electron and ion fluids, so $f_{ce} = -f_{ci}$.

Equation (8.7) is sometimes written

$$\rho_m \frac{dv}{dt} = \rho E + (J \times B) - \nabla p \qquad (8.7a)$$

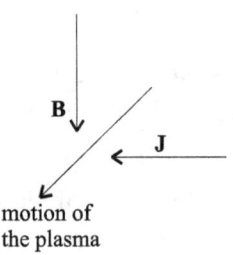

B

J

motion of
the plasma

**Fig. 8.2   Effect of $J \times B$
on plasma**

The ρE term is often omitted when ρ, as defined by equation (4.4): $\rho = \varepsilon_0 (\nabla \cdot E)$, is too small to be significant. This is the quasi-neutrality situation described in Section 5.1 and is the basis on which the MHD model is built. The $J \times B$ term is the magnetic force on a current-carrying conductor, $F = nqv \times B$ which, in a plasma, produces acceleration perpendicular to both J and B (Fig. 8.2). It is the $J \times B$ term in equation (8.7a) which Alfvén assumed to be the dominant term when he devised the MHD equations in the 1940s, and which provides the "magneto" in the name magnetohydrodynamics.

8.2.2   Continuity equation

The continuity equation, describes particle behaviour and the conservation of mass mathematically. It states what happens to the number density of a fluid as it flows from one region to another. If we assume that there are no sources or sinks (losses) of charged particles from ionisation or recombination within a given region then mass must be conserved. If more of a fluid flows into a region than flows out of it, there will be a build-up of fluid mass — an increase in density — in that region over a period of time. The continuity equation therefore expresses the fact that if fluid is not being created or destroyed, the only way the fluid density can change is through a source or sink.

Returning to our small cube of plasma (Fig. 8.1), the number of particles flowing into or out of the box depends on the particle density, n, and the average velocity, v, of the particles. If there is no net change in particle density within the volume (no sources or sinks), then the rate of change in the number of particles, $\partial n/\partial t$, is equal to the net flux of particles across all the faces of the box. By analogy with the derivation of equation (8.1), the change in the number of particles caused by movement in the x-direction is the difference between the number of particles entering the volume in the x-direction and the number leaving it, in a time interval dt. Thus

$$\left(\frac{\partial n}{\partial t} \right)_x = -\frac{\partial n v_x}{\partial x}$$

and so, in three dimensions:
$$\frac{\partial n}{\partial t} = -\nabla \cdot (n\mathbf{v})$$

If we again separate the three types of particle into their respective "fluids", the continuity equations for the three "fluids" can be written:

electrons:
$$\frac{\partial n_e}{\partial t} + \nabla \cdot (n_e \mathbf{v}_e) = 0 \tag{8.8}$$

ions:
$$\frac{\partial n_i}{\partial t} + \nabla \cdot (n_i \mathbf{v}_i) = 0 \tag{8.9}$$

neutrals:
$$\frac{\partial n_n}{\partial t} + \nabla \cdot (n_n \mathbf{v}_n) = 0 \tag{8.10}$$

These equations, if multiplied by their respective electron and ion masses, can be combined to produce the *total mass continuity equation*:

$$\frac{\partial \rho_m}{\partial t} + \nabla \cdot (\rho_m \mathbf{v}) = 0 \tag{8.11}$$

Any sources or sinks of particles will appear on the right-hand side of the continuity equation, thus the zero in equation (8.11) indicates no ionisation, recombination, etc.

The electron and ion continuity equations, (8.8) and (8.9), can be subtracted from one another to form the *charge continuity equation*:

$$\frac{\partial \rho}{\partial t} + \nabla \cdot \mathbf{J} = 0 \tag{8.12}$$

The $\partial \rho / \partial t$ term can often be ignored for the same reasons that the $\rho \mathbf{E}$ term in equation (8.7a) is often omitted. Equation (8.12) then becomes $\nabla \cdot \mathbf{J} = 0$, indicating once again the absence of any sources or sinks of current — all currents in MHD systems must close on themselves.

8.2.3   The MHD equations

The six hydrodynamic equations, (8.4 - 8.6) and (8.8 - 8.10) contain nine unknown quantities: n_e, n_i, n_n, $\mathbf{v}_e$, $\mathbf{v}_i$, $\mathbf{v}_n$, p_e, p_i, p_n. To be able to solve for these nine variables we need three more equations which describe the plasma and relate the nine unknowns to each other. A useful relation for this purpose is the plasma version of the ideal gas law which relates pressure, density and temperature within the plasma:

electrons:
$$p_e = n_e k_B T_e \tag{8.13}$$

ions:
$$p_i = n_i k_B T_i \tag{8.14}$$

neutrals:
$$p_n = n_n k_B T_n \tag{8.15}$$

So we have the full set of equations which together describe the plasma as an electrical fluid: the nine individual equations (8.4 - 8.6), (8.8 - 8.10) and (8.13 - 8.15) plus Maxwell's equations (4.4 - 4.6) and (4.8). Solutions can often be obtained by assuming that the plasma is surrounded by a rigid perfectly-conducting wall, as is the case in laboratory experiments. There can then be no electric field, $\mathbf{E}$, parallel to the wall; no velocity $\mathbf{v}$ normal to it; and no motion of the magnetic field lines in the wall. So, if n is the unit vector normal to the wall, the *boundary conditions* become:

$$n \times E = 0 \tag{8.16}$$
$$n \cdot v = 0 \tag{8.17}$$
$$n \cdot \frac{\partial B}{\partial t} = 0 \tag{8.18}$$

If there is a surface current density, then $n \cdot B = 0$ (8.19)

Finally, we need Ohm's law to relate current and field. Equation (5.16) defined resistivity as $\eta = E/J$ — the simple form of Ohm's law. For MHD plasmas, where the plasma is treated as a single fluid, this becomes

$$E + v \times B = \eta J \tag{8.20}$$

The left-hand side of this equation represents the effect of an electric field, E, which is felt by a fluid element moving with velocity v across a magnetic field B. In a high-temperature (collisionless) plasma, resistivity is very small and equation (8.20) becomes the perfect conductivity equation

$$E + v \times B = 0 \tag{8.21}$$

This form of Ohm's law is used in the so-called "ideal" MHD model for investigating plasma instabilities in which the effects of plasma resistivity can be ignored. It also has significance for plasma motion. If the plasma is at rest, $v = 0$ and equation (8.21) reduces to $E = 0$ — the electric field vanishes. Charged particles will move in response to a perturbing electric field, inducing another electric field to cancel out the perturbing field. If v is not zero, equation (8.21) states that there must be an electric field. Equally, if there is an electric field, the plasma must flow since v cannot be zero.

8.3 Magnetic pressure

The single-fluid equation of motion was given in equation (8.7a):

$$\rho_m \frac{dv}{dt} = \rho E + (J \times B) - \nabla p$$

Under steady-state conditions, no forces act on the plasma and the left-hand side of this equation becomes zero. In most plasmas, ion density is assumed equal to electron density and so $\rho \ (= e(n_i - n_e))$ is zero. Equation (8.7a) thus reduces to

$$\nabla p = J \times B \tag{8.22}$$

This simply states that the plasma pressure gradient and the Lorentz force must be balanced. Fig. 8.3 shows this equilibrium situation in a plasma column,[a] with ∇p radially directed towards the axis.

---

[a]   Plasma behaviour is generally investigated in laboratories, where the plasma, in gas discharge tubes, etc., forms a cylindrical column.

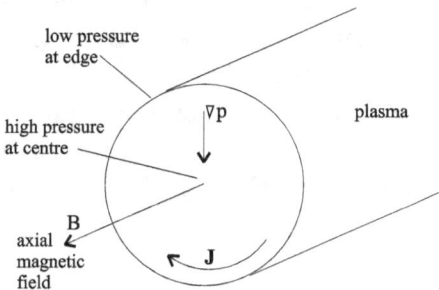

low pressure
at edge

∇p

plasma

high pressure
at centre

axial
magnetic
field

B

J

Fig. 8.3 Equilibrium in a plasma column

The inward $J \times B$ force balances the outward force produced by the pressure gradient. Ampère's law, $\nabla \times B = \mu_0 J$ (equation (4.8)), provides an expression for J which can be substituted into equation (8.22):

$$\nabla p = \left[\frac{1}{\mu_0} (\nabla \times B) \right] \times B$$

By rearranging the vector identity: $(\mathbf{A} \cdot \nabla)\mathbf{A} = \nabla(\frac{1}{2}|\mathbf{A}|^2) - \mathbf{A} \times (\nabla \times \mathbf{A})$ we obtain:

$$(\mathbf{A} \cdot \nabla)\mathbf{A} - \nabla(\frac{1}{2}|\mathbf{A}|^2) = (\nabla \times \mathbf{A}) \times \mathbf{A}$$

(Note the change of sign on the right hand side: $-\mathbf{A} \times (\nabla \times \mathbf{A}) = (\nabla \times \mathbf{A}) \times \mathbf{A}$.) Similarly, our expression for ∇p becomes

$$\nabla p = \frac{1}{\mu_0} \left[(B \cdot \nabla)B - \nabla \left(\frac{1}{2} B^2 \right) \right]$$

Rearranging once more, we obtain:

$$\nabla \left(p + \frac{B^2}{2\mu_0} \right) = \frac{(B \cdot \nabla)B}{\mu_0} \tag{8.23}$$

This is the *pressure balance condition*. The term on the right represents distortion and compression of the magnetic field. In a straight cylinder with an axial magnetic field as shown in Fig. 8.3, the field lines are straight and parallel to the axis of the cylinder and there is no variation along B. Under these conditions, the term on the right-hand side of (8.23) disappears and the pressure balance becomes:

$$p + \frac{B^2}{2\mu_0} = constant \tag{8.24}$$

Here, p is the plasma kinetic pressure from the motion of the particles: the higher the temperature, the greater is the particle velocity and the higher is the pressure. The term $B^2/2\mu_0$ is the *magnetic field pressure*, often called the *magnetic energy density* because it measures the energy per unit volume contained within the magnetic field. If the magnetic field is in matter, μ_0 must be replaced by the magnetic permeability, μ, of the material. When the magnetic field is curved, the curvature is often small and equation (8.24) still applies.

This equation demonstrates the diamagnetic nature of plasmas (Sections 4.5.1 and 4.8): as plasma pressure increases, any externally-applied magnetic field decreases in strength. So, at the centre of the plasma, where pressure is greatest, the strength of the external magnetic field is at its lowest. The maximum plasma pressure that can be confined by an external magnetic field is $B^2/2\mu_0$. For a field strength of 0.5 Tesla, this is one atmosphere.[b]

8.3.1   Plasma beta, β

The ratio of the outward plasma pressure to the inward magnetic-field pressure provides a measure of the diamagnetic effect. This pressure ratio, known as β (beta), is given by

$$\beta = \frac{plasma\ pressure}{magnetic\ field\ pressure} = \frac{2\mu_0 p}{|\boldsymbol{B}|^2} \tag{8.25}$$

The higher the value of β, the greater is the depression of the magnetic field by the plasma pressure and therefore the greater is the effect of diamagnetic currents on plasma confinement. If β is small, plasma pressure is not important and magnetic effects dominate.

An alternative definition of β which is sometimes used is:

$$\beta = 1 - \frac{B^2}{B_0^{\,2}} \tag{8.26}$$

where B_0 is the magnetic field strength at the plasma surface. An estimate is provided by the ratio of the ion sound speed to the Alfvén speed:

$$\beta = v_s^{\,2}/v_A^{\,2}$$

In practice, the value of beta varies throughout the plasma. The external magnetic field permeates the plasma, so field strength and plasma pressure will be different at different points within the volume. A plasma is described as low-beta when β is very much less than 1, and high-beta when β is approximately equal to 1. With a low-β plasma, the magnetic field can contain the plasma providing there are no instabilities, but if β is greater than 1 there is no possibility of magnetic confinement. In astrophysical plasmas β may be close to 1, an example being the interstellar medium. In stars it may even be greater than 1. In the laboratory, β is usually small.

8.4   Plasma instabilities

So far, we have assumed a state of equilibrium where all forces are balanced, but the point was made in Chapter 3 that complete thermodynamic equilibrium is unlikely in plasma. A magnetically-confined plasma is not in thermodynamic equilibrium because radiation can escape even if the particles are constrained by the magnetic field. In many naturally-occurring

---

[b]   1 atmosphere = 101325 N m^{-2}. $(0.5)^2/2\mu_0$ = 99472 N m^{-2} ≈ 1 atm.

plasmas there may be spatial variations in density, temperature, magnetic field strength, etc. which put the plasma a long way from thermodynamic equilibrium.

Even when all forces are in balance there may still be free energy available to disturb the plasma. Such disturbances come not from slow diffusion of plasma across the magnetic field, or loss from magnetic mirrors, etc., but from a sudden and potentially catastrophic large-scale movement of the plasma. A slight disturbance in the motion of the plasma may deform the magnetic field, producing magnetic forces which amplify the original disturbance. Such an unstable situation can drastically alter the plasma properties.

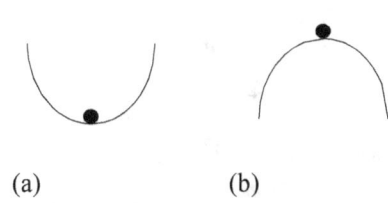

(a)   (b)

Fig. 8.4   Stable/unstable equilibrium

Any equilibrium state is unstable if a slight disturbance grows in time; and stable if it decays. Fig. 8.4 shows the two situations. If the ball in (a) is displaced slightly it will return to its original equilibrium position: its equilibrium is stable. The ball in (b), however, if displaced slightly will roll away. Its equilibrium is unstable.

Different types of plasma disturbance are possible, and an equilibrium can be described as stable or unstable to a particular mode or type of instability. The growth of instabilities can produce random fluctuations and, ultimately, plasma turbulence. This in turn causes movement of plasma and energy across the confining magnetic field, resulting in loss of energy and sometimes loss of plasma. Instabilities are not always disastrous for plasma confinement and often occur in laboratory experiments which still work satisfactorily despite the disturbance. However, understanding the mechanisms involved in their growth is essential to the control of fusion plasmas and is the subject of much research. The theory is beyond the scope of this book and we will confine ourselves here to just a brief description of a few instabilities.

Plasma instabilities are generally classified under two headings: *macro-instabilities*, which are associated with large-scale disturbances of the plasma and can lead to complete loss of confinement; and *micro-instabilities*, involving small disturbances such as fluctuations in the electric or magnetic fields within the plasma. The turbulence resulting from micro-instabilities causes diffusion of plasma across the magnetic field, at speeds much lower than the sound speed. They are usually small-scale and slow-growing, often just causing an increase in electron drift.

A further distinction can be made on the basis of whether resistivity is present in the plasma. Under ideal conditions, there is no resistivity, and the plasma is perfectly-conducting. We can therefore identify "ideal" instabilities and resistive instabilities.

8.4.1   "Ideal" instabilities

Perfectly-conducting magnetically-confined plasmas can develop instabilities with scale lengths comparable with the dimensions of the plasma. Ohm's law for "ideal" MHD, where ηJ is zero (equation (8.21)), is used to model these instabilities. The plasma is treated as a conducting fluid "tied" or *frozen* to the magnetic field — any fluid element located on any particular magnetic field line will still be located on the same field line after an arbitrary motion of the plasma, no matter how complicated that motion may be. This is a property of any high-conductivity MHD fluid.

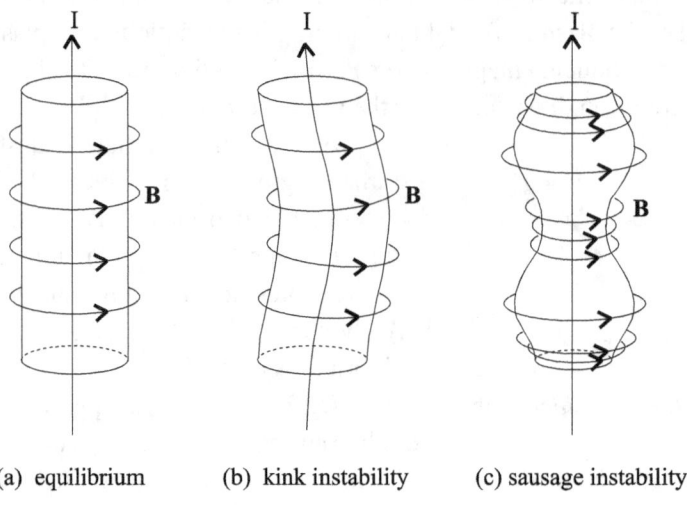

(a) equilibrium   (b) kink instability   (c) sausage instability

Fig. 8.5   Plasma instabilities

In reality the situation is complicated by geometry, for example the shape of the containment vessel, and by the nonlinear development of instabilities. Ideal instabilities draw on the free energy associated with pressure gradients or plasma currents. The unstable modes grow rapidly and can result in a sudden loss of stored plasma energy. Many instabilities with highly descriptive names — flute, kink, sausage, etc. — have been identified. Examples are shown in Fig. 8.5. In (a), the column of plasma with an axial current, I, is in equilibrium. The current produces a magnetic field which encircles the column. For stable equilibrium, the particle pressure inside the column must equal the magnetic field pressure on the outside of the column and this pressure balance must be uniform along the field.

In an unstable equilibrium, the column may be displaced by, for example, an unequal pressure balance. The resulting *kink instability* (Fig. 8.5(b)) continues to develop because the magnetic pressure is increased on the inside of the bend and reduced on the outside. The pressure balance inequality grows because there is nothing to counteract the original displacement. Slight variations in magnetic field strength around the column can produce the *sausage instability* shown in Fig. 8.5(c), so-called because it can look like a string of sausages. Here, increased magnetic field strength (the field lines group together) compresses the plasma at intervals, squeezing it into regions of weaker magnetic field. As there are no adequate restoring forces, the discharge is eventually interrupted by being broken apart.

In hydrodynamics, if a heavy fluid is supported by a less dense fluid, ripples at the interface allow the heavier fluid to sink through the less dense fluid. This is the *Rayleigh-*

Taylor instability. In astrophysical plasmas, this situation can arise at the boundary between plasmas of different densities, as when a dense plasma is supported against gravity by the pressure of a magnetic field. In a laboratory plasma, the *flute* or *interchange instability* is a pressure-driven version of the Rayleigh-Taylor instability.

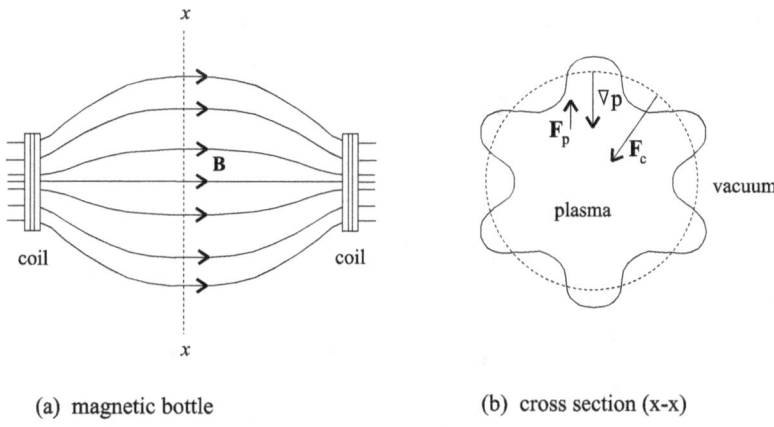

(a) magnetic bottle   (b) cross section (x-x)

Fig. 8.6   Flute or interchange instability

Fig. 8.6 shows a plasma confined within a magnetic "bottle". In curved magnetic fields, particle motion along the curved field lines produces a centripetal force on the plasma (Section 4.9.1c) which acts like a gravitational force, shown as F_c in Fig. 8.6(b). This force is in the opposite direction to the pressure force, F_p. The dotted line in Fig. 8.6(b) marks the stable situation in cross-section. Ripples may develop on the outer surface of the plasma, at the interface with the surrounding vacuum, beginning in the middle section (marked as *x-x* in (a)) and extending rapidly along the length of the bottle towards the mirror points, which are regions of greater stability. The flute instability is driven by the non-uniform plasma pressure and causes the plasma to break through the magnetic confinement to reach the walls of the containment vessel. This large-scale movement of plasma at very high speed, in the region of 10^5 m s^{-1} in hot plasmas, destroys the plasma column.

8.4.2   Resistive instabilities

Plasma is seldom perfectly-conducting so resistivity must be included when considering phenomena which develop over a long timescale. Produced by collisions between the particles, resistivity frees the plasma from being tied to the magnetic field lines by scattering the particles and allowing a slow diffusion across the magnetic field. The presence of resistivity may damp disturbances, reducing their effect, or it may amplify a perturbation, drawing on the magnetic energy generated by currents in the plasma and leading to the development of resistive instabilities. Resistivity allows magnetic fields which were absent at the formation of the plasma to diffuse into it. A perfectly-conducting plasma acts to exclude such fields.

144

With resistivity, Ohm's law takes the form given by equation (8.20):

$$E + v \times B = \eta J$$

To find out what happens if the magnetic field varies over time, we rearrange this to give: $E = \eta J - v \times B$. Substituting for E in Faraday's law (equation 4.5) we obtain

$$\nabla \times (\eta J - v \times B) = -\frac{\partial B}{\partial t}$$

So

$$\frac{\partial B}{\partial t} = \nabla \times (v \times B) - \nabla \times (\eta J) \qquad (8.27)$$

Ampère's law, equation (4.8):

$$J = \frac{1}{\mu_0}(\nabla \times B)$$

when substituted in (8.27) gives

$$\frac{\partial B}{\partial t} = \nabla \times (v \times B) - \nabla \times \left(\frac{\eta}{\mu_0} \nabla \times B \right)$$

Using the vector identity $\nabla^2 B = \nabla(\nabla \cdot B) - \nabla \times \nabla \times B$

with $\nabla \cdot B = 0$ (equation 4.6), we arrive at a final description for $\partial B/\partial t$:

$$\frac{\partial B}{\partial t} = \nabla \times (v \times B) + \frac{\eta}{\mu_0} \nabla^2 B \qquad (8.28)$$

Equation (8.28) describes a magnetic field that changes in time partly by convection or movement of the plasma — because the field is tied, or coupled, to the plasma (the first term on the right-hand side) — and partly by diffusion (the final term). The diffusion of the magnetic field is caused by the resistivity. In the simple case shown in Fig. 8.7, plasma resistivity allows the oppositely-directed magnetic fields on either side of the dotted line to diffuse into one another and "cancel out". The scale length of this diffusion is generally very small compared with the dimensions of the plasma. The ratio of the convection term to the diffusion term is a dimensionless quantity which, if sufficiently large, indicates that the plasma is perfectly-conducting. This ratio is sometimes called the *magnetic Reynolds number* because it is similar to the Reynolds number used in fluid mechanics to determine whether a fluid flow will be smooth or turbulent.

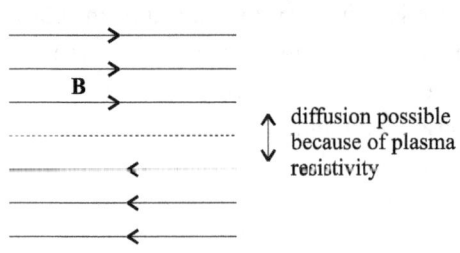

Fig. 8.7   Magnetic diffusion

An important instability, known as the *resistive tearing instability*, can develop when wave-like disturbances cause the magnetic field lines to bend or ripple. This "tearing mode" instability is of particular importance in astrophysical plasmas where it can develop in current sheets in which magnetic field lines are anti-parallel, as in Fig. 8.7. Similar instabilities occur in magnetically-confined laboratory plasmas. The situation is shown in Fig. 8.8.

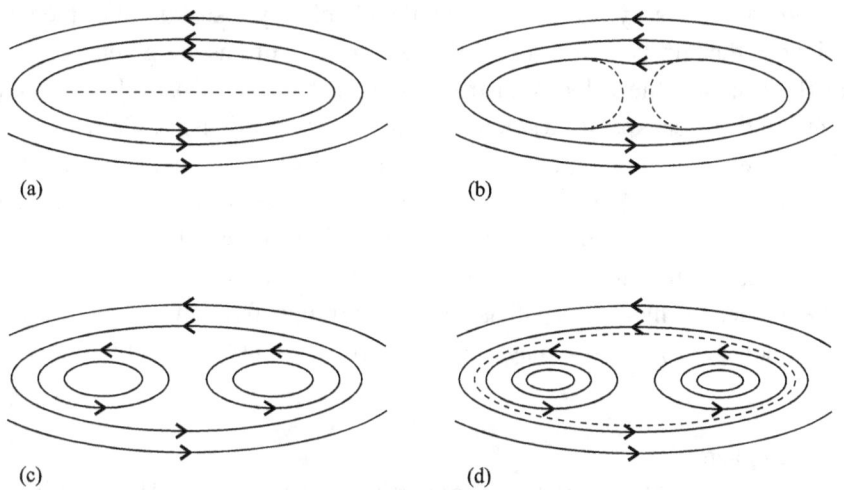

Fig. 8.8 Magnetic reconnection and magnetic island development

Perturbation of the parallel field lines in (a) produces ripples, as shown in (b). This ultimately leads to breaking and reconnection of the field lines, shown as dotted lines in (b), with the end result as shown in (c). The closed-field-line regions produced by this breaking and reconnection of the field lines are called *magnetic islands*. These islands may amplify the disturbance and grow in size, causing further breakage and reconnection of adjacent field lines. In three dimensions, the magnetic field lines will form magnetic flux surfaces. The surface separating the closed-field-line surfaces, or islands, from the open-field-line surfaces is called the (magnetic) *separatrix*, and is marked as a dotted line in Fig. 8.8(d). Equation (8.28) can be used to model the evolution of this magnetic field disturbance, with *v* representing the disturbance in the velocity field.

This process of *magnetic reconnection* was first proposed by James Dungey in 1953[2] but not accepted until the 1960s. It may be responsible for reduced confinement times in laboratory and fusion plasmas, since non-uniformities and weaknesses in the magnetic field can allow the pent-up magnetic energy to escape as the field lines break and reconnect. This energy may be explosively converted into plasma kinetic energy, as happens in solar flares.

8.5 Kinetic theory

The fluid model provides the simplest description of a plasma, describing the average properties of a system of particles but still managing to account for most of the large-scale phenomena. Unfortunately, some aspects of plasma behaviour cannot be dealt with satisfactorily using a fluid approximation. For these, we must return to kinetic theory and the velocity distribution function, $f(v)$, encountered in Chapter 3. The kinetic model accounts for the bulk properties of plasmas in terms of the motion of the constituent particles and thus provides a more detailed description of particle behaviour. The solution to the equation which forms the basis of the theory reveals the spatial and velocity distribution of the particles — electrons, ions and neutrals — if initial and boundary conditions are given.

In fluid theory, the velocity distribution of each type of particle is assumed to be Maxwellian and therefore specified solely by temperature and density. Collision frequencies are usually high enough to assume that a Maxwellian distribution is maintained throughout the plasma. As plasma temperature increases, or density decreases, collisions become less frequent. Deviations from local thermodynamic equilibrium can become long-lasting and these assumptions are then no longer valid. Kinetic theory is used for particle flow along the magnetic field when the mean free path is long, and for high-frequency flow (greater than the cyclotron frequency, ω_c) across a magnetic field. It is also used when there is no magnetic field.

8.5.1   Phase space

The properties of large collections of particles can be predicted by applying statistical principles and reasoned assumptions about the probable behaviour of the particles to reduce the amount of information to be handled. This is the basis of statistical mechanics.

To specify the state of a system of particles completely from a classical (non-quantum-mechanical) point of view, the position and momentum of each particle in the system at any instant must be known. Each particle will have a position (in three dimensions), and may also be moving at a given speed in any direction, so three coordinates in velocity space will specify its motion (or momentum) at that instant. In velocity space the velocity vector can be thought of as a position vector. A particle moving with constant velocity is therefore represented in velocity space by a stationary point. The density of these points is described by a velocity distribution such as that encountered in Chapter 3. Each particle's position coordinates (x,y,z) are combined with its velocity-space coordinates (v_x, v_y, v_z), using coordinate axes (r, v). So each is represented by a single point in a six-dimensional phase space[c] which has coordinates (x, y, z, v_x, v_y, v_z).

Statistical mechanics assumes that every macroscopic state of a system, such as a gas, results from a particular distribution of the constituent particles among the various "cells" of phase space, each cell representing both a small volume of space and a particular amount of momentum. At any instant, the state of a system of N moving particles can be represented by N points in phase space. Each particle, because it has velocity, will change its position coordinates with time, and any change in velocity, due to collisions, etc., will result in a

---

[c]   Phase space is a multi-dimensional space in which coordinates represent the variables required to specify the state of a system of particles.

corresponding movement in phase space. With plasma, rather than deal with each individual particle, we can take an average over distances which are greater than the particle separation distance but less than a Debye length. In effect, we do not distinguish between particles with similar velocities and with a similar location, and by considering averaged densities we can obtain a complete description of the whole system. The averaged density of these representative points in phase space is known as the *phase space distribution function, f(r,v,t)*. This is the probability that a particle will be found in a certain phase space volume element *dr dv* at time *t*.

8.6   Boltzmann equation

The evolution of the phase space distribution function *f(r,v,t)* is described by the Boltzmann equation, which is fundamental to kinetic theory and is not specific to plasmas. It gives a more detailed description of the behaviour of a system of particles than do the MHD equations. MHD describes the average properties of the system, the Boltzmann equation reveals the distribution of the particles in space and in velocity. If the initial conditions and the boundary conditions are given, and a solution can be found to the equation, then all the properties of the system of particles being modelled can be obtained from the solution. Unfortunately, it is much more difficult to solve the Boltzmann equation than the MHD equations.

The Boltzmann equation can be used in its various forms to obtain dispersion relations for plasma waves and oscillations, etc., and to investigate the transport processes in a plasma. These include particle diffusion and heat conduction across the magnetic field, viscosity, resistivity and energy transfer, all of which are affected by collisions between particles or the collective particle-scattering effects of plasma waves.

To develop the basic ideas of kinetic theory in Chapter 3, we used a small box, or "differential volume" (since $dV = dx\,dy\,dz$). If we imagine a differential volume in six dimensions, $dV = dx\,dy\,dz\,dv_x\,dv_y\,dv_z$, then the number of points representing (say) electrons in this volume can be expressed as a function of *r*, *v*, and *t*:

$$n_e = f_e(\boldsymbol{r}_e, \boldsymbol{v}_e, t)$$

Any change with time in the number of these points is caused either by a drift of points in position coordinates, (x,y,z), due to the inherent velocity of the particles, or by a drift of points in velocity coordinates, (v_x, v_y, v_z), due to electric and magnetic forces. In six dimensions, the total time derivative of the distribution function *f(r,v,t)* — the way the state of the system changes with time — is given by:

$$\frac{df}{dt} = \frac{\partial f}{\partial t} + \frac{\partial f}{\partial x}\frac{\partial x}{\partial t} + \frac{\partial f}{\partial y}\frac{\partial y}{\partial t} + \frac{\partial f}{\partial z}\frac{\partial z}{\partial t} + \frac{\partial f}{\partial v_x}\frac{\partial v_x}{\partial t} + \frac{\partial f}{\partial v_y}\frac{\partial v_y}{\partial t} + \frac{\partial f}{\partial v_z}\frac{\partial v_z}{\partial t}$$

$$= \frac{\partial f}{\partial t} + \frac{\partial f}{\partial x}v_x + \frac{\partial f}{\partial y}v_y + \frac{\partial f}{\partial z}v_z + \frac{\partial f}{\partial v_x}a_x + \frac{\partial f}{\partial v_y}a_y + \frac{\partial f}{\partial v_z}a_z$$

where a_x, a_y and a_z are the acceleration components ($a_x = \partial v_x/\partial t$, etc.).

Thus

$$\frac{df}{dt} = \frac{\partial f}{\partial t} + \boldsymbol{v} \cdot \frac{\partial f}{\partial \boldsymbol{r}} + \frac{\boldsymbol{F}}{m} \cdot \frac{\partial f}{\partial \boldsymbol{v}} \qquad (8.29)$$

which can be written

$$\frac{df}{dt} = \frac{\partial f}{\partial t} + \boldsymbol{v} \cdot \nabla_r f + \frac{\boldsymbol{F}}{m} \cdot \nabla_v f \qquad (8.29a)$$

The first term on the right in equation (8.29), $\partial f/\partial t$, represents the rate of change of f at a fixed point. The second term, $\boldsymbol{v} \cdot \partial f/\partial \boldsymbol{r}$, represents the drift of points in position coordinates (x,y,z) due to the inherent velocity of the particles. The final term, $(\boldsymbol{F}/m) \cdot \partial f/\partial \boldsymbol{v}$ is the drift of points in velocity coordinates (v_x, v_y, v_z). This is the general form of the Boltzmann equation. The nature of the force, $\boldsymbol{F}$, and whether the distribution can change in time determine the forms of the equation specific to plasmas.

In plasma, the drift of points in velocity coordinates is due to electric and magnetic forces, so the Lorentz force (equation 4.3) is substituted for $\boldsymbol{F}$:

$$\frac{\boldsymbol{F}}{m} = \frac{q}{m}(\boldsymbol{E} + \boldsymbol{v} \times \boldsymbol{B}) \qquad (8.30)$$

The field $\boldsymbol{B}$ is the static external magnetic field, while $\boldsymbol{E}$ includes fields arising from the plasma particles as well as any externally-applied field.

8.6.1 Collisionless Boltzmann, or Vlasov equation

Plasmas are often sufficiently tenuous for collisions to be ignored if the time between collisions, or the mean free path, is long compared to times or distances of interest. This is the case in the majority of astrophysical plasmas. It may also be so in laboratory or industrial plasmas operating at high temperatures or low pressures.

The density of points in motion does not change when a plasma is collisionless. In the absence of short-range collisions, the representative points move in continuous curves and df/dt in equation (8.29) is zero. Thus:

$$\frac{\partial f}{\partial t} + \boldsymbol{v} \cdot \frac{\partial f}{\partial \boldsymbol{r}} + \frac{\boldsymbol{F}}{m} \cdot \frac{\partial f}{\partial \boldsymbol{v}} = 0 \qquad (8.31)$$

or, using equation (8.30):

$$\frac{\partial f}{\partial t} + \boldsymbol{v} \cdot \frac{\partial f}{\partial \boldsymbol{r}} + \frac{q}{m}(\boldsymbol{E} + \boldsymbol{v} \times \boldsymbol{B}) \cdot \frac{\partial f}{\partial \boldsymbol{v}} = 0 \qquad (8.31a)$$

This is the plasma form of the collisionless Boltzmann equation, known as the *Vlasov equation* after its Russian originator. A different equation is required for each type of particle in the plasma with a different mass and charge:

electrons: $\quad \dfrac{\partial f_e}{\partial t} + \boldsymbol{v}_e \cdot \dfrac{\partial f_e}{\partial \boldsymbol{r}} - \dfrac{e}{m_e}(\boldsymbol{E} + \boldsymbol{v}_e \times \boldsymbol{B}) \cdot \dfrac{\partial f_e}{\partial \boldsymbol{v}} = 0 \qquad (8.32)$

ions: $\quad \dfrac{\partial f_i}{\partial t} + v_i \cdot \dfrac{\partial f_i}{\partial r} + \dfrac{Z_i\,|e|}{m_i}\,(E + v_i \times B) \cdot \dfrac{\partial f_i}{\partial v} = 0$ (8.33)

neutrals: $\quad \dfrac{\partial f_n}{\partial t} + v_n \cdot \dfrac{\partial f_n}{\partial v} = 0$ (8.34)

As with the fluid model, the Lorentz force does not apply to neutral particles.

The Vlasov equation describes the evolution of the distribution function $f(r,v,t)$ in six-dimensional space. Combining the set of equations (8.32 - 8.34) with Maxwell's equations, provides a complete description of the behaviour of a collisionless plasma. Analytical solutions to the Vlasov equation are usually difficult to find because the fields E and B are assumed to be either external or due to the rest of the plasma. In reality, both can be influenced by the distribution function, f, itself.

8.6.2 The effect of collisions on the Boltzmann equation: the Fokker-Planck equation

Collisions between particles will cause small changes in velocity, which in turn alter the distribution function and df/dt is no longer zero. These collisional effects are incorporated in the collision term:

$$\left(\dfrac{\partial f}{\partial t} \right)_{col}$$

Equation (8.31) now becomes

$$\dfrac{\partial f}{\partial t} + v \cdot \dfrac{\partial f}{\partial r} + \dfrac{q}{m}(E + v \times B) \cdot \dfrac{\partial f}{\partial v} = \left(\dfrac{\partial f}{\partial t} \right)_{col}$$ (8.35)

Once again, a different equation is needed for each type of particle. Calculating the collision term requires knowledge of the type of interactions taking place in the plasma. The equations for the different types of particles — electrons, ions and neutrals — are generally linked through their collision terms. For example, the collision term in the electron equation may contain the distribution functions of the ions and neutrals as well as the electrons. The effects of all relevant interactions between the particles must be included and the form the collision term takes will vary depending on which particular version of the kinetic model is used. Information is needed about the type of collisions taking place in the system before the collision term can be calculated. A Coulomb collision, because of the electric forces involved, causes a drift of particles in velocity coordinates and also involves a change in position coordinates, and so can only be approximately accounted for by the collision term.

In fully-ionised plasmas, changes in the distribution function are caused by small changes in the velocity of the particles which result from the many small-angle long-range Coulomb collisions. The collision term therefore represents only the cumulative effect of these collisions. Large-angle scattering is ignored. For this reason a form of the *Fokker-Planck equation* is usually used for the collision term. The Fokker-Planck equation originated in the study of Brownian motion and is an important equation in statistical physics generally.

150

It describes the situation in velocity space only (i.e. not phase space) and is used when a probability distribution changes slowly in time due to a very large number of small changes, such as, in the case of plasmas, small-angle collisions. The Fokker-Planck equation for plasma was initially derived by Landau in 1937, but there are several derivations and modifications in the literature.[3] Maxwell's equations for the electromagnetic field are once again required to complete the picture.

8.7   Summary

1.   Maxwell's equations, the momentum transfer equations, continuity equations and the plasma version of the ideal gas law together describe the plasma as an electric fluid — the MHD model. Maxwellian distributions are assumed. Neglecting the displacement current means that MHD deals with low frequency phenomena   ($v \ll c$).

2.   Ideal MHD applies to perfectly-conducting fluids. Non-ideal MHD applies to plasma with resistivity. When the fluid is not perfectly-conducting or if viscous effects are present, MHD oscillations will be damped.

3.   Plasma beta, the ratio of plasma pressure to magnetic field pressure, measures the effect of diamagnetic currents on plasma confinement. If $\beta > 1$ confinement by magnetic fields alone becomes impossible.

4.   Plasma instabilities may be small- or large-scale. The turbulence produced may simply cause diffusion across the magnetic field, or may result in catastrophic loss of plasma energy and confinement.

5.   Kinetic theory is important in the absence of a magnetic field; when collision frequency is low (and the mean free path is long) and particle flow is along the magnetic field; or for high frequency flow across the magnetic field. A statistical approach is used to specify position and velocity of the system of particles in 6D phase space. The equations used for computing transport processes, etc., in plasmas are the Boltzmann equation and, in fully-ionised plasmas, the Fokker-Planck equation.

References

1.   Cowling, T.G.: *Rep. Prog. Phys.,* **25**, p.244-86 (1962)

2.   Dungey, J.W.: *Phil. Mag.,* **44**, p.725-38 (1953)

3.   *Collected papers of L.D. Landau.* (ed.) D. Ter Haar. Pergamon Press (1965) p.163-70
Chandrasekhar, S.: *Rev. Mod. Phys.,* **15**, p.2-89 (1943) [abstr.]
Rosenbluth, M.N., W.M. MacDonald & D.L. Judd: *Phys. Rev.,* **107**, p.1-6 (1957)

Chapter 9

Gas Discharges

9.1   Introduction

Electrical discharges through gases developed from the electrostatic experiments of the 1700s, via the invention of the voltaic cell and electric arc, to the now familiar fluorescent light tube and "neon" advertising signs first introduced in the early 1900s. Exciting developments elsewhere in physics in the 1920s caused gas discharge research to decline. It was thought that, with limited applications, the field had little more to offer. Interest revived in the 1950s with the advent of nuclear fusion research and the need to understand and control the behaviour of hot plasmas. The arrival of space exploration and an increasing number of technical and industrial applications added further stimulus to research.

9.2   What is a gas discharge?

The early discharge tube was a type of capacitor: two conductors separated by an insulating (dielectric) layer — either a partial vacuum, gas or air. The maximum potential difference tolerated depends on the electrical breakdown characteristics of the dielectric. When breakdown occurs, the circuit is closed and the capacitor discharges.

The term "gas discharge" is now used to describe any flow of electric current through a gas. When an electric field is applied across the gas, collisions between the atoms in the gas lead to ionisation and the creation of free electrons and ions. The electrons are accelerated by the electric field and collide with other atoms and ions, producing an avalanche of electrons. In many discharges the degree of ionisation is below 10%. All that is needed for a current to flow through the gas is that some of the particles are ionised, and there is an electric field to drive the charged particles to form a current. No electrodes are necessary. Lightning is a naturally-occurring gas discharge, as is the aurora. Man-made examples include fluorescent and neon tubes, arc lamps for welding, and glow discharges.

Gas pressure and electrical circuitry determine the behaviour of a discharge. Some are short-lived and transient, others are steady-state, self-sustaining processes. The currents they carry range from less than 10^{-6}A to more than 10^6A. Electron temperature is usually much higher than positive-ion and neutral-atom temperatures, especially at low pressures, because the energy producing the discharge comes from the applied electric field. This heats the electrons more efficiently than the heavier ions. Energy transfer from the fast-moving electrons to the heavier particles is by means of collisions and is very slow, so electrons in the discharge can reach temperatures of about 1 eV while ions and neutrals remain around room temperature. Even in welding arcs, where electron temperatures of several eV are produced, positive ion and neutral particle temperatures are only about 0.5 eV.

The temperatures may seem high, but low density means that very little heat is transferred from the electrons to the gas or the walls of the tube. As pressure or density increases, more energy is transferred from the electrons to the neutrals and ions. The gas temperature rises and electron temperature falls. At pressures between 10 and 100 torr the gas and electron temperatures begin to converge. In arcs at atmospheric pressure the two temperatures are equal.

9.3   Breakdown

A gas completely shielded from all forms of radiation would be a good insulator. In reality, this is unlikely — any gas contains a few charged particles produced by ultraviolet or cosmic radiation. This naturally-occurring process of photo-ionisation ensures that two electrodes a short distance apart, in air, will each emit a few electrons even if no voltage is applied. No current flows between them because the rate of electron production equals that of recombination.

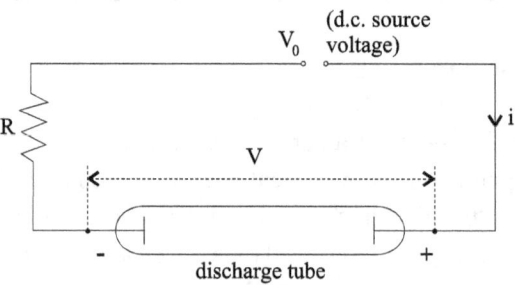

Fig. 9.1   Discharge circuit

Fig. 9.1 shows two cold plane electrodes in a discharge tube filled with gas at reduced pressure. The potential difference between the electrodes can be varied by shining ultra-violet light on the cathode. The discharge is started by applying a small variable voltage across the electrodes, which creates a very weak electric field between them. The photo-ionised charged particles drift along this field, producing a similarly very weak current, i, of around 10^{-11} A. Current strength is proportional to particle speed. Gas conductivity depends on charged particle mobility and their rates of production and recombination. This stage is sometimes called the *background ionisation regime*.

The electric field between the electrodes increases as the voltage across them increases. Current density increases as more charged particles are produced, the current increasing with the voltage. As more ions and electrons reach the electrodes and are neutralised, the effective recombination coefficient increases and the rate of increase of current with voltage slows. When the voltage across the electrodes is about 15 V, and provided there is no change in the rate of ionisation as the electric field increases, all the charged particles will reach the electrodes before they have had chance to recombine. They have insufficient energy to create additional ions in collisions with other particles, so the number of charges arriving at the electrodes will equal the number being produced. Saturation is reached at i_0 in Fig. 9.2. Current density, J, becomes a function of the distance d between the electrodes and the rate of production of (singly) charged particles per unit volume, dn/dt:

$$J = d\, q\frac{dn}{dt} \qquad (9.1)$$

where q is the charge on the electron. Unlike equation (5.18) — $J = nqv_e$ — neither the electric field between the electrodes nor the particle velocity has any effect on current density at this point. Equation (9.1) defines the *saturation current density* which is usually less than $10^{-5}\,\text{A m}^{-2}$. Excitation and emission rates are too low for this discharge to be visible. It is known as a *dark discharge*.

Fig. 9.2 shows how the discharge current varies with voltage. The curve is known as the discharge, or I-V characteristic. Once the saturation zone is reached the current increases very slowly, remaining almost constant for quite large increases in voltage, with the rate of increase in current depending on the gas pressure. This region of the characteristic, marked as T_1 in Fig. 9.2, is known as a *Townsend discharge* and is the simplest form of gas amplification of current.

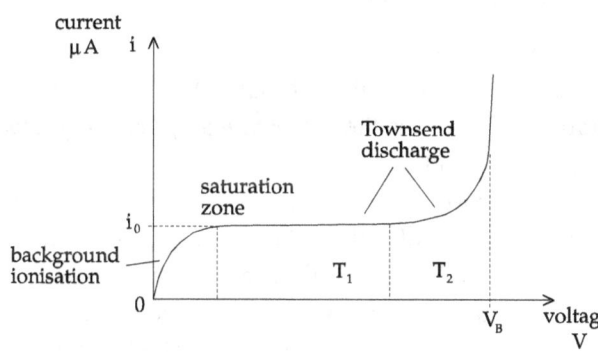

Fig. 9.2   Pre-spark conditions

As the voltage across the discharge tube increases, each electron emitted by the cathode is increasingly accelerated by the electric field between the electrodes, until it gains enough energy to be involved in ionising collisions with atoms of the gas. The newly-freed electrons are themselves accelerated by the electric field and further ionising collisions may occur. This primary and secondary ionisation means that the number of electrons arriving per second at the anode is greater than the number leaving the cathode, ultimately producing an effect known as a *Townsend avalanche* of electrons. The symbol α, known as the *first Townsend ionisation coefficient* or *coefficient of primary ionisation*, represents the number of new electrons created, per metre, by this electron multiplication process. The value of α depends on the type of gas, its pressure and the electric field strength, which must be uniform. An approximate value is given by

$$\alpha = pA\exp\left[-B\left(\frac{p}{E}\right)\right] \qquad (9.2)$$

where A and B are constants specific to the gas and p is the gas pressure at $T = 0°C$. B is equivalent to A times the effective ionisation potential, which in turn is greater than the actual ionisation potential, V_i, because of the effect of collisions. Equation (9.2) was first derived by Townsend and is only accurate over a limited range of E/p, since early measurements of α, before the 1960s, were inaccurate due to contamination by mercury vapour from vacuum pumps and gauges.

Meanwhile, the much heavier positive ions move slowly towards the cathode and accumulate there, outnumbering the electrons. At the anode, ion concentration is zero and electron concentration is at its greatest, although still less than that of the ions at the cathode. This uneven distribution of charge distorts the otherwise uniform electric field and therefore the value of α.

In a uniform electric field, the number of electrons produced from each initial electron increases exponentially with distance from the cathode. In travelling a distance dx, n electrons produce $\alpha\, n\, dx$ others. The change in the number of electrons, dn, is therefore given by $dn = \alpha\, n\, dx$, so:

$$n_x = n_0\, e^{\,\alpha x} \qquad (9.3)$$

For simplicity, electron losses through recombination or diffusion to the walls of the discharge tube have been ignored. If the spacing between the electrodes is d and no electrons are lost, then the current between the electrodes in the tube is given by

$$i = n_0\, q e^{\,\alpha d} = i_0\, e^{\,\alpha d} \qquad (9.4)$$

where i_0 is the primary current — the electron current at the cathode. As d increases, the current increases exponentially. The ratio i/i_0 ($= e^{\,\alpha d}$) is the *electron multiplication factor*. Equation (9.4) can be used to measure the coefficient α.

As the voltage across the electrodes increases still further, the current increases more rapidly than is possible through simple electron multiplication. At higher values of i and d, equation (9.4) is no longer valid. The electric field has become sufficiently strong for some positive ions to arrive at the cathode with enough energy to eject an electron from the electrode. This is the region marked T_2 in Fig. 9.2, and sometimes called a "type 2" Townsend discharge. Equation (9.3) now becomes

$$n_{tot} = (n_0 + n_+)\, e^{\,\alpha d} \qquad (9.5)$$

where n_0 is the number of electrons produced at the cathode by external radiation, and n_+ is the number produced by secondary emission at the cathode. The current between the electrodes is given by

$$i = i_0 \frac{e^{\,\alpha d}}{1 - \gamma(e^{\,\alpha d} - 1)} \qquad (9.6)$$

The factor γ is often called the *cathode yield* (in electrons per incident ion) although all secondary effects, such as ionisation en route, can be included. The value of γ depends on the nature of the gas and on the electrode material. It is always less than 1, and is usually in the region of 10^{-2}.

The characteristic shown in Fig. 9.2 shows a sharp increase in current at voltage V_B, where di/dV becomes infinite. This is the *breakdown point* and V_B, which has values of 200V upwards, is the breakdown voltage. It is a function of gas pressure (p) and the distance (d) between the electrodes and, for uniform fields, is given by:

$$V_B = f(pd) = \frac{B}{\ln(pd) + \ln\left(\dfrac{A}{\ln\left(1 + \dfrac{1}{\gamma}\right)}\right)} \qquad (9.7)$$

If pd is less than 1 cm torr, a typical value for V_B is 300 V. At the breakdown point, the gas no longer acts as an insulator. The current rises uncontrollably, its value determined by the resistance of the external circuit, and a spark results in the transition to a self-sustaining discharge. Up to this point, the discharge can be stopped by switching off the UV light. Once V (in Fig. 9.1) equals V_B, the discharge will continue even if the light source is removed.

For low pressure discharges, the Townsend current, as given by equation (9.6), becomes (theoretically) infinite if

$$1 - \gamma(e^{\alpha d} - 1) = 0 \qquad (9.8)$$

This can be written $\gamma e^{\alpha d} = \gamma + 1$ but, since γ is usually very much less than 1, we can define the *Townsend criterion* or *sparking criterion* as

$$\gamma e^{\alpha d} = 1 \qquad (9.9)$$

When $\gamma e^{\alpha d}$ is less than 1 the discharge depends on the initial current, i_0 (equation 9.6), and ceases with i_0 — it is not self-sustaining. The sparking threshold (the threshold for a self-sustaining discharge independent of i_0) is $\gamma e^{\alpha d} = 1$. For values of $\gamma e^{\alpha d}$ greater than 1, the ionisation of successive avalanches is cumulative. The space charge will grow and the spark will appear sooner the more that $\gamma e^{\alpha d}$ exceeds 1.

In effect, each primary electron produces one secondary electron to continue the process. Electron production no longer depends on external radiation and the discharge becomes self-sustaining. Experimental observations support the criterion as given in equation (9.9), despite its assumption of a steady-state situation where the electric field is uniform and where the current can be maintained at any value. In reality, breakdown tends to occur as a rapid increase in current which is difficult to control externally, and whose distortion of space-charge affects the electric field. Once breakdown has occurred, the discharge becomes self-sustaining and takes the form of a glow or arc discharge, depending on the gas and circuit conditions.

The breakdown voltage depends on the pressure and nature of the gas, the spacing between the electrodes, their shape and the materials used. Impurities can reduce the value of V_B. If the pressure is low or the electrode spacing is small, very few collisions will occur between primary electrons from the cathode and atoms of the gas. If the pressure is too high the primary electrons will gain insufficient energy between collisions to be able to release secondary electrons in ionising collisions with atoms. If the electrodes are too far apart, the ions and electrons will recombine before the ions reach the cathode to release more electrons. In all these cases, there will be very few secondary electrons and the breakdown voltage needed to sustain the discharge will be high.

Breakdown at high pressure differs slightly from the low-pressure situation. When positive ions near the anode take a long time to reach the cathode, and especially if the gap is large, the breakdown develops as a luminous streamer from the anode, which effectively moves the anode potential closer to the electron avalanche originating at the cathode. The two eventually join, resulting in breakdown.

9.4   Types of discharge

Electric discharges through gases can be classified as stable or transient. Stable discharges require a continuous supply of charged particles. Transient discharges are usually initiated by breakdown of the insulating spark gap and last until the source of energy is removed.

Whether stable discharges are self-sustaining depends on the current they carry. Below about 10^{-11} A the discharge current produces insufficient ionisation and an external agent such as photo-ionisation is required to excite the particles. When this is removed the discharge ceases. Examples include: Geiger counters, ionisation gauges and gas-filled photo-electric cells. Currents larger than 10^{-10} A can produce sufficient ionisation for the discharge to become self-sustaining: once started, the current continues to flow and depends only on the circuit and discharge parameters.

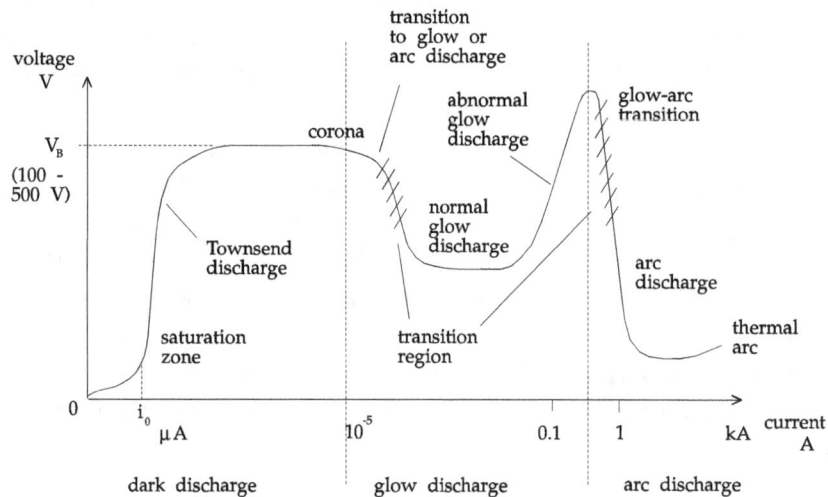

Fig. 9.3   D.c. discharge current-voltage characteristic

If we take a low pressure discharge and measure the current flowing through it as the voltage is gradually increased, the resulting current-voltage curve will be as shown in Fig. 9.3. The far left of the curve is that shown in Fig. 9.2: the pre-spark conditions.

9.4.1 Dark discharge or Townsend discharge

The name Townsend, or dark, discharge is usually applied to a low-pressure discharge in which the electric field is made uniform by having large plane (flat) electrodes. It is not self-sustaining. The avalanche of electrons needed to raise the current above the saturation value is initiated by external radiation — UV light, X-rays or cosmic rays — or a negative electrode. A dark discharge is invisible simply because the number of excited atoms emitting light in the visible spectrum is very small — the radiation is too weak to be seen. The currents carried are also very small, up to 10^{-1} A but usually in the region of 10^{-8} A, so the cathode remains cold. This discharge is the main cause of leakage from charged conductors and is the starting point of other types of discharge.

9.4.2 Corona discharge

Meaning "crown-like", the corona discharge takes its name from mariners' descriptions of the St. Elmo's fire discharge which forms a glow around ships masts during electrical storms. If one electrode in a discharge has an uneven surface, with sharp points, edges or wires, then a localised concentration of electric field develops which is greater than the breakdown field. This produces an incomplete breakdown called a corona, or unipolar, discharge. Corona occurs near the point of transition from the Townsend discharge to a glow or arc (see Fig. 9.3) and should not be confused with sparks, which are transient. Ionisation is confined to a small region of intense field strength near the electrode and a low-current self-sustaining luminous glow forms around the electrode. Currents of the order of 1 mA are possible. At very low currents, the corona is dark. The brush discharge, first described by Faraday in 1838, is a luminous discharge in a non-uniform field in which many simultaneous corona discharges form streamers around the electrode.

Corona is important in high-voltage apparatus where it can cause significant power loss, for example in high-voltage transmission systems. By facilitating flashover[a], it can destroy insulating materials. Industrial applications of corona discharges include removing surface charge from plastic or photographic film, and improving the adhesive properties of polymers to take paints or printing inks.

9.4.3 D.c. low-pressure glow discharge

The direct current (d.c.) glow discharge is a self-sustaining luminous discharge which is easy to produce and maintain at pressures less than 0.1 atm, once breakdown has occurred,

---

[a]   an electric discharge over the surface of an insulator

158

provided that secondary electron emission is sufficient. Currents carried range from 10^{-6} A to about 0.1 A. Electrodes, inserted into each end of the tube, are connected to a d.c. power supply in series with a means of controlling the discharge current. The minimum voltage needed to produce and sustain the discharge is the breakdown voltage, V_B. Once extensive breakdown occurs in the gas the discharge starts to glow, noticeable at first just at the cathode. The voltage drops and the current increases (Fig. 9.3). The colour of the diffuse glow depends on the gas used.

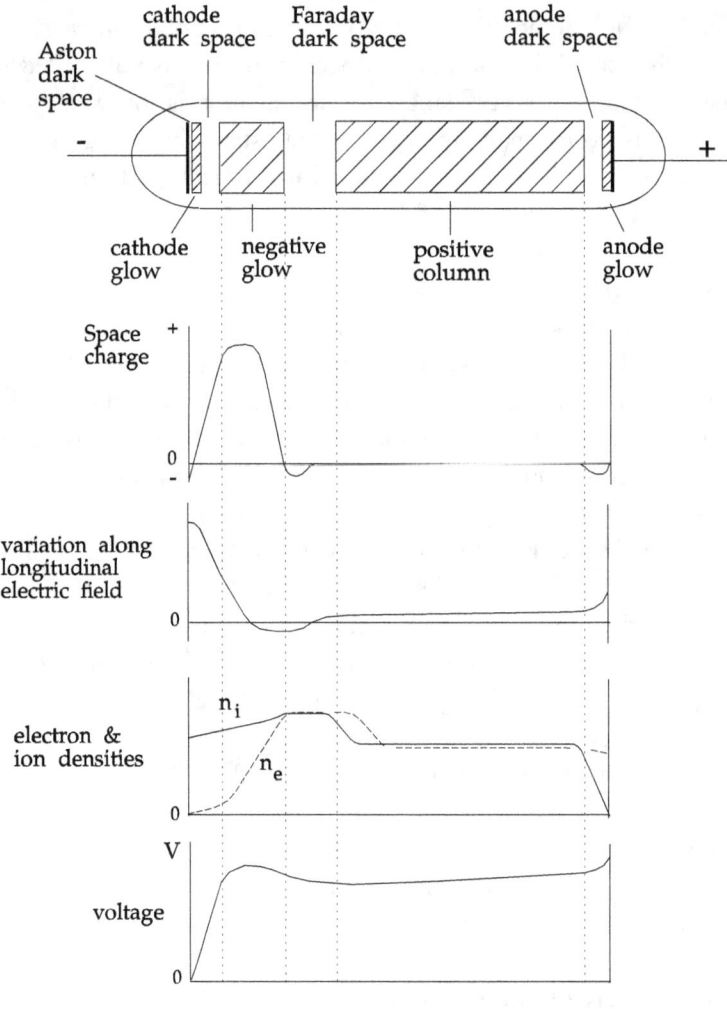

Fig. 9.4   d.c. glow discharge

Fig. 9.4 shows the typical structure of a glow discharge, its actual appearance determined by the pressure within the tube. Along its length, dark regions are interspersed with bright glows resulting from radiative recombination (Section 7.4.3). Working from left to right in Fig. 9.4: photo-ionisation at the cathode plus positive ion bombardment of the cathode produces the electrons needed to maintain the discharge. The electrons leave the surface and form a negative space charge region close to the cathode (the *Aston dark space*) before being accelerated by the electric field. The Aston dark space is often obscured by the cathode glow. Slower-moving positive ions accumulate near the cathode until they greatly outnumber the electrons. The current in this part of the tube is therefore carried mainly by positive ions originating in the cathode dark space and negative glow regions of the tube. As a result, the space charge becomes positive only a short distance from the cathode and remains so through the cathode dark space. There is, in effect, a plasma sheath at the cathode.

The *cathode glow* is produced by slow electrons recombining with incoming positive ions. Its length depends on the type and pressure of the gas. In air it is often a reddish-orange colour. The cathode glow of a normal discharge usually only covers part of the cathode surface, the area varying with the current. As the abnormal discharge stage (see Fig. 9.3) approaches, the whole surface becomes covered by the glow.

The length of the *cathode dark space*, sometimes called the cathode fall, varies inversely with pressure, filling the tube when the pressure is low. Under normal glow discharge pressures, it is around 10 mm in length. The large positive space charge of the cathode dark space creates a voltage gradient to balance the ion current hitting the cathode with the flow of electrons from the negative glow to the anode. The voltage drop can be of the order of 100 V and depends on the gas and electrode material. This accelerates the electrons, producing intense ionisation. At the anode end of the cathode dark space, electron density begins to increase through secondary ionisation, and the current is increasingly carried by the electrons.

In the *negative glow*, the voltage gradient decreases slightly as the net space charge becomes negative. Most of the ions which are accelerated to the cathode originate in the cathode dark space and the first part of the negative glow. There is no further acceleration of the electrons emerging from the cathode dark space and their energy is absorbed by collisions with other particles. The resulting excitation and ionisation produces the bright negative glow, at its most intense on the cathode side. The *Faraday dark space* marks the end of this energy absorption. Electron energy has become too low to produce further excitation. Recombination and diffusion in the dark space reduces electron density. By the end of the Faraday dark space, the net space charge has become zero and the electric field along the axis of the tube has acquired a small positive value. At this point, the *positive column* begins.

The positive column forms a plasma extending almost to the anode, and occupying as much of the tube as it is allowed to. It maintains the conduction of current by electrons which was established in the cathode region. The voltage gradient is usually small, decreasing with increasing current at fixed pressure, and increasing with pressure at fixed current. The electric field is just sufficient to ensure that loss of electrons to the wall of the tube is balanced by production of electrons by ionisation throughout the length of the positive column. The column is luminous because of the energy of the electrons. Variations in electron density from the axis of the tube to the wall produces a radial electric field which reduces the flow of the more mobile electrons and concentrates plasma density along the centre of the tube. Unlike in the cathode dark space and the negative glow, where particle flow along the tube determines the behaviour of the discharge, radial processes dominate in the positive column. To maintain quasi-neutrality, electrons and ions in this region must flow towards the wall of the tube with equal current densities. A sheath therefore forms between the plasma column and the wall of the tube, ensuring that the wall floats at a constant potential. Luminous bands called *striations*, linked to fluctuations in electron and ion densities, often appear along the length of the positive column. In some low-pressure discharges, the Faraday dark space and positive column disappear because the anode is brought close to the negative glow. The minimum separation distance of the electrodes is twice the length of the cathode dark space — any less and the glow discharge is extinguished.

At the anode, electrons from the positive column form a thin boundary layer or sheath with a negative space charge. There is frequently an accompanying *anode dark space* and *anode glow*. The anode glow, when present, is often brighter than the positive column. The anode attracts electrons from the positive column, while the sheath at the anode adjusts the current densities of the ions and electrons to those required to carry the discharge current to the anode surface. No ions are produced at the anode. The small ion current in the positive column comes from electron collisions close to the anode.

Electrons generally have very close to a Maxwellian velocity distribution in all luminous parts of the discharge and in the Faraday dark space, but not in the regions of cathode and anode falls of potential or, as a rule, in the dark spaces between striations. It is therefore possible to define an electron temperature in the main part of the discharge.

Glow discharge tubes, with cold cathodes, have for many years been used as voltage regulators in electronic circuits. More recent applications include nitrogen bombardment of steel to harden it and protect items such as engine parts against wear. Low-pressure glow discharges may also be used to "pump" gas lasers into their lasing state.

9.4.4   Arc discharges

In the glow discharge, electrons are removed from the surface of the cold metal of the cathode by the impact of positive ions. In the arc discharge, a high current heats up the cathode producing thermionic emission — electron emission from a heated material as a result of thermal energy. Arc discharges are often started in open air because the heat released can destroy the glass tube.

When the gas pressure is close to atmospheric pressure and the resistance of the external circuit is relatively low, breakdown will result directly in an arc discharge. The gas becomes intensely luminous with apparently violent turbulence. Alternatively, if pressure or current increase, an established low-pressure glow discharge will become less stable and degenerate into an arc discharge. The glow-to-arc transition is triggered by an increase in electron emission from the cathode, usually from positive-ion bombardment of the cathode, i.e. the action of the current itself. The transition takes place quite suddenly, and often unpredictably (see Fig. 9.3). As the pressure increases in the glow discharge, the cathode dark space region contracts. Current density in this region increases noticeably, accompanied by an increase in the voltage gradient in the positive column. This is the "abnormal glow" region of the discharge characteristic. As current density increases, so too does the field at the cathode. The cathode is heated by positive ion impact, the heating increasing as current density increases. The increasing energy which the ions deposit at the cathode may raise its temperature enough to produce thermal emission of electrons. This adds to the current, increasing the current density still further. At this point, the voltage decreases sharply and the glow discharge becomes an arc discharge.

The discharge is then sustained by the intense production of positive ions over a small distance due to the high current density, which maintains the temperature of the cathode. The luminous area at the cathode reduces to a very small *cathode spot*. The high temperature of the spot may cause the electrode metal to evaporate while the rest of the cathode remains at a much lower temperature. The temperature of the cathode spot depends on current density, electrode material and the type of gas. Atmospheric arcs generally have cathode temperatures between 2200 K and 3300 K. Unlike the glow discharge, the cathode sheath of the arc discharge is very thin, often much less than 1 mm and close to the Debye length.

The arc discharge has a lower voltage than the glow discharge (less than 50 V compared to 300-500V in a glow discharge) and a higher current (greater than 0.1 A), determined mainly by the external circuit. The positive column or *arc column*, if present, is intensely luminous because of the higher current density and the cathode region is no longer distinguishable. In very short arcs there may be little evidence of a positive column. At low pressure the positive column is diffused and similar to that in a glow discharge. At high pressure, around one atmosphere, the arc appears to have an intense core with no noticeable structure; the potential gradient along the positive column and the apparent diameter of the column remain constant. At low pressures, below about one torr (equivalent to 10^{-3} atm), the temperature of the gas in the positive column is much less than the electron temperature, which may reach 5 eV. As pressure increases, electron temperature falls and the gas temperature rises until the two coincide. At atmospheric pressure the temperature of the whole gas in an arc discharge is usually over 5000 K (around 0.5 eV). As with the glow discharge, a very thin negative space charge sheath forms at the anode. Positive ions produced in this region are repelled from the anode and add to the drift current in the positive column.

The regions of the arc close to the electrodes are often called the *cathode jet* and the *anode jet* because of the high velocities and temperatures which occur in these regions. Arcs require a copious emission of electrons from the cathode, provided by thermionic emission

162

or field emission[b] since the cathode sheath is too small for secondary electron emission to be significant. Thermionic emission is used to explain the electron emission from electrodes of high boiling-point materials such as tungsten and carbon. Arcs in which the boiling point of the electrode material is too low for adequate thermionic emission are described as cold-cathode arcs. These include mercury arcs and arcs with copper electrodes. In such cases, field emission may be responsible.

An arc discharge can also be produced by taking two touching carbon or metal current-carrying electrodes and moving them a small distance apart. This is a *drawn arc* and, once formed, is similar in all respects to the arc discharge obtained from the glow-to-arc transition described above. Just before the rods separate, so much electrical resistance develops at their boundary that the tips begin to glow through the emission of electrons from the cathode. The electrons ionise the air immediately surrounding the electrode tips so the current continues to flow when the electrodes are separated. The situation is shown in Fig. 9.5. The anode becomes white hot under the bombardment of electrons to which it is exposed and a plasma forms in the arc itself. The temperature in the arc can be between 2 and 5 eV.

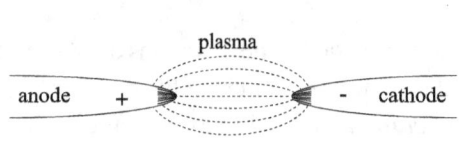

Fig. 9.5   D.c. drawn arc

No breakdown voltage is required to produce a drawn arc, so its life depends on the separation distance of the electrodes and the available voltage in the circuit. Unless the electrodes are fed forward to maintain the spacing as they burn, the arc will be extinguished if the gap between the electrodes becomes too great.

9.4.5   Radio-frequency (r.f.) discharges

Conventional d.c. glow discharges require electrically-conductive electrodes to be inserted into the tube in direct contact with the plasma, risking contamination of the plasma by electrode material, or of the electrodes by the plasma. In industrial processing, for example, the principle of high-energy ions hitting the cathode is used to deposit thin-film coatings on surfaces (or *substrates*) placed on or close to the electrodes. If the film being deposited is a semi-conductor or an insulator, the electrodes exposed to the plasma will soon become covered with an insulating film and the discharge will be extinguished.

Radio frequency (r.f.) discharges can produce and maintain a plasma using either conductive or non-conductive electrodes placed outside the discharge tube, which must be made of a dielectric material such as quartz. A low-frequency alternating electric field is applied between the two electrodes so that, in one complete cycle, each electrode acts

---

[b]   A high voltage gradient produces an intense electric field at the electrode which strips electrons from the surface.

alternately as a cathode and an anode. With external electrodes, the power is transferred to the plasma either via inductive coupling, in which a coil is wrapped around the tube, or via capacitative coupling, where two separated electrodes are fixed on the outside of the tube (Fig. 9.6).

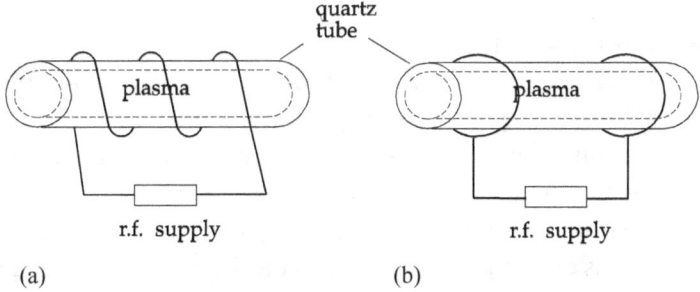

Fig. 9.6   r.f. discharges: (a) inductive coupling (b) capacitative coupling

Once the breakdown voltage, V_B, is reached on each half-cycle, a temporary d.c. glow discharge is initiated. When the voltage drops below V_B during the cycle, the discharge ends and the space charge decays before the discharge is restarted with the reverse polarity. The plasma in an alternating current discharge can be thought of as a double-ended negative glow with sheaths between the plasma and the electrodes.

As the frequency of the alternating electric field is increased beyond a critical frequency, f_c, given by

$$f_c = \frac{\langle v_e \rangle}{2d} \tag{9.10}$$

where $\langle v_e \rangle$ is the average drift velocity of the electrons and d is the distance between the two electrodes, charged particles do not recombine between cycles. At frequencies higher than 50 kHz the oscillating electrons gain sufficient energy to cause ionising collisions. The voltages needed to start and maintain the a.c. discharge are therefore much less than those required for a d.c. discharge. The frequencies used in these high-frequency discharges are in the range of radio transmission (typically 13.56 MHz), hence the name *r.f. discharge*. Most capacitative r.f. discharges used in industry are unmagnetised and operate below the plasma frequency at between 1 and 100 MHz.

Ions produced from collisions of atoms with electrons flow into the sheath regions and are accelerated by the alternating field, taking several cycles to reach the electrode. Light ions take fewer cycles and reach the highest energies. In low-pressure discharges, collisions in the sheath are rare and the ion energy is used to etch metals, semiconductors, etc. As the pressure increases, collisions in the sheath become more frequent and ion energies are lower. Discharges operating at higher pressure are used for depositing thin films of metal, glass or polymer on to a substrate.

R.f. discharges allow greater control over the energy of the ions bombarding the substrate than is possible in the d.c. discharge. The high-energy ions bombarding the cathode of a d.c. discharge are accelerated at voltages that must be above the minimum breakdown voltage. The resulting ion velocity can damage sensitive substrates placed on the cathode. Such damage is less likely in r.f. discharges.

9.4.6   Microwave gas discharges

The microwave gas discharge was first developed in the late 1940s following wartime investigations of microwave equipment and radar technology.[1]   It served as the basic mechanism of the transmit-receive tube, a switch to protect a radar receiver from damage during the high-energy transmitter pulse, when both transmitter and receiver are used with the same scanning aerial through a common waveguide system.

A microwave gas discharge is simpler than the d.c. glow discharge in that it has just one bright region, a plasma similar to the positive column of the d.c. discharge. Most microwave-generated plasmas are produced in a waveguide structure or resonant cavity. There are no internal electrodes, so no erosion of electrode material to contaminate the plasma. It is an alternating-current discharge and as a result, the electrons can only move a very small distance during a half-cycle of the electric field. In the d.c. glow discharge, the electrons gain energy only by moving towards the anode. In the microwave discharge, the electron motion is more random, with energy being gained from collisions with other particles.

Radiation at microwave frequencies, generally 2.45 GHz or 915 MHz, interacts with the plasma as a dielectric medium rather than with individual electrons, as is the case with r.f. discharges. Electron temperature ranges from 5 eV to 15 eV, higher than the 1 or 2 eV of d.c. and low-frequency r.f. discharges, and pressures are generally lower: the critical number density (Section 6.3.2) is 7.4×10^{16} m^{-3} at a frequency of 2.45 GHz. Microwave discharges, because of their higher temperatures and lower operating pressures, can provide a greater degree of ionisation and dissociation than d.c. or low-frequency r.f. discharges, which can be important in plasma chemical applications.

Since the 1980s, microwave-generated plasmas have been increasingly used in microelectronic plasma processing, continuous-flow plasma torches and chemical reactors.

9.4.7   Electron cyclotron resonance (ECR) discharges

Certain industrial applications of plasma, for example some etching and deposition processes, require low-energy ions to act perpendicularly on a surface. Microwave discharges, especially ECR discharges, are preferable to r.f. discharges where, at low pressure, perpendicular incidence is often associated with high-energy ions. Electron cyclotron resonance was originally used to heat plasmas for fusion research but since the 1980s has been increasingly used in plasma processing.

ECR plasmas can operate at lower pressures and produce higher ionisation and plasma density than r.f. plasmas. As we saw in Chapter 4, if a magnetic field B is applied to a plasma, the charged particles rotate around the field lines at the cyclotron frequency, $\omega_c = eB/m$. For electrons in a magnetic field where $B \approx 8.75 \times 10^{-2}$ tesla, the cyclotron frequency becomes 2.45 GHz (using $f = \omega/2\pi$). If microwaves of this frequency are used to produce and maintain a plasma, the gyromotion of the electrons resonates with the microwaves, and the plasma is called an ECR plasma.[c] The plasma flows out of the source chamber along a divergent magnetic field towards a substrate in a separate vacuum chamber which is usually at a lower pressure than the source. The greater speed of the electrons streaming out of the source chamber creates an electric field which accelerates the slower ions towards the substrate. Around the resonance frequency, the average power absorbed by the electrons from the electric field is at a maximum. The magnetic confinement enhances the degree of ionisation, allowing low-pressure operation, usually between 10^{-5} and 10^{-3} torr. Electron temperatures, are in the region of 5 eV — higher than in unmagnetised plasmas. Ion energy is generally about four or five times the electron temperature.

Reference

1.   Herlin, M.A., & S.C. Brown: *Phys. Rev.*, **74**, p.291-6 (1948)
          Biondi, M.A.: *Elec. Engng. (NY)*, **69**, p.806-9 (1950)
          Cobine, J.D.: *Elec. Engng. (NY)*, **69**, p.499-504 (1950)

---

[c]   The resonance is often closer to the upper hybrid frequency (equation 6.33) than to the cyclotron frequency, but the difference is generally small. If $n_e = 10^{17} \, \text{m}^{-3}$, $\omega_{pe} = 1.78 \times 10^{10} \, \text{rad s}^{-1}$.
           If $f = 2.45 \times 10^{9}$ Hz (so $\omega_{ce} = 1.54 \times 10^{10} \, \text{rad s}^{-1}$), $\omega_{UH} = 2.35 \times 10^{10} \, \text{rad s}^{-1}$.

Chapter 10

Industrial and environmental applications of gas discharges

10.1 Background

When a plasma interacts with a neutral gas, the higher kinetic temperature of the plasma particles causes collisions with the gas particles which produce ionisation, dissociation of molecules into their constituent atoms, and chemical reactions to form new compounds. Investigation of the chemical effects of electrical discharges through gases began in the mid-19th century, but no general theory could be established until the processes in the discharge tube were understood. By the 1870s, arc lamps were being used in experimental lighting in the United States. Gas-filled valves and switchgear based on discharge tube technology soon followed, and by the turn of the century nitric oxide was being synthesised in an arc in the first industrial use of plasma chemical processing.[1] The demands of radio technology for discharge tubes during the First World War (1914-18) further accelerated their development.

In the 1920s, attention turned once again to the chemical changes produced by electrical discharges. The rate of chemical formation could not yet be determined, either experimentally or theoretically, preventing the development of a satisfactory theory of the processes involved. By 1930 it was known that chemical action in the glow discharge is initiated primarily by positive ions in the negative glow of the discharge. The following year, saw the decomposition of methane at liquid air temperatures (less than -196°C) into ethylene and hydrogen ($2CH_4 \rightarrow C_2H_4 + 2H_2$) in the negative glow. This was regarded as significant since it showed that the reaction process was very simple.[2] The late 1940s saw the introduction of microwave and r.f. discharges, while MHD power-generation, an idea which had been around since the 1930s, began to receive attention in the 1960s.

There was limited industrial use of low-temperature, weakly-ionised discharge tube plasmas until the development of surface modification by ion implantation in the early 1970s. Metals, semiconductors, ceramics and polymers could now have their surfaces modified or coated using plasma-based processes such as sputtering and chemical vapour deposition in low-pressure gas discharges. By the 1980s, plasma was being used to manufacture integrated circuits for the electronics industry. Cold processing plasmas, with temperatures below 10^4K replaced earlier chemical etching of semiconductors for the computer industry. Plasma reactors gave greater control of the pattern transfer and cleaner, finer lines, reducing component size and facilitating vast improvements in computer memory. The more

168

environmentally-conscious 1990s saw the development of waste processing, in particular of toxic materials, using plasma technology. Other applications include "cold" pasteurisation of food and sterilisation of medical products. Plasma propulsion for space travel is also under investigation.

Many processing plasmas are low-temperature (around 10 eV) non-equilibrium plasmas with electron densities around $10^{16}\,\mathrm{m}^{-3}$. Glow discharge plasmas are popular because they operate with low voltages and currents. Other methods of generating the plasma include positive or negative corona discharges, silent discharges, r.f. and microwave discharges. Each has its own particular advantages and disadvantages. Corona discharges tend to become unstable at high pressure, turning into an arc if the power supply provides enough current. The silent discharge has one or both metal electrodes covered with a dielectric layer, such as glass, which separates them from the discharge gas. Sometimes called a dielectric-barrier discharge, the design was first used in 1857 to produce ozone and is now used in toxic waste treatments. Radio frequency discharges, with no electrodes exposed to the plasma, are widely used in optical emission spectroscopy and for etching in semiconductor manufacture.

Although the large-scale use of plasma for industrial processing is still fairly new, many applications and techniques are already in operation and other potential uses await the development of the necessary technology. It is only possible here to provide a brief outline.

10.2   Arc applications

The high temperature of arc discharges has been used for more than a century for cutting, melting and welding metals. Electrons emerging as a beam from the curved or hollow cathode of an arc or glow discharge are used for welding. The geometry of the cathode determines the structure of the beam. Its energy density depends on discharge voltage, distance between the electrodes, the nature and pressure of the gas, and any focusing via electric or magnetic "lenses". An arc-furnace is a welding arc on a grand scale, in which the metal being melted forms one electrode.

Other applications of electric arcs include rectifiers and control tubes such as thyratrons. A rectifier allows electric current to pass in one direction only and is used to convert a.c. into d.c. It was one of the earliest applications of electric arcs. A thyratron is a grid-controlled rectifier — a thermionic gas-filled vacuum tube containing a heated (and therefore emitting) cathode, an anode and a control electrode or grid whose potential controls the flow of electrons from cathode to anode.[3]

The plasma torch, or plasma jet, is an arc burning between a central tungsten electrode and a water-cooled copper tube which forms the nozzle. Solids, liquids or more often gases are forced through the nozzle and emerge as the plasma of the arc column in a jet rather like a flame. The high temperatures of the plasma torch have many applications in melting or cutting materials, and in surface treatments and deposition.

Arc discharges can also be used as circuit breakers since an arc inevitably forms when a current-carrying circuit is interrupted. Arc discharge lamps are highly efficient light sources, the carbon arc being one of the first practical electric lamps.

10.3 Gas discharges as light sources

Two types of light source are in general use: tungsten-filament incandescent lamps — the conventional light bulb — and plasma lamps based on the discharge tube. These may be either: (a) low-pressure, low-power-density, weakly-ionised discharges that are far from thermal equilibrium, such as fluorescent lamps and low-pressure sodium lamps; or (b) high-pressure, high input power density, weakly-ionised discharges that are close to local thermal equilibrium. Here, the plasma is contained in a much smaller volume and operating temperatures are much higher. Examples include the high-pressure sodium street lamp and high-intensity halogen lamps used in commercial, industrial and recreational environments.

The *low-pressure (mercury vapour) fluorescent lamp* consists of a long tube in which an arc burns in mercury vapour at very low pressure: between 0.1 and 2.5 torr (10^{-5} - 0.003 atm). Over 95% of the radiation from the plasma lies in the ultraviolet, mainly at the mercury resonance wavelength of 253.7 nm. The inside of the glass tube is coated with a phosphor layer, the atoms of which absorb ultraviolet-wavelength photons from the mercury and are excited. As they return to the ground state, each atom emits two or more photons of lower energy and longer wavelength, producing visible light, the colour produced depending primarily on the choice of fluorescent substance used to coat the walls of the tube. The mercury vapour pressure in the tube is very low, so argon gas is added to enable the discharge to start. Many lamps, because of their tube length, use a starter circuit, or choke, to provide enough electrons to ensure rapid breakdown.

The so-called "neon" lights used in electric advertising signs are cold cathode discharge tubes, with currents of less than 0.1 A, containing neon (which gives a fiery red glow) and other rare gases. Tinted glass or phosphors which produce line radiation from the excited atoms of these rare gases provide the wide range of colours.

One of the earliest discharge light sources was the *low-pressure sodium lamp*, an arc burning in sodium vapour and producing a yellow light. Electrical energy from the discharge produces radiation in the visible spectrum, almost all of it at 589.0 nm and 589.6 nm: the sodium resonant doublet, or "yellow D-lines". Since these resonance wavelengths are visible and strong, there is no need for a phosphor coating on the tube walls. To minimise radiation absorption by the plasma in the positive column, a pressure of about 5 mtorr is required. This would require a very high voltage for breakdown to occur, so a mixture of neon and argon is added to the sodium vapour to enable the discharge to start. The neon produces the characteristic red glow seen in the first few minutes of operation.

The *high-pressure mercury vapour lamp* once again produces radiation in the ultraviolet range, chiefly at 253.7 nm, but by operating at high pressure — up to 10 atm — the mercury vapour absorbs and re-emits the radiation, becoming a good source of visible light. Argon is added to reduce the breakdown voltage. A phosphor coating is often applied to the outer "envelope" of the tube to improve the colour of the light produced.

The *high-pressure sodium lamp*, which replaced the blue-white high-pressure mercury street lamps, operates with a vapour pressure of about 100 torr. At this pressure, less light is produced at the resonant frequencies and more comes from neighbouring frequencies, producing a more golden glow. Xenon is added to reduce the breakdown voltage. Mercury may also be included in the mixture to control electrical and thermal properties. It has no effect on the colour of the light produced as the sodium plasma is too cold (around 700°C or 0.085 eV in the cooler parts) to excite the mercury atoms.

10.4   Synthesis of chemical compounds

In the early 1900s, the Rev. P.J. Kirkby, working at New College, Oxford, investigated whether Townsend's ideas on electron multiplication applied to chemical changes. Using a glow discharge containing a mixture of hydrogen and oxygen at low pressure, and with differing electric currents, he found that the gases began to combine, producing water vapour ($2H_2 + O_2 \rightarrow 2H_2O$). Unlike electrolysis, the quantity of vapour was proportional to the strength of current passing through the gases in the tube. It was independent of the nature of the electrodes and occurred throughout the discharge, not just around the electrodes.[4]

While the synthesis of water is of purely academic interest, ozone production has important practical applications, especially in sterilising fluids, where relatively small amounts are required to destroy bacteria. In 1857, German inventor Werner von Siemens (1816-92) produced ozone from oxygen flowing through an annular discharge space between two coaxial glass tubes — the first use of the dielectric-barrier, or silent, discharge. He described the process as "electrolysis of the gas phase". In 1904, with a better understanding of electrical discharge processes, E. Warburg ascribed the formation of ozone from oxygen in a discharge tube not to electrolysis but to the effect of cathode rays and ultraviolet radiation.[5] Ozone has subsequently been produced in ozonisers in which oxygen is subjected to an electrical brush or corona discharge: electrons collide with O_2 molecules, removing some oxygen atoms which partially associate with other O_2 molecules to form ozone (O_3).

Initially thought to be beneficial — in the late 1800s it was thought there was not enough of it — ozone is now known to be toxic in the lower atmosphere. It is a stronger oxidising agent than all other industrial chemicals except fluorine, which makes it very effective at destroying bacteria. It is a more potent disinfectant than the chlorine gas often used for sterilising water but the sterilisation process is more complicated. In chlorination, chlorine gas is simply bubbled through water to the required concentration, whereas in ozonization a solution of ozone-in-water is added to the water, which is then filtered through charcoal. A little chlorine is needed to keep the water sterile since, unlike ozone, it remains active for a long time. Ozone has been used to treat domestic water supplies in France since

1907. A similar system began in St. Petersburg in Russia in 1910. Ozone can also be used to sterilise waste water from sewage and remove the bacteria which inhibit heat transfer from the cooling water in power stations. It is also a strong bleaching agent and is often used to treat kaolin, textiles and pulp for making paper.

Apart from ozone, only acetylene, formed from hydrocarbons in an electric arc, is produced on an industrial scale using gas discharges. Attempts have been made to synthesise compounds such as hydrogen peroxide, ammonia and hydrazine using discharge tubes but there is little commercial interest because alternative methods already exist. The synthesis of hydrazine from ammonia in a silent electric discharge was first reported in 1911 by A. Besson, but further investigation showed that only low yields were possible and the product quickly began to decompose in the discharge.[6]

10.5   Sputtering: Surface cleaning and thin-film deposition

Sputtering is the removal of atoms from a surface, especially a cathode, by positive-ion bombardment. This disintegration of an electrode, although often a problem, is frequently used deliberately for surface-cleaning and to deposit thin films of metal onto glass, plastic or other metals. The effect was first reported by William Grove in 1852[7] but even in 1930, Karl Compton and Irving Langmuir were describing its behaviour and explanation as being "rather obscure".[8]

Plasma sputtering exploits the fact that any surface exposed to a plasma is negative relative to the plasma because of the formation of a sheath at the surface. This accelerates positive ions towards the surface. These have greater momentum than electrons and are more effective in producing surface reactions. If an ion arrives with sufficient energy it can remove, or "sputter", atoms from the surface. The number of atoms removed per incident ion — the sputtering yield — depends on the type of ion and its kinetic energy, the yield increasing steadily from a threshold at between 10 and 50 eV, to several atoms per ion at energies in the keV range.

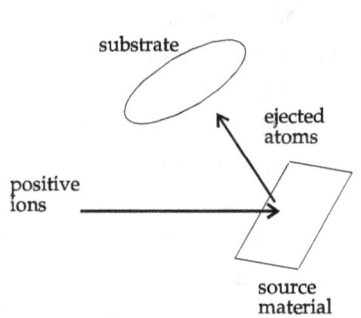

Fig. 10.1   Sputtering

The deliberate use of sputtering to deposit a thin-film coating on a surface can be achieved in various ways. The general principle is shown in Fig. 10.1. The surface to be treated, the *substrate*, is contained within a low-pressure glow discharge and placed on or near the anode. The source material — that required for treating the substrate — is often the cathode, but may be an independent, negatively-biased target. Energetic ions from the discharge bombard the source material ejecting atoms from it which are deposited on adjacent surfaces. The mean free path is often short, so, to minimise diffusion, the source-substrate distance is kept small. This technique can also be used to clean a surface by removing surface atoms and impurities.

R.f. discharges are frequently used to avoid contamination of electrodes by film particles. Conductive or non-conductive electrodes can be placed inside or outside the containment vessel, allowing a much wider range of coating materials to be used. When the discharge gases are corrosive or when the electrode material might contaminate the plasma, external electrodes are essential. Capacitatively-coupled r.f. discharges also allow greater control over the energy of the ions bombarding the substrate. In a d.c. discharge the high-energy ions are accelerated at voltages which are, by definition, above the minimum breakdown voltage. The impact of these ions can damage sensitive substrates placed on the cathode. This is less likely in r.f. discharges, which operate over a wide range of pressures. As the pressure increases, collisions increase, reducing the energies of the ions. This enables a wide range of applications which do not require high energy ions. Examples include the deposition of metals, polymers, ceramics such as aluminium nitride, glasses such as silica, and diamond-like carbon.

Sputtering is also used in *secondary ion mass spectrometry* (SIMS) to analyze the composition of a surface. Here, energetic ions eject secondary ions and atoms from the surface for mass analysis, to identify the sputtered particles.

10.6 Surface modification

The growing demand for materials with resistance to corrosion or wear, with decorative finishes, or with specific electrical, magnetic or optical properties is being met increasingly by plasma processing. Much of the surface treatment takes the form of coatings. Plasma treatment ensures that the particular coating binds to the substrate, which is generally chosen for its thermal or mechanical properties rather than its ability to bond with the coating.

The use of plasma for surface modification resulted from observations that metallic electrodes became discoloured, and solid polymers or compounds were often produced in electrical discharges.[9] Thin-film coatings, only a few micrometres thick, can be "grown" using chemical vapour deposition techniques which exploit the chemical reactions between a substrate and a plasma. The type of film produced will depend on the chemical composition of the plasma and on the substrate. Non-conducting materials, such as ceramics, can be coated in preparation for metallising processes and subsequent use in electronics and appliance-manufacturing industries; anti-reflective and scratch-resistant coatings can be applied to optical equipment. While plasma processing may be more expensive than traditional chemical or thermal methods using a simple oven, the time taken is less and the end result is usually better.

One of the oldest techniques for surface modification with cold plasmas, dating back to the 1950s and 1960s, is that of growing an oxide film on metal or semiconductor surfaces using oxygen plasmas.[10] When the substrate immersed in the plasma is at floating potential the process is called *plasma oxidation*. No current flows through the floating substrate and the oxide growth is generally thin — around 10^{-8}m at low temperatures. Oxides of 10^{-7} m thickness can be formed on some materials at temperatures below 100°C. When the substrate is positively-biased above the floating potential, a current flows through the oxide, accelerating its growth to thicknesses of a few micrometres. This process is called *plasma anodization*, and can be carried out in low pressure d.c., r.f. and microwave plasmas.

The surface of ceramics, polymers such as PVC, and metals can be modified using cold plasmas of gases such as argon, nitrogen or oxygen. Once again, ion bombardment produces chemical reactions and changes in the surface of the substrate. Other materials often need to be deposited on the surface of polymers used in food packaging and decorative products, and as insulation for electrical devices, but not all polymers have the necessary physical or chemical properties for good adhesion. Plasma surface treatment uses nitrogen or oxygen plasmas to incorporate plasma-based compounds into the polymer surface, improving adhesion while maintaining the desired properties of the bulk material.

Plasma nitriding is used to harden metals, especially steel, and improve resistance to corrosion by modifying the surface. Nitrides are compounds of metals with nitrogen, prepared conventionally by passing nitrogen or ammonia (NH_3) over heated metal. Plasma nitriding is a much quicker process than conventional thermal nitriding and provides greater control over the structure and composition of the nitrided layer. It is used particularly in the manufacture of vehicle engine parts and for implanting nitrogen into the titanium alloy of replacement hip joints, etc. Titanium nitride is used for coating cutting tools. The metal parts to be treated are used as the cathode in a d.c. glow discharge or abnormal glow, in which the applied voltage is pulsed to prevent the discharge becoming an arc. High-energy nitrogen-ion bombardment heats the metal to 500 - 600°C. The nitriding results from absorption of nitrogen into the surface, and the deposition of products from chemical reactions between sputtered surface atoms and the plasma. The end result is a hardened wear-resistant layer about 1 mm thick in the surface of the metal part.

Plasma processing of textiles can make them impervious to stains or chemicals. The fabric is placed in a gas discharge and ions from the gas bond to the fabric, producing a breathable non-stick molecular structure on the surface of the material. Even delicate fabrics such as silks, which may be adversely affected by conventional liquid protection methods, can be made dirt-resistant without affecting the handling qualities of the material.

10.6.1 Surface modification by ion implantation

Ions from a glow discharge plasma are accelerated to high energies (typically 10 keV to 500 keV) through a set of accelerating grids (Fig. 10.2, overleaf). The pressure of the source plasma is less than 1 mtorr, so the ion mean free path is longer than the source-target distance. The resulting beam of ions, all flowing in the same direction and with the same kinetic energy, is then injected into materials such as metals, plastics or ceramics to modify the surface.

174

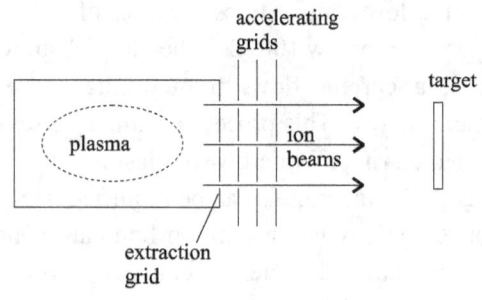

accelerating
grids

target

plasma

ion
beams

extraction
grid

Fig. 10.2   Ion source

By controlling the energy of the ions, and thus the depth of penetration, materials can be produced with finely defined compositions. New materials with new properties can be produced and surface properties can be altered without affecting any desirable features of the bulk material. There are none of the problems of bonding failure or delamination which are associated with other (non-plasma-based) surface-modification techniques. The material's surface is modified, not coated, and, since it is a low temperature process, there are no changes in dimensions or in the sharpness of cutting edges.

Conventional ion implantation is a line-of-sight process. To ensure uniform implantation, the target material is rotated in the ion beam, which limits the size of target which can be processed. A major problem is that of sputtering due to ions grazing the surface rather than being implanted as the target rotates.

An alternative technique developed in the mid-1980s places the target material directly in the plasma source. If the target is biased to a high negative potential (up to -50 kV) relative to the containment vessel walls, the plasma ions are accelerated across the sheath which forms around the target, bombarding it on all surfaces simultaneously. Implantation depths are in the tens of nanometres and sputtering is greatly reduced because the ions approach the target at close to 90°. The density of the plasma now becomes significant, since the sheath width decreases with increasing density. Complex targets in a higher-density plasma can therefore be implanted more uniformly in a shorter time, although there is an increased risk of arcing.[11]

10.7   Paint-spraying

Corona discharges are frequently used in spray-painting. A high negative voltage (with respect to earth) is applied to a point electrode of a spray gun, producing a corona discharge around the tip. The spray gun releases a mist of (initially neutral) liquid paint droplets. The droplets acquire a negative charge from the ionised air produced by the corona and are driven from the negative electrode towards the surface requiring painting. This surface is earthed and therefore appears as a positive electrode, attracting the mist of negatively-charged paint droplets to it. Coatings of dry powder or short fibres can be applied in a similar way.

10.8   Xerographic processes

Small corona discharges act as surface chargers in the non-chemical photographic process of xerography (the basis of many photocopiers). The dielectric surface of the copier's drum is given a uniform charge before being exposed to the light image. Exposure to light discharges the dielectric surface, leaving patterns of charge on the drum. The initial charge on the drum depends on the requirements of the toner, but is usually negative. Particles of toner, carrying opposite charge, are attracted to the patterns on the drum and then transferred to the copy paper using a corona charger behind the paper. This corona must obviously be of the same polarity as the drum. The resulting toner image is then fused to the paper by heat. Corona chargers remove any residual charge from the drum before it is cleaned for the next exposure. The corona device used for final cleaning is of opposite charge to that of the drum.

10.9 Removal of compounds from flue gases: electrostatic precipitators

Whilst on a visit to Montreal in August 1884, Oliver Lodge demonstrated a new method of removing dust particles from air by passing an electric current through a wire in a bell jar of illuminated magnesium smoke.[12] The smoke particles quickly collected into long filaments aligned along the electric field and dropped under their own weight when the current was removed. A higher potential broke up the collections of particles, driving them against the side of the jar. Lodge suggested that this prototype electrostatic precipitator — so-called because an electrostatic field is used to precipitate solid or liquid particles in a gas — could be used for clearing smoke-rooms or disinfecting hospital air.

Modern devices use corona discharges to collect and remove particles from the atmosphere in a wide variety of industries, including iron and steel, cement, and ore-processing; and removing dust particles in power stations. They can also be found in domestic and commercial building-ventilation systems. Ions from the corona charge individual dust particles as they are carried through the precipitator. The charged particles move along the electric field of the precipitator towards flat collecting plates which are oppositely-charged, or earthed, electrodes. Once collected, the particles can be easily removed. Similar processes can be used to remove carbon, sulphurous and nitrous oxides from diesel-engine exhaust and other waste gases.[13]

10.10 Plasma etching

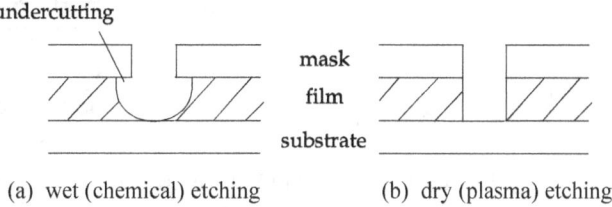

(a) wet (chemical) etching   (b) dry (plasma) etching

Fig. 10.3 Thin-film etching

With the growing demand for larger and faster computer memory and smaller components in the early 1980s, chemical methods of manufacturing integrated circuits began

176

to be replaced by plasma ion implantation and etching techniques. Chemical, or "wet etching" processes can eat sideways as well as downwards into the film, as shown in Fig. 10.3(a). This undercutting of the mask becomes a serious problem when the thickness of the film is comparable with the dimensions of the pattern and the lines to be etched.

Plasma etching, or "dry etching", is used to transfer patterns on to semiconductors to make integrated circuits (silicon chips). The wafer of silicon (the substrate) is mounted on a grounded electrode with the r.f. power supply connected to the opposite-powered electrode. The pattern, defined by a mask or stencil placed on a layer of *photo-resist* (photosensitive film), is cut through the photo-resist and etched on to the underlying substrate (Fig. 10.3). The discharge operates at between 0.1 and 1 torr and contains a gas which reacts with the film to form volatile compounds. These evaporate, leaving a cleanly-etched surface with deep straight-sided channels (Fig. 10.3(b)). Oxygen is used for etching photo-sensitive material and carbon-based films, while carbon tetrafluoride (CF_4) is more often used to etch silicon. The etching is therefore a combination of chemical processes and ion bombardment. Undercutting is controlled by varying the chemistry and parameters of the plasma. The micro-electronic processing industry is now a major employer of gas discharges, especially r.f. discharges, since energetic ions are undesirable.

10.11   Gas laser

The first gas laser to be developed was the helium-neon (He-Ne) laser shown in Fig. 10.4. Visible and infra-red high-power lasers of various types are now used for cutting, drilling, welding and annealing[a] metals and other materials such as ceramics, paper and textiles.

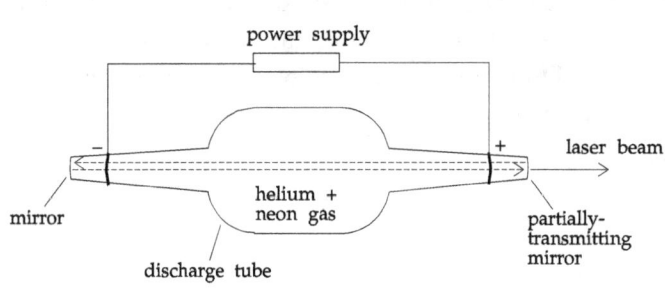

Fig. 10.4   He-Ne laser

A combination of several gases is used at pressures between 1 and 100 torr. In the He-Ne laser, a mixture of helium and neon, each at a pressure of about 1 torr, is contained within a glass discharge tube. In the carbon dioxide laser, the active medium is a mixture of carbon dioxide and gases such as nitrogen. Applying the necessary voltage produces a glow discharge within the tube, exciting the gas mixture. Radiation from the resulting plasma is reflected back and forth between two parallel mirrors, one at each end of the tube, stimulating the emission of more

---

[a]   heating glass or metal, followed by slow cooling to temper or toughen the material

coherent light. One of the mirrors allows a portion of the beam to emerge from the tube as an intense beam of coherent monochromatic light: a laser beam.

The infra-red CO_2 laser operates at a wavelength of 10.6×10^{-6} m, while the output of the red helium-neon (He-Ne) laser, at 632.8 nm, is in the visible region of the spectrum. In the case of X-ray lasers, since X-ray photons have energies approaching 1 keV, the lasing medium in the discharge tube must have a similar temperature. Laser output at 20.6 nm and 20.9 nm have been obtained using highly ionised selenium plasma as the lasing medium.

10.12   MHD power generators

Commercial research into magnetohydrodynamic power generation began in the US in the 1930s and large-scale generators began to appear in the 1960s.[14] MHD generators are based on the principle that, when a conductor, whether solid, liquid or gas, moves across a magnetic field, an electromotive force is generated. The MHD generator — in its simplest form it can be just a pair of electrodes in a fast-moving ionised gas — produces electrical energy from a plasma flowing through a transverse magnetic field. The resulting electrical power comes from the energy of the moving gas stream. MHD generators use plasmas which are only partially ionised — less than 1 part in 1000 — and while they might be relatively cheap to build, efficiency is low and they have received little interest as stand-alone power sources.

10.13   Particle accelerators

Particle accelerators have been used since the 1930s to investigate nuclear reactions and their method of operation has remained the same: the particles are accelerated by strong electric fields and guided by magnetic fields. The strength of the electric field is limited by breakdown so, as particle-energy requirements increase, the particles are accelerated over larger and larger distances to gain more speed. Accelerator size and cost have increased as a result. Existing technology may therefore be reaching practical limits and the possibility of using plasma to accelerate the particles has been under investigation since the 1980s.

A plasma has already broken down electrically and can support a much larger electric field, but the field must be set up in such a way that relativistic particles experience a constant field. This can be achieved by exciting an electron plasma wave, or Langmuir wave (Section 6.3), in the plasma. Oscillating electrons produce longitudinal plasma waves of positive and negative regions travelling through the plasma. If charged particles are injected into the plasma at roughly the same velocity as the plasma waves they will stay in phase with the wave, absorbing energy from it and accelerating steadily. The frequency of the plasma wave in any given plasma is determined by the electron density, but its wavelength can be varied, so the velocity of the wave can be made almost equal to the speed of light.

The oscillations producing the plasma wave can be created in various ways. A cluster of electrons can be sent through a plasma to produce a wake of plasma waves, like a boat

178

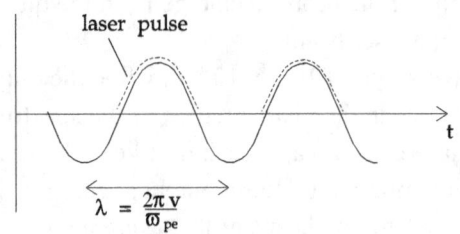

Fig. 10.5 Laser-induced Langmuir wave

moving through water. A similar effect can be achieved using laser beam pulses. If the pulse length is half the period of the Langmuir wave, i.e. π/ω_{pe}, a Langmuir oscillation is induced (Fig. 10.5). The pattern of the wake from the oscillation moves through the plasma at the group velocity of the laser pulse, which is slightly less than the speed of light. Particles injected into the plasma at relativistic speeds can move in phase with the wave, experiencing a constant electric field as they follow the laser pulse through the plasma.

Another method combines two laser beams of different frequencies to create interference patterns — regions where the waves are in phase (and reinforce each other) plus regions where the waves are out of phase and cancel out. This produces a beat wave oscillating at a frequency which is the difference between the frequencies of the two laser beams. If the beat wave frequency equals the natural oscillation frequency of the plasma and the wave is focused into the plasma, creating regions of high and low radiation pressure, the plasma electrons will oscillate resonantly with the wave, producing strong plasma waves.[15]

10.14   Environmental applications and waste processing

Plasma technology has widespread potential for use in environmental applications, since hazardous waste, chemical weapons, and toxic industrial, hospital or municipal waste can be dealt with even at room temperature. The silent electrical, or dielectric-barrier, discharge is frequently used in this context since it can operate at atmospheric pressure and above, producing a low-temperature ("cold") non-equilibrium plasma. The electrons, ions and neutrals in these plasmas all have different kinetic temperatures, with ions and neutrals close to room temperature, while the electrons typically have temperatures of 1 to 10 eV.

Free radicals[b] are easily produced by these low-temperature plasmas and are used, with the plasma electrons, to selectively decompose weak concentrations (1 - 1000 ppm) of toxic particles in air and other gases. By attaching themselves to the toxic molecules, they produce atoms or compounds which are more readily processed, either chemically or by further plasma treatment. Chemicals such as hydrocarbons and those used in chemical weapons can be processed and rendered harmless in this way. Hazardous organic chemicals, bacteria and viruses are quickly attacked by free radicals such as atomic oxygen, nitric oxide and hydroxyls, converting them to less harmful substances: acids, carbon dioxide, water. Surgical instruments are conventionally sterilised by heat treatment in a steam chamber

---

[b]   Atoms or molecules with an unpaired electron which can therefore react rapidly with other radicals or other molecules.

known as an autoclave, but plastic containers and other items that would melt with heat sterilisation have to be excluded. Ultraviolet light can kill bacteria but cannot deal with cracked surfaces which provide hiding places for micro-organisms. Cold plasma, containing large quantities of free radicals, can deal with situations such as these, by disrupting the cellular membranes of both bacteria and viruses. Air conditioning systems can be cleared of bacteria and fungi, while micro-electronic equipment can be safely sterilised.

The actual temperatures and energies required for electron and free radical production depends on the particular product being decomposed. Reaction rates are high, enabling large quantities to be processed. Unfortunately, decomposition normally occurs at atmospheric pressure, making it difficult to keep the plasma at the required temperature — frequent collisions between the charged particles tend to equalise electron and ion temperatures, reducing the effectiveness of the treatment. Low-temperature plasma processing is therefore often used in the more difficult initial-decomposition stage, with conventional methods being employed for subsequent decompositions.

An electron-beam with an energy of up to 1 MeV could be used to generate an avalanche of secondary electrons by ionising the atoms of the gas containing the toxic molecules. This would supply sufficient energetic electrons for decomposition to proceed despite the collisions. Electron beams can also provide continuous large volume non-equilibrium plasmas very efficiently and have been used to reduce pollutants such as vinyl chlorides from air.[16] Electron beam irradiation produces ions in the host gas which react with the pollutant to destroy it, even at low concentrations. The host gas is unaffected, since, in the case of air, the ionisation potential of the normal constituents is higher than that of most pollutants. Conventional methods involve thermal incineration of the pollutants which is only efficient at high concentrations.

High temperature ("hot" or "thermal") plasmas are generally arc plasmas in thermal equilibrium — all particle species have the same temperature — and can reach temperatures up to 10^4 K. They can be used at atmospheric pressure, either as an arc formed between two electrodes, such as the plasma torch; or as an arc between an electrode and the material to be processed. Two electrodes allows for greater flexibility, since the toxic material can be placed between them. Making the material one of the electrodes, assuming it is electrically conducting, is more effective in heating solids.

Thermal plasmas are used to treat solids, liquids and gases by melting waste products, particularly those with high melting temperatures, to form stable compounds. Hazardous waste can be processed by high-temperature arc plasmas operating in a furnace at atmospheric pressure, with particle temperatures in the region of 10^4 K. At this temperature, the plasma melts the waste to form stable, glass-like products which can be recycled or used in land-fill sites. Emission of toxic pollutants is much reduced because the high temperature ensures total combustion and decomposition. Medical and radioactive waste products, used tyres, oil and municipal waste are all suitable candidates for treatment.

180

10.15   Space propulsion

In December 2001 the first ion-engine spacecraft finally ran out of fuel, lost power and drifted off into space. Launched in October 1998, *Deep Space 1* was designed to test, amongst other things, the idea that electrically-charged particles could be used for rocket propulsion. In a conventional rocket, several tons of fuel burns with tremendous force for a short time. As the exhaust gases stream out in one direction, the rocket is propelled in the opposite direction. In an ion-drive system, plasma provides the propulsion. *Deep Space 1* carried about 81 kg of xenon as its propellant. A hollow cathode, powered by a solar array, produced electrons to ionise the xenon. The ions were then electrically accelerated to a speed of about 3×10^4 m s^{-1} before being expelled as exhaust.

Ion propulsion does not have the power of a conventional rocket — *Deep Space 1* would have taken about four days to accelerate from 0 to 27 m s^{-1} (97 kph or 60 mph) — but, unlike a conventional rocket which burns through its entire fuel supply in a matter of minutes, plasma propulsion continues for a long time. Acceleration lasts for years: *Deep Space 1* had been accelerating for two years when its fuel supply ran out, making it the fastest spacecraft ever.[17] It was followed by the European Space Agency's *Smart-1* (Small Missions for Advanced Research and Technology) which was launched into Earth orbit by a conventional Ariane-5 rocket on 27th September 2003. Once in orbit, the ion-drive propulsion took over (the propellant once again being xenon), driving the probe with a thrust of just 0.07 Newtons — equivalent to blowing lightly on a piece of paper — towards its ultimate destination: lunar orbit. Here, it would spend the next two and a half years scanning the lunar surface. Accelerating steadily at around 2×10^{-4} m s^{-2}, it finally arrived in lunar orbit on 15th November 2004, almost fourteen months after launch.

The idea of ion propulsion for spacecraft has been around for many years, with the first ion engine being built by NASA in 1960. Various ideas have been investigated, including expanding electrically-heated plasmas, laser propulsion, and plasma guns or thrusters, but despite numerous tests in the laboratory, no space mission was willing to gamble on an ion drive system for its primary propulsion in case it did not work. The advantage of a plasma system over the conventional fuel system is that plasma propulsion requires less fuel mass than chemical systems and could therefore make launching a spacecraft, especially from an orbiting space station, less expensive. Conventional rockets would still be needed to launch spacecraft from Earth and provide sufficient acceleration to escape Earth's gravity, but then plasma propulsion would take over, dramatically reducing journey times, particularly on deep-space missions.

An alternative propulsion system, proposed in 1999 by a team from the University of Washington in Seattle, is the mini-magnetospheric plasma-propulsion (M2P2) system. This would create a magnetic plasma "balloon" between 30 and 60 km in diameter which would be propelled by the solar wind. The team's calculations indicated that there is enough power in the solar wind to accelerate a 136 kg spacecraft to speeds up to 8×10^4 m s^{-1}. It would take about nine or ten years to reach the heliopause, more than 150 AU from Earth. (*Voyager 1*, launched in 1977, should arrive there in 2019.) M2P2 would use an r.f. discharge surrounded by solenoid coils to produce a magnetic field. The magnetic field would contain the plasma, possibly hydrogen or helium, while being "inflated" by it to form a mini-magnetosphere around the spacecraft. This remains attached to the spacecraft and acts like a sail, interacting with the solar wind particles, and propelling the spacecraft through space. Course changes would be made by adjusting the angle between the electromagnetic field of the "sail" and the solar wind. Power would come from solar panels or, for deep-space missions, a source such as a radio isotope.

References

1.   Eliasson, B., & U. Kogelschatz: *IEEE Trans. Plasma Sci.*, **19**, p.1013-77 (1991) [review paper]

2.   Linder, E.G., & A.P. Davis: *J. Phys. Chem.*, **35**, p.3649-72 (1931)
      Brewer, A.K., & P.D. Kueck: *J. Phys. Chem.*, **35**, p.1293-1302 (1931)

3.   Langmuir, I.: *Phys. Rev.* 2nd ser. **2**, p.450-86 (1913)
      Mackay, G.M.J., & C.V. Ferguson: *Frank. Inst. J.*, **181**, p.209-16 (1916)

4.   Kirkby, Rev. P.J.: *Phil. Mag.*, **7**, p.223-32 (1904); *Phil. Mag.*, **9**, p.171-85 (1905); *Phil. Mag.*, **13**, p.289-312 (1907); *Proc. Roy. Soc.*, **85**, p.151-74 (1911)

5.   Siemens, W.: *Ann. Phys. & Chem.*, **102**, p.66-122 (1857)
      Warburg, E.: *Ann. Phys.*, **13**, p.464-76 (1904)

6.   Besson, A.: *Comptes Rendus*, **152**, p.1850-2 (1911)
      Thornton, J.D., & P.L. Spedding: *Nature*, **213**, p.1118-9 (1967)
      Ibberson, V.J.: *High temperatures — High Pressures*, **1**, (no. 3), p.243-68 (1969)

7.   Grove, W.R.: *Phil. Trans.*, **142** (1), p.87-101 (1852)

8.   Compton, K.T., & I. Langmuir: *Rev. Mod. Phys.*, **2**, p.123-242 (Apr. 1930)

9.   Stewart, R.L.: *Phys. Rev.*, **45**, p.488-90 (1934)

10.   O'Hanlon, J.F.: *J. Vac. Sci. Technol.*, **7** (2), p.330-8 (1970) [review paper]

11.   Conrad, J.R., et al: *J. Appl. Phys.*, **62**, p.4591-6 (Dec. 1987)
      Tendys, J., et al: *Appl. Phys. Lett.*, **53**, p.2143-5 (Nov. 1988)

12.   Lodge, O.J.: *Nature*, **31**, p.265-9 (Jan. 1885).

13.   Higashi, M., et al: *IEEE Trans. Plasma Sci.*, **20**, p.1-12 (1992)

14.   Rosa, R.J.: *Phys. of Fluids*, **4**, p.182 (1961)
      Thring, M.W.: *J. IEEE*, **8**, p.237-41 (1962)
      Nucl. Fusn., **11** (5), p.535-40 (1971) [Conference report]

15.   Dawson, J.M.: *Sci. Am.*, **260**, p.34-41 (March 1989)

16.   Slater, R.C., & D.H. Douglas-Hamilton: *J. Appl. Phys.*, **52**, (9), p.5820-8 (1981)

17.   Reichhardt, T.: *Nature*, **400**, p.392 (29 July 1999)

Chapter 11

Nuclear fusion

11.1   "A lot of nonsense"

In 1932, John Cockcroft and Ernest Walton split the atom for the first time by bombarding lithium with protons in an accelerator.[1]   A year later, Ernest Rutherford discounted the possibility that useful power could result from the process: "These transformations of the atom are of extraordinary interest to scientists," he said, "but we cannot control atomic energy to an extent which would be of any value commercially, and I believe we are not likely ever to be able to do so.   A lot of nonsense has been talked about transmutations.   Our interest in the matter is purely scientific."[2]   On this occasion Rutherford's forecast would prove to be wrong, but in the early 1930s many people shared his view, and it is easy to see why.   Using Einstein's 1905 equation for the equivalence of mass and energy, $E = mc^2$, the mass energy of a hydrogen atom ($m = 1.673 \times 10^{-27}$ kg) is roughly 1.5×10^{-10} J, and this very small amount is the energy released if the entire mass of the hydrogen atom is converted into energy.

Atomic structure had been worked out by the 1920s and theories about nuclear energy began to develop soon afterwards.   Rutherford remarked in 1923 that helium was probably a very close combination of four hydrogen nuclei and two electrons.   Arthur Eddington had recently suggested that solar energy might result from building up complex elements out of simpler ones, and Rutherford confirmed that mass losses in synthesising a helium nucleus from hydrogen nuclei did indeed indicate that the process released large quantities of energy as radiation.[3]

In 1928, George Gamow (1904-68) in Copenhagen and Edward Condon (1902-74) at Princeton independently used quantum theory, and in particular Schrödinger's wave equation, to explain how an α-particle can escape from an atomic nucleus by appearing to have "tunnelled" out.[4]   This tunnelling effect could, it was suggested, work in reverse, enabling an α-particle to penetrate the nucleus of an atom with which it collides.[a]   In Berlin,

---

[a]   When two protons approach one another, their like charge prevents them from touching; but if they get sufficiently close, their quantum wave functions (which describe the particle in wave terms) can overlap, enabling them to interact.   It is as if one proton tunnels through the potential barrier between them, to join the other.

Robert D'Escourt Atkinson (1898-1982) and Fritz Houtermans quickly adapted Gamow's ideas to their stellar energy investigations. They suggested that, in the case of proton-proton impacts, a proton entering a nucleus would "anchor itself there by radiating" — they had not yet worked out the details. The necessary conditions for sustained thermonuclear fusion reactions — high temperatures and high particle densities — existed at the centre of stars, so heavier elements could be gradually built up from lighter ones as Eddington had suggested.[5] By 1930, therefore, the broad principles of nuclear fusion were established; all that remained was for the detail to be filled in.

In the mid-1920s Francis William Aston, having switched his attention from gas discharges to the hydrogen atom, made the first accurate measurements of the masses of individual atoms. Further investigations revealed minute variations in the atomic masses of hydrogen when determined by different methods. Birge and Menzel, in California, concluded in 1931 that there must therefore be a heavy isotope of hydrogen "of mass 2, present to the extent of 1 part in 4500 parts of hydrogen of mass 1". Harold Clayton Urey (b.1893), with his colleagues at Columbia University in New York, searched for atomic spectra of both ^2H and ^3H in a hydrogen discharge tube and later that year, on 5th December, reported finding faint lines at the calculated positions for the lines of ^2H accompanying H_β, H_γ and H_δ. As R.H. Fowler later explained, they had found the isotope's Balmer lines displaced by the correct distance from the Balmer lines of hydrogen. These lines did not agree in wavelength with any known molecular lines and the relative abundance was about 1:4000, in agreement with Birge and Menzel's estimate. No evidence for ^3H was found.[6]

The importance of this new isotope was recognised immediately and the choice of a name was raised at a Royal Society discussion on heavy hydrogen early in 1934. Urey had proposed the name "deuterium" for the ^2H atom and his fellow-countryman Gilbert Lewis had suggested "deuton" or "deuteron" for the nucleus. In the UK, Rutherford and his team had already adopted the names "diplogen" (from the Greek for "double") for the atom and "diplon" for its nucleus.[7] It was several years before the more widely-accepted *deuterium* (for the atom) and *deuteron* (for the nucleus) were used in the UK.

In the spring of 1934, Mark Oliphant, P. Harteck and Ernest Rutherford announced the results of their deuterium experiments at the Cavendish Laboratory.[8] Compounds such as ammonium chloride (NH_4Cl), in which the hydrogen had been largely replaced by deuterium, had been bombarded by high velocity deuterium nuclei. An "enormous emission" of fast protons had been detected, together with a similar quantity of singly-charged particles of much shorter range (16 mm compared to 143 mm for the fast protons). They thought that this second group might be nuclei of another new isotope of hydrogen, of mass 3.0151. A large number of neutrons with energies up to 3 MeV had also been observed. Rough estimates of their number indicated that the reaction which produced them was less frequent than that producing the protons. They suggested that two deuterons combined to form an (as yet) undetected high-energy helium nucleus which was unstable and immediately broke up into two components: either a proton and a ^{3}H nucleus or a helium isotope of mass 3 and a neutron. The first, based on their respective ranges, would account for the equal numbers of fast protons ($^1_1$H) and singly charged particles they had detected. The possibility of a $^3_1$H isotope was accepted but none had yet been found. The second would account for the observed high energy neutrons. Another Cavendish Laboratory researcher, P.I. Dee, confirmed their results using a cloud chamber to examine the particles produced in the reaction. His results showed that $^3_1$H (later called *tritium*) and protons were produced and recoiled in opposite directions, as expected of two positively-charged particles.[9]

Three years later, Otto Hahn, Lise Meitner and F. Strassmann bombarded uranium atoms with neutrons, splitting the uranium nuclei into two and releasing energy and more neutrons. Meitner and Otto Frisch later named the splitting process *nuclear fission*. In June 1939, Niels Bohr and John Archibald Wheeler, at Princeton, explained that fission had occurred not in the stable nucleus of uranium-238 but in that of the much rarer isotope uranium-235. Their calculations showed that the force holding the nucleus together (now called the strong nuclear force) was dominant in all elements with an atomic number less than that of silver ($Z = 47$). Beyond that, the repulsive electrostatic force within the nucleus took over and strong particle bombardment would break up the nuclei of the heavier elements, many of which were inherently unstable, with the emission of large amounts of nuclear energy. They predicted that a spontaneous fusion process should therefore result when two nuclei with a combined atomic weight less than that of silver are brought together. The two nuclei would combine to form a heavier element, releasing energy.[10]

The outbreak of war in Europe disrupted the research and many scientists fled to the United States. One of these, Enrico Fermi (1901-54), oversaw the building and testing of the first nuclear fission reactor in a squash court at Chicago University in 1942. The following year he moved to Los Alamos, in New Mexico, as a member of the top-secret Manhattan Project to develop the first atomic bomb.[b] Alongside the Manhattan Project was an equally secret Project Matterhorn led by Edward Teller (1908-2003). Its purpose was to develop a "super" bomb — a thermonuclear hydrogen bomb. This would use a uranium fission explosion (a small atomic bomb) to trigger a thermonuclear reaction in a quantity of deuterium and tritium. Unlike the atomic bomb, a hydrogen bomb has no chain reaction and no critical mass. All that is needed is extremely high temperatures for the nuclei to fuse. The small fission bomb was designed to produce these temperatures. In the early years of the

---

[b]   The atom bomb depends on the fission of uranium-235 or plutonium-239 for its energy. In a small quantity of fissionable material, any neutrons generated will easily escape. Very few will cause secondary fissions. As the size of the material increases, proportionately fewer neutrons escape and more will cause secondary fissions. A chain reaction begins and the material becomes "critical" — explosion becomes imminent.

project, there were fears that a thermonuclear reaction such as this might ignite the atmosphere.

The first atomic (fission) bomb was dropped on Hiroshima on 6th August 1945. Seven years later, on 1st November 1952, the first hydrogen bomb was detonated by the United States at Eniwetok Atoll in the Pacific Ocean. The Russians, led by Andrei Dmitrievich Sakharov (1921-89), exploded a similar device a few months later on 12th August 1953. A hydrogen bomb is essentially an uncontrolled thermonuclear reaction. Researchers now began to consider the possibility of controlling the process and extracting energy from it to satisfy the world's growing energy needs.

11.2   The binding energy of the nucleus

Just as the energy of electrons in atoms is quantized, so too is the energy of the particles making up the nucleus. While the energy levels of the electrons in the atom are measured in electronvolts, those of the protons and neutrons in the nucleus are measured in MeV. This potential energy of the protons and neutrons is associated with the strong nuclear force which holds the particles together in the nucleus and is called the *binding energy*. If the binding energy of a nucleus is said to be 10 MeV it means that 10 MeV of energy must be transferred to the nucleus to free its constituent protons and neutrons, while that same amount of energy is released when free protons and neutrons are brought together to form a nucleus. Put simply, the 10 MeV is the energy required to break the nucleus apart.

To find out how this energy is calculated, let us take the helium nucleus, $^4_2$He (an α-particle), as an example. The mass of the helium nucleus is approximately 6.645×10^{-27} kg. It consists of two protons and two neutrons tightly bound together (mass number, $A = 4$, atomic number, $Z = 2$). The equation describing the "unbinding" of the helium nucleus into its constituent protons and neutrons can be written:

helium nucleus + binding energy → 2 free protons + 2 free neutrons

The mass of a free proton is 1.6726×10^{-27} kg and that of a free neutron is 1.675×10^{-27} kg. The total rest mass of two free protons plus two free neutrons, is therefore 6.6952×10^{-27} kg. This is greater than the mass of the helium nucleus by 5.02×10^{-29} kg. From relativity theory, the *principle of mass-energy conservation* states that, in any isolated system, a change in the sum of the rest masses is accompanied by an equal and opposite change in the total energy of the system. This change in energy is accounted for by the binding energy.

For energy to be conserved, the energy of the helium nucleus plus the binding energy must equal the energy of the free protons and neutrons. The energy needed to free the particles from the nucleus can be calculated using Einstein's mass-energy equation, $E = mc^2$:

$$\begin{array}{ccc} \text{mass energy of} & + \text{ binding} & \rightarrow \text{ total mass energy of} \\ \text{helium nucleus} & \text{energy} & \text{2 protons + 2 neutrons} \\ (m_{\text{He}}\,c^2) & \text{of }^4_2\text{He} & (2\,m_{\text{p}}\,c^2 + 2\,m_{\text{n}}\,c^2) \end{array}$$

Rearranging this:

$$\text{binding energy of } {}^{4}_{2}\text{He} = 2\, m_{p}c^{2} + 2\, m_{n}c^{2} - m_{He}c^{2}$$
$$= c^{2}\,(2\, m_{p} + 2\, m_{n} - m_{He})$$
$$= (2.998 \times 10^{8})^{2}\,(5.02 \times 10^{-29}\,\text{kg})$$
$$= 4.512 \times 10^{-12}\,\text{kg m}^{2}\,\text{s}^{-2}$$
$$= 28.16\,\text{MeV}$$

since 1 kg m^2 s^{-2} = 1J and 1 MeV = 1.602 x 10^{-13} J. The energy needed to free each proton or neutron from the helium nucleus is one quarter of this amount: approximately 7 MeV.

This method can be applied to any nucleus: the measured difference between the mass of the nucleus and the total mass of its free protons and neutrons, when multiplied by c^2 gives the total binding energy of the nucleus. Dividing by the mass number, A, gives the average binding energy per proton or neutron. This is approximately the same (8 MeV) for values of A greater than 10 (see Fig. 11.1).

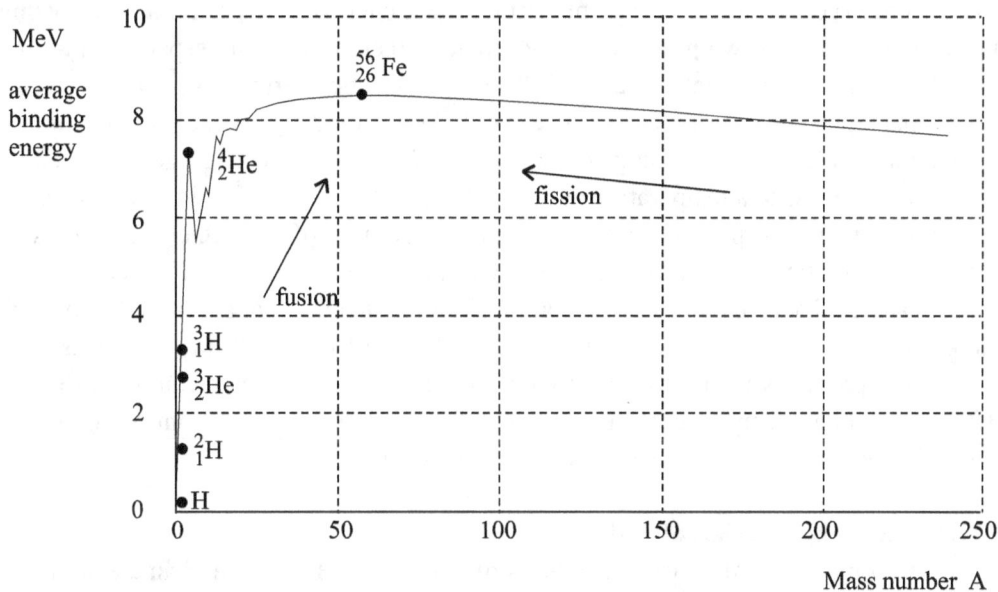

Fig. 11.1 Average binding energy of the nucleus

To overcome the strong force and break up the nucleus requires an input of energy. To put the pieces of the nucleus back together again releases energy. This reconstruction process is known as *nuclear fusion*. The general increase in average binding energy up to mass number $A = 50$ means that fusion reactions between light nuclei are usually exothermic — they release heat energy in large quantities. Once we reach the region of iron, with 26 protons and $A = 56$, the Coulomb repulsion becomes so large that energy would need to be supplied to force the nuclei to fuse. Iron is the most stable of all the elements of the periodic table, and one of the most abundant. A simplistic way of interpreting Fig. 11.1 is that lighter elements fuse, and heavier elements split, to form more iron-like elements. In practice, the only reactions of any interest are those between the isotopes of hydrogen and between them and helium.

11.3   Nuclear fusion

When two nuclei fuse they release energy and form a heavier nucleus, whose constituent protons and neutrons are more tightly bound, and whose mass is slightly less than the total mass of the initial particles. This nuclear fusion reaction occurs when two light positively-charged nuclei get close enough to each other — typically 2×10^{-15} m apart — for the attractive strong nuclear force to hold them together. For this to happen, the particles must approach one another with a kinetic energy at least equal to their electrostatic potential energy in order to overcome (or "tunnel" through) the potential energy, or Coulomb, barrier caused by their mutual repulsion. Tunnelling only becomes effective after most of the Coulomb barrier has been conventionally overcome by the kinetic energy of the particles. The potential energy, $q_1 q_2 / 4 \pi \varepsilon_0 r$, of two protons 2×10^{-15} m apart (centre-to-centre separation) is about 1.15×10^{-13} J, or approximately 0.72 MeV. This is the kinetic energy they need to overcome their mutual repulsion. This much energy can only be acquired at very high temperatures — hence *thermo*nuclear fusion. Using equation (3.6), $\langle E_{kin} \rangle = \frac{3}{2} k_B T$, an average kinetic energy of 0.72 MeV equates to a temperature of 5.57×10^9 K or 480 keV. Fusion can, in fact, take place at temperatures some 50 times lower than this because of the tunnelling process and because many particles will have higher energies that the average. At these temperatures, charged particles have velocities of several thousand km per second: a proton, for example, with kinetic energy equal to 1.15×10^{-13} J has a velocity of 1.17×10^7 m s^{-1}. This is where many of the problems facing fusion research programmes lie. It is technologically very difficult to maintain temperatures of even a few thousand kelvin while controlling, for any length of time, the forces produced by such energetic particles.

11.3.1   The raw materials of fusion

A normal atom of hydrogen consists of a proton ($_1^1$H or $_1^1$p) and an electron. The nuclei of the two heavier isotopes of hydrogen: *deuterium* and *tritium*, are known as deuterons ($_1^2$D or $_1^2$H) and tritons ($_1^3$T). Deuterium occurs naturally: 1 m^3 of water contains 34 g of deuterium, equivalent to one part in 6500. With one proton and one neutron in the nucleus, it is twice as heavy as ordinary hydrogen and has an atomic mass number $A = 2$. It behaves exactly like hydrogen, combining with oxygen to form D_2O or "heavy water".

Tritium has one proton and two neutrons, and therefore has an atomic mass of 3. It only occurs naturally in very small quantities (about 1 atom in 10^{17} in natural hydrogen) but could be manufactured in the fusion process. Tritium is radioactive, with a half-life of 12.3

years.[c] While its use in power generation might present some problems, they would be small compared to the handling of radioactive waste from nuclear fission reactors.

11.3.2   Fusion reactions

Exothermic fusion reactions are possible between a variety of light nuclei; the most important are detailed below. Energy generated in fusion reactions appears in three forms: kinetic energy of the charged particles produced in the reaction; kinetic energy of emitted neutrons; and electromagnetic radiation. Only the first remains within the plasma. The other two cannot be contained by magnetic fields, and would be lost from the process without the addition of a lithium "jacket" around the reactor.[d] This would absorb escaping energy, raising the temperature of the jacket to about 500° C. Heat can then be extracted, via a heat exchanger, to produce steam and hence electricity by conventional methods. A major drawback is that the neutrons will damage the vacuum vessel walls and make the entire reactor structure radioactive.

a)   Deuterium (D-D) reaction

As Oliphant and his co-workers proposed in 1934, there are two possible outcomes when two deuterium nuclei fuse. In one reaction, two deuterium nuclei form a helium nucleus, emitting a neutron and releasing energy:

$$\,_1^2D + \,_1^2D \; \rightarrow \; \,_2^3He + \,_0^1n + \text{?energy} \tag{11.1}$$

The protons and neutrons in the helium nucleus are more tightly bound than those in the deuterium nucleus, as shown by Fig. 11.1 — more energy is needed to separate the $\,_2^3He$ nucleus than the $\,_1^2H$ nucleus. Taking the left-hand side of equation (11.1), the mass of the deuterium nucleus is approximately 3.3436×10^{-27} kg. Following the $\,_2^4He$ example shown earlier, the total energy required to separate the proton and neutron from one deuterium nucleus is therefore 2.24 MeV:

$$2\{\,_1^2D + \text{binding energy (2.24 MeV)}\} \; = \; 2\,_1^1p + 2\,_0^1n$$

The right-hand side of equation (11.1) represents the situation after fusion. The mass of a $\,_2^3He$ nucleus is approximately 5.0064×10^{-27} kg. The total energy required to separate the two protons and one neutron from the $\,_2^3He$ nucleus is therefore 7.74 MeV. Thus:

---

[c]   This is the time required for the strength of a radioactive source to decay (releasing energy) to half its original value. Decay is exponential, so in the case of tritium, 50% of the isotope will have become stable after 12.3 years, half of the remaining 50% after another 12.3 years, and so on.

[d]   Lithium, like other elements in the Earth's crust, is in finite supply, although known resources could last about 1000 years.

$$_1^2D + _1^2D + 4.48 \text{ MeV} = 2\,_1^1p + 2\,_0^1n = (_2^3He + 7.74 \text{ MeV}) + _0^1n$$

If two free protons and one neutron are brought together to form a helium nucleus, 7.74 MeV of energy is *released*. The net energy release in this D-D reaction is therefore 7.74 - 4.48 = 3.26 MeV. Equation (11.1) becomes:

$$_1^2D + _1^2D \rightarrow _2^3He + _0^1n + 3.26 \text{ MeV} \tag{11.2}$$

From momentum conservation, 2.45 MeV of the total 3.26 MeV is carried away as kinetic energy by the neutron; the helium nucleus takes the remaining 0.81 MeV.

An equally possible D-D reaction combines one proton and two neutrons to form a tritium nucleus with an approximate mass of 5.0074×10^{-27} kg, the final equation being:

$$_1^2D + _1^2D \rightarrow _1^3T + _1^1p + 4.03 \text{ MeV} \tag{11.3}$$

Here, the proton takes 3.02 MeV of the kinetic energy, the triton the remaining 1.01 MeV.

b)   Deuterium-tritium (D-T) reaction

The D-T reaction is more complicated. The fusion reaction itself is straightforward:

$$_1^2D + _1^3T \rightarrow _2^4He + _0^1n + 17.6 \text{ MeV} \tag{11.4}$$

Most of the energy from D-T reactions is carried by high-speed neutrons. By the law of momentum conservation, of the 17.6 MeV of energy released, 14.1 MeV appears as kinetic energy of the neutron and just 3.5 MeV as kinetic energy of the $_2^4He$ nucleus. Only the energy of the helium nucleus would remain in the plasma to heat and ionise other particles and keep the fusion process going. The neutron energy would be absorbed by the jacket surrounding the reactor, increasing the temperature of the jacket and allowing generation of electricity. The neutron produced in the reaction can be "recycled", free neutrons reacting with one or other of the two stable isotopes of lithium to "breed" more tritium:

$$_3^6Li + _0^1n \rightarrow _2^4He + _1^3T + 4.8 \text{ MeV} \tag{11.5}$$

$$_3^7Li + _0^1n \rightarrow _2^4He + _1^3T + _0^1n - 2.5 \text{ MeV} \tag{11.6}$$

The raw materials consumed in the fusion reaction would therefore be deuterium and lithium.

c)   Deuterium-helium 3 reaction

The conventional fusion reaction, using isotopes of hydrogen, produces large numbers of energetic neutrons which can seriously damage the walls of the containment vessel, requiring its periodic replacement. In the late 1980s, it was suggested by a team from Wisconsin University, led by Gerald Kulcinski, that a fusion reactor fuelled with helium-3 instead of hydrogen would be a less-destructive alternative.[11]

Normal helium ($_2^4He$) has two protons and two neutrons in its nucleus. Helium-3 has one less neutron and occurs naturally only rarely on Earth. It is produced in D-D fusion reactions (equations (11.1) and (11.2)) and as a by-product of the maintenance and decommissioning of nuclear weapons. However, rocks brought back from the Moon in the late 1960s and early 1970s by the Apollo astronauts have been found to contain significant quantities. In their 1988 paper, Kulcinski and his co-workers suggested that large quantities produced in the Sun had been carried away by the solar wind. The Earth's atmosphere prevented them from reaching the surface here, but huge deposits are believed to have accumulated in the lunar surface and the feasibility of mining these deposits is under consideration.

Helium-3 reacts with deuterium to produce a proton and helium-4. The equation of the reaction is given by:

$$_1^2D + {}_2^3He \rightarrow {}_2^4He + {}_1^1p + 18.3 \text{ MeV}$$

The helium-4 nucleus takes 3.6 MeV of the kinetic energy, the proton 14.7 MeV. More energy is released than in the D-T reaction of equation (11.3) and no highly-energetic neutrons are produced.

c) Some other useful reactions

$$_1^3T + {}_1^3T \rightarrow {}_2^4He + {}_0^1n + {}_0^1n + 11.3 \text{ MeV}$$

$$_1^3T + {}_2^3He \rightarrow {}_2^4He + {}_1^1p + {}_0^1n + 12.1 \text{ MeV}$$

$$_1^3T + {}_2^3He \rightarrow {}_2^4He + {}_1^2D + 14.3 \text{ MeV}$$

11.3.3   Fusion cross-section, σ_f

A D-D reaction is achieved by ionising deuterium atoms and providing the nuclei with sufficient energy to fuse when they collide. Firing a beam of deuterons at a target of deuterium is not energy-productive because too much input energy is required, and because the probability that the reaction will take place — the *cross-section* — is so small that most of the nuclei would lose their energy in collisions with atoms of the target before they could fuse with other deuterons.

The fusion cross-section, σ_f, is the probability that one particle will fuse with another, measured as an effective area. Put simply, it is the size of the area of the nucleus which the incident particle must "hit" for the reaction to occur. The unit of measurement is the *barn* (10^{-28} m^2). The term was adopted in the 1940s because to an incoming nucleus, the size of the atom that could capture it appears to be "as big as a barn door".

The value of the cross-section depends on the nature and energy of the particles involved, their relative speeds and the angle of collision. It can be much smaller or larger than the actual size of the nuclei involved. Tritium has a larger cross-section than deuterium at the same energy so it should be possible to get a reaction going at a lower temperature. At 100 keV bombarding energy the combined cross-section for the D-D reactions (equations 11.2 and 11.3) is approximately 0.03 barn while for the D-T reaction (equation 11.4) it is about 5 barns.[12]

11.3.4   Reaction rates

The rate of reaction at any given density and temperature will affect the amount of energy produced in the fusion process. The reaction rate is the number of fusions per second in a cubic metre of plasma. For the D-T reaction this is given by

$$R_{DT} = n_D n_T \langle \sigma_f v \rangle \text{ m}^{-3} \text{ s}^{-1} \qquad (11.7)$$

and for the D-D reaction

$$R_{DD} = \tfrac{1}{2} n_D^2 \langle \sigma_f v \rangle \text{ m}^{-3} \text{ s}^{-1} \qquad (11.8)$$

Here n is the number of deuterons or tritons; v is the magnitude of the relative velocity of two interacting nuclei; σ_f is the cross section for the fusion reaction at velocity v; and $\langle \sigma_f v \rangle$, sometimes called the *rate constant*, is the value of $\sigma_f v$ averaged over the Maxwellian distribution appropriate to the kinetic temperature of the plasma.[e]   Since high particle velocities are required to bring the nuclei close enough together for fusion to occur, $\langle \sigma_f v \rangle$ is highly sensitive to plasma temperature. Temperatures in excess of 1 keV are required to produce the D-T reaction. For a D-D reaction, approximately 26 keV (about 3×10^8 K) is required, and for a D-He 3 reaction, a minimum of 10 keV.

A simple numerical example will illustrate the importance of plasma temperature in power generation: From published tables, $\langle \sigma_f v_{DT} \rangle$ is 2.6×10^{-25} m^3 s^{-1} when the temperature is 2 keV. The energy generated by the D-T reaction of equation (11.4) is obtained by multiplying R_{DT} (equation 11.7) by the energy released in a D-T reaction:

$$E = n_D n_T \langle \sigma_f v_{DT} \rangle \times 17.6 \text{ MeV m}^{-3} \text{ s}^{-1}$$

Since 1 MeV = 1.602×10^{-13} J and 1 W = 1 J s^{-1}, the power per cubic metre is

$$P = n_D n_T (2.6 \times 10^{-25})(17.6 \times 1.602 \times 10^{-13})$$
$$= 7.33 \times 10^{-37} n_D n_T \text{ W m}^{-3}$$

If we assume a particle density of $n = 5 \times 10^{19}$ m^{-3} for both deuterium and tritium,[f] the power produced in this example is 1.83 kW m^{-3}. At a temperature of 30 keV, $\langle \sigma_f v_{DT} \rangle$ becomes 8×10^{-22} m^3 s^{-1} and the power generated is approximately 5.6 MW m^{-3}.

The D-D reaction, at similar density ($n_D = 10^{20}$ m^{-3}) and temperatures, produces much less power. Using the reaction given in equation (11.2):

$$E = \tfrac{1}{2} n_D^2 \langle \sigma_f v_{DD} \rangle \times 3.3 \text{ MeV m}^{-3} \text{ s}^{-1}$$

When the temperature is 2 keV, $\langle \sigma_f v_{DD} \rangle$ is 5.4×10^{-27} m^3 s^{-1} and power generation is about 14.3 W m^{-3}. At 30 keV, $\langle \sigma_f v_{DD} \rangle$ is 8×10^{-23} m^3 s^{-1} and power generation has increased to about 0.21 MW m^{-3}.

---

[e]   Values for σ_f, $\langle \sigma_f v \rangle$, R_{DD}, and R_{DT} can be obtained from the literature; e.g. R.F. Post, *Rev. Mod. Phys*, **28**, 338-362 (1956)

[f]   Total particle density is therefore $n = 10^{20}$ m^{-3}, as previous examples.

For a thermonuclear fusion reactor to be economically viable as an energy generator it is essential that the power output exceeds that required to sustain the plasma. Although deuterium is more readily available than tritium, the higher temperatures that are required for the D-D reaction and the much-reduced energy output from this reaction would indicate that the deuterium-tritium reaction is the more economically desirable of the two.

References

1. Cockcroft, J.D., & E.T.S. Walton: *Proc. Roy. Soc.,* **137**, p.229-42 (1932)

2. Snow, C.P.: *Variety of Men.* Curtis Brown Ltd., London (1967)

3. Eddington, A.S.: *Nature*, **111**, p.5-12 (12 May 1923)
Rutherford, E.: *Science*, **58**, p.209-21 (1923); *Nature*, **111**, p.644 (1923)

4. Gamow, G.: *Nature*, **122**, p.805-6 (24 Nov. 1928)
Gamow, G., & F.G. Houtermans: *Zeits. f. Phys.*, **52**, p.496-509 (1928)
Gurney, R.W., & E.U. Condon: *Phys. Rev.*, **33**, p.127-40 (Feb. 1929); *Nature*, **122**, p.439 (22 Sept. 1928)

5. Gamow, G.: *Zeits. f. Phys.*, **52**, p.510-15 (1928)
Atkinson, R.D'E., & F.G. Houtermans: *Nature*, **123**, p.567-8 (1929)
Atkinson, R.d'E.: *Astrophys. J.*, **73**, p.250-95, 309-47 (1931)

6. Birge, R.T., & D.H. Menzel: *Phys. Rev.* **37**, p.1669 (1931)
Urey, H.C., F.G. Brickwedde & G.M. Murphy: *Phys. Rev.* **39**, pp.164-5 (1932)
Fowler, R.H.: *Proc. Camb. Phil. Soc.*, **30**, p.225-41 (1934)

7. Heavy hydrogen discussion. *Proc. Roy. Soc.*, **144**, p.1-28 (1934)

8. Oliphant, M.L.E., P. Harteck & Rutherford: *Nature*, **133**, p.413 (1934); *Roy. Soc. Proc.* ser. A., **144**, p.692-703 (1934)

9. Dee, P.I.: *Nature*, **133**, p.564 (1934)

10. Meitner, L., O. Hahn & F. Strassmann: *Zeits. f. Phys.*, **106**, p.249-70 (1937)
Meitner, L., & O.R. Frisch: *Nature*, **143**, p.239-40 (Feb. 1939)
Bohr, N., & J.A. Wheeler: *Phys. Rev.*, **56**, p.426-50 (1939)

11. Kulcinski, Cameron, Santarius, Sviatoslavski & Wittenberg: *Fusion energy from the moon for the 21st century* (1988), Fusion Technology Institute, University of Wisconsin

12. Thompson, W.B.: *Nature*, **179**, p.886-9 (1957)

194

Chapter 12

Controlled nuclear fusion: the confinement problem

12.1   Introduction

To be usable, energy from nuclear fusion must be released at a controlled rate. Herein lies one of the fundamental problems confronting science: how to contain a high-energy plasma for sufficient time to enable a large enough number of fusions to take place and generate useful (and usable) power. The aim is not to recreate the Sun in a laboratory but to build a machine which is small enough to be economically viable while still functioning as a reactor. The energy produced must exceed that needed to heat the fuel and maintain the plasma. In a laboratory situation, we cannot rely on gravity to hold the plasma together, as happens in astrophysical plasmas. Other means of containing the plasma must be used instead. The two principal methods currently under investigation are inertial confinement and magnetic confinement. In inertial confinement, small high-density plasmas are produced for a very short time, of the order of nanoseconds; in magnetic confinement, lower density plasmas are constrained by magnetic fields for a much longer time: maybe one or two seconds.

12.2   Historical background

Until the 1940s, nuclear fusion research was largely concerned with sources of stellar energy while plasma physics focused on gas discharge tubes, for which a knowledge of nuclear physics was not essential. The search for usable power from controlled thermonuclear fusion reactions following the explosion of the hydrogen bomb brought the two disciplines together.

Before the Second World War both ideas and personnel were freely exchanged between laboratories around the world. During the war, the atomic bomb was developed in the USA by an international team with major contributions from Canada, Britain and France. The advent of the Cold War in 1945 brought secrecy and military influence, with a desire to safeguard national interests. The USSR became the enemy and the USA classified its weapons research, excluding even its allies in the West from access to information. Fusion research, with its links to the hydrogen bomb, was included in the ban. Members of the war-time international team were summoned home from the USA to recreate the whole technology and science from memory.

196

In the UK, work began on linear and ring-shaped (toroidal) discharges in 1946 at Imperial College, London, led by George Paget Thomson (1892-1975), son of J.J. It was well known that in high-current gaseous discharges, the self-magnetic field could cause a constriction of the discharge. Bennett, Tonks and Alfvén had all considered the theoretical aspects of this "pinch effect" in linear discharges in the 1930s and 1940s but no experimental observations had as yet been recorded.

The ring discharge, first produced by J.J. Thomson in 1891, was often used as a light source in spectroscopy but otherwise had received little attention. Its time had now come. Thomson's device had been contained within a solenoid through which flowed rapidly alternating currents. It was suggested that coils wound around a ring discharge, or *torus*, could produce a magnetic field which would totally enclose the plasma, holding it firmly in place. At Imperial College, S.W. Cousins and A.A. Ware used a rotating mirror camera to study the toroidal ring discharge and in 1951 reported seeing a noticeable constriction in argon discharges at low pressures.[1] This produced a lateral oscillation of the discharge as the current filament expanded and contracted. The oscillations were thought to be due to plasma ion waves excited by the pinch effect. Cousins and Ware concluded that large currents produce strong "pinch forces" in the gas which initially compress the outer edges of the plasma where the pinch forces are strongest. The compression then travels inwards as a shock wave in the plasma, producing the observed oscillations. In the UK, attention now focused on the pinch discharge — in secret because, prompted by spy scandals, defections and court cases, fusion research had become a classified activity here too.

That same year, the scientific world was briefly stunned by an announcement from Argentina that fusion had been achieved, but claims that a prototype reactor was working soon proved to be false. Whether it was political trickery to flatter the Argentinian dictator Juan Perón, or naive optimism on the part of the scientists is hard to tell, but it added impetus to research projects elsewhere.

While the obsession with secrecy was understandable given the global political situation and mounting fears about nuclear war, it hindered scientific progress. In the UK there was much theoretical interest in fusion in the early 1950s but the practical problems of confinement meant that little progress was made. In the USSR, spurred on by the Argentinian claims, Andrei Sakharov and Lev Artsimovich began working on an alternative toroidal device which they named a *tokamak*.[a] They believed that a strong electric current passed through the plasma would simultaneously heat it and produce a second magnetic field, creating a twisted magnetic "cage" around the plasma. The first (top-secret) tokamak, TMB, was built in the USSR in 1955. It suffered from very low temperatures due to impurities.

In the USA, several projects were under way. In Los Alamos, nuclear fusion research

---

[a] A Russian acronym from *to*roidalnaya *ka*mera *mag*nitnym polem (toroidal chamber with a magnetic field).

continued under the code-name "Project Sherwood". Its official start-date was given as 1951, but much of the basic theory had been formulated by the international team working there during the War. One member of that team, James Tuck, had been recalled to the UK at the start of the Cold War. He now returned to Los Alamos with knowledge of British views on controlled fusion. Some of these ideas were used in his first pinch machine for the Americans: a simple ring-shaped vacuum vessel with a strong current within the plasma, induced by transformer action, to create the magnetic confinement.

At the Livermore Laboratory in California, work began on developing a magnetic mirror machine: a long tube encased in magnetic coils, the strength of which increased at each end. On the east coast, unaware of the classified work at Los Alamos, the Princeton astrophysicist Lyman Spitzer Jnr. put forward his own design for a reactor early in 1951. Spitzer believed that a ring-shaped vacuum vessel would make it difficult to initiate the plasma, but if the two ends of a linear discharge were twisted to form a figure-of-eight, the plasma should form easily. The twist would also cancel the outward drift of the plasma since "outward" on one half of the figure-of-eight was "inward" on the other. The model-A Stellarator was built at Princeton in 1953.

By the end of 1955, the obsession with secrecy was becoming a hindrance. Very little thermonuclear fusion work had been published since the end of the War. Each research group was working in isolation and the need to share ideas and discuss problems led to the first tentative steps towards resuming international collaboration. The USA and the UK began sharing classified information on fusion and the first exchange visits took place. The Russian fusion expert Igor Kurchatov visited the UK and outlined his country's previously secret efforts since 1950 on the pinch-effect of an arc discharge. The Russians had found that, immediately after breakdown, the discharge current was carried in a thin cylindrical layer, adjacent to the tube walls, which contracted radially, like a shock-wave, ionising and compressing the gas. The plasma then expanded and contracted again, this time emitting short bursts of neutrons and X-rays as it contracted. Neutron emission was supposed to indicate thermonuclear temperatures, but the Russians attributed it to unstable behaviour and sudden movement of the plasma. They, like the Americans and the British, had encountered problems with plasma instabilities in their machines.

In June 1956, the UK Atomic Energy Authority introduced British industry and universities to the concepts and problems of fusion research. A comment made at that symposium reveals the state of knowledge at that time: "Our vision is of a power station sited perhaps on the coast, with a pipe bringing water from the sea, helium leaving by the chimney and electrical power flowing into the Grid. We do not know what to put inside the power station."[2] The hope was, that by involving industry and academia, that problem might soon be solved.

The UK's first large-scale toroidal pinch discharge, known as ZETA (Zero Energy Toroidal Assembly), began operating at Harwell in August 1957, initially with a plasma current of 200 kA lasting for four milliseconds. It was the first attempt to produce high temperatures and fusion reactions in a toroidal discharge at what was then the largest fusion research facility in the West and it showed that a toroidal configuration could confine a plasma for many milliseconds. Early claims in 1958 that neutrons from fusion reactions had been detected were soon found to be due to the heating process rather than high temperatures. The machine was branded a failure by the British media who did not understand the enormous problems involved and thought that the generation of limitless electricity was imminent. ZETA was, in fact, extremely successful in achieving what it was designed to do and remained in operation until 1968.

The 1958 Atoms for Peace Conference in Geneva heralded the modern era of international cooperation in plasma physics. As the delegates, chiefly from the USA, USSR and UK, outlined their projects and the progress made towards fusion, the complexity of the problems facing them became clear. Lev Artsimovich later described it as "essentially a display of ideas, only thinly draped with rough experimental data".[3] Japan and Europe soon joined the research efforts. Costs were escalating and projects became smaller, with the aim of understanding what was going on within the plasma, rather than trying to build a large device with the dimensions and densities calculated necessary for fusion. It was now realised that vacuum conditions must be the best possible, and impurities kept to the absolute minimum, if meaningful results and progress were to be achieved. Since 1960, much of the research has centred on overcoming instabilities within the plasma, improving containment, and developing diagnostic techniques — equipment to measure basic parameters such as temperature and density.

Stability problems meant that work on pinches declined in favour of magnetic mirrors and stellarators as the emphasis switched to more basic studies. An historic Anglo-Soviet collaboration in 1968-9, in which a British team measured the temperatures in the Soviet tokamak T3, showed that Russian claims of stability, high temperatures and thermonuclear neutrons in tokamaks were underestimated.[4] The tokamak now became the favoured design. Small-scale tokamak experiments in Europe and elsewhere in the 1970s indicated that a tokamak fusion reactor would have large dimensions and a large plasma current.

The 1973 oil crisis exposed the dependence of the industrialised nations in the West on overseas fossil fuel resources. In Europe, the enlargement of the European Community with the inclusion of Britain, Denmark and Ireland led to the idea of a Community-wide research project into fusion power from tokamaks as an alternative fuel source. Coordination of European fusion research had started in 1957 with the formation of Euratom (*Eur*opean *Atom*ic Energy Community), although at the time fusion formed a very small part of its programme. Euratom organised research in the national laboratories of the member states, encouraging the exchange of staff and information, providing financial aid and avoiding unnecessary duplication of effort.

In America, work began on the design and construction of the Tokamak Fusion Test

Reactor (TFTR) at Princeton. Across the Atlantic, plans for the Joint European Torus (JET), based near Oxford, took shape. This would become the largest single project of the Euratom fusion research programme. Both machines were specifically designed to run eventually with a mixture of deuterium and tritium, and establish conditions approaching those in a fusion reactor. Both were operational by mid-1983.

Although work continued on reversed-field pinch machines as an alternative line of research in a few centres, notably the UK and Italy, the number of tokamaks and similar devices grew. Their performance improved through the 1970s and 1980s as knowledge accumulated, despite variations in size, design and operating conditions producing differing results. By the late 1980s support for magnetic confinement designs other than the tokamak was declining and many of the alternative devices, such as magnetic mirrors and reversed-field pinches, shut down as funding was cut.

Improved plasma heating greatly increased electron and ion temperatures, bringing the goal of fusion power a little nearer. Finally, on 10 November 1991, JET produced a 1.7 megawatt pulse of power while operating with a mixture of deuterium plus 10% tritium in an experiment that lasted 2 seconds. This was the first time that a substantial amount of energy had been produced by nuclear fusion in a controlled experiment. The amount of tritium was deliberately kept small to minimise radioactive contamination of the machine. Two years later, on 9th December 1993, a similar experiment on TFTR, this time with a 50:50 D-T mixture, yielded 10.7 MW of fusion power with a peak of 3 MW. TFTR shut down in April 1997 because of budget cuts and a change in the focus of US fusion research, leaving JET as the only tokamak where tritium experiments could be performed. By the end of the 1990s it had produced 16 MW of fusion power and 21.7 MJ of energy.[5]

An alternative method of confinement involves focusing laser pulses onto tiny pellets of deuterium gas. Known as inertial confinement, it compresses the pellet for the fraction of a second necessary for the deuterium nuclei to fuse and produce a miniature thermonuclear explosion. Following the development of the laser in 1960, experimental laser fusion programmes began in the USSR, at Limeil in France and at the Livermore Laboratory in California (which became the major centre for inertial confinement research). Gas breakdown and the formation of a spark in the focus of a laser beam, similar to a conventional spark discharge, was observed in Russia in 1962. At Princeton, J.M. Dawson calculated that a laser pulse delivering power of the order of 10^{10} W to a liquid or solid particle 0.1 mm in diameter would produce a plasma with temperatures of several hundred eV. By 1967, temperatures in excess of 10^5 K (8.6 eV) and shockwave velocities as high as 2×10^5 m s^{-1} in the imploding pellets had been measured in Russia.[6] American research was largely conducted in secret, as part of US defence systems research, and benefitted initially from defence budget spending. When the programme was declassified in 1972, John Nuckolls reported that laser light had been focused to intensities greater than 10^{15} W m^{-2}, producing radiation pressure, or momentum flux, of almost 10^8 atm.[b]

Particle beam accelerators have also been tried as an alternative means of imploding the fusion target. As with magnetic confinement, inertial methods have their own inherent problems, such as focusing, as well as those of size and cost to be overcome and by 2000 the target of ignition was still some way off.

Underlying both magnetic and inertial confinement is the conviction that power extraction from nuclear fusion requires extremely high temperatures. There was therefore considerable scepticism and some alarm in the field of fusion research in March 1989 when two electrochemists, Professor Martin Fleischmann and Dr. Stanley Pons, claimed to have achieved so-called "cold" fusion at room temperature. Fleischmann was a distinguished British scientist, well-established in the field of electrochemistry at Southampton University, while Pons, having obtained his PhD from Southampton, was working at the University of Utah. Their experiment, in Utah, involved electrolysis using heavy water (D_2O) and a platinum anode surrounding a cathode made of palladium, a metal which can absorb large amounts of deuterium. They had observed bursts of heat emerging from the cells, too large to be explained by chemistry: a return of four watts of heat from one watt of electricity when current was passed through the electrode for 100 hours. This was attributed to the palladium absorbing so much deuterium that some of it fused to form helium-3 or tritium simply because the nuclei were so close together. Neutron emission had also been detected, which made the two men think that the excess heat came from D-D fusion. A team from Brigham Young University confirmed that they too had observed neutron generation in similar equipment.

Fleischmann and Pons were rushed into publishing their results without proper scrutiny. They were also claiming a major discovery outside their own field — nuclear research being regarded as the domain of physicists, not chemists — and they were threatening the well-established thermonuclear fusion effort. Nevertheless, because the two men were highly regarded in their field, attempts were made to repeat the experiment, with differing results since the electrolytic conditions were hard to reproduce. Neutron counts proved too low to be due to fusion and it was generally agreed that Fleischmann and Pons might have discovered a thermal reaction, but it was not nuclear. Cold fusion disappeared from mainstream research, although small-scale investigations continued throughout the 1990s. By early 2002, several research groups were beginning to hint that some form of low-energy fusion reactions might be possible.

12.3   Energy confinement — Lawson's criterion ($n\tau_E$)

Output energy must exceed input energy. This determines the minimum temperature — the *ignition temperature* — required for any fusion reactor to be self-sustaining. Above this temperature, more energy is produced by fusion than is lost through processes such as

---

[b]   Radiation pressure in the centre of the Sun is believed to be 10^{11} atm.

radiation. At low temperatures, losses through bremsstrahlung can greatly exceed energy production.

If a hot plasma touches the walls of its containment vessel it immediately loses energy and is cooled through recombination and impurity ionisation processes. To keep power losses to a minimum, therefore, the plasma must be isolated from its surroundings. The effectiveness of this confinement is measured by the *energy confinement time*, τ_E. This is the time needed for the total energy to be lost from the plasma; in other words, the time it takes for the system to cool down once all external forms of heating are switched off.

In 1957, at the UK's Harwell research site, J.D. Lawson calculated the theoretical power balance required for various types of thermonuclear reactor. He found that temperatures must be high enough, and the reaction sustained for long enough to allow a definite fraction of the fuel to be burnt.[7] Later calculations have revealed that, in a fusion reactor, the following typical values of ion temperature, density and energy confinement time must be achieved simultaneously:

$$T_i = 10 - 20 \text{ keV } (\approx 1 - 2 \times 10^8 \text{ K})$$
$$n_i = 2.5 \times 10^{20} \text{ m}^{-3}$$
$$\tau_E = 1 - 2 \text{ sec.}$$

so that the *fusion triple product* $(n_i T_i \tau_E)$ is greater than 5×10^{21} keV s m^{-3}. The requirement of high temperature and good thermal insulation $(n\tau_E)$, known as *Lawson's criterion*, is very difficult to achieve. For a deuterium-tritium plasma, $n\tau_E$ must be about 5×10^{20} s m^{-3}, with ion temperatures in excess of 900 eV. For a deuterium-only plasma the estimate for $n\tau_E$ is even higher: in the region of 10^{22} s m^{-3}, with ion temperatures greater than 10 keV.

ZETA, in 1957, had a confinement time, τ_E, of around 10^{-4} s, a temperature of about 50 eV and density of 10^{20} m^{-3}; $n_i T_i \tau_E$ was therefore $10^{20} \times 5 \times 10^{-2} \times 10^{-4} = 5 \times 10^{14}$ keV s m^{-3}. By the early 1980s, confinement times of about 0.15 s had been achieved on some machines. Temperatures were in the region of 6 keV. By 1995, the simultaneous measurement of a fusion triple product in the region of 1.1×10^{21} keV s m^{-3} had been accomplished. Ignition parameters had been achieved on the largest machines (JET, TFTR, JT-60): maximum central ion temperatures of about 40keV, particle densities of about 3×10^{20} m^{-3} and confinement times of about one second, but not at the same time. The first 50:50 D-T experiment on TFTR in December 1993 produced ion temperatures of 35 keV and τ_E equal to 0.19 s, compared to 27 keV and 0.16 s in corresponding deuterium-only shots.[8] Adding tritium clearly enhanced the results.

12.4 Magnetic confinement and plasma β

Since plasma can be controlled by magnetic fields, most laboratory confinement devices consist of a vacuum chamber surrounded by external current-carrying coils or electromagnets to create the magnetic fields. Field lines must not touch the vacuum vessel walls otherwise charged particles orbiting the field lines would be lost through recombination

202

at the wall, and impurities would be introduced into the plasma. Keeping the hot plasma away from the walls introduces gradients in plasma temperature and density.

A magnetic field exerts pressure on the plasma because of the energy it contains, and the ratio of outward plasma pressure, p, to inward magnetic field pressure — the plasma β (equation 8.25) — indicates the likelihood of confinement. If p is measured at a point within the plasma and the magnetic field strength, B, is measured at the plasma surface (B_0), β measures the efficiency of the magnetic confinement. For long-term stable confinement, the outward plasma pressure must be less than the inward magnetic field pressure, $B_0^2/2\mu_0$, at the plasma surface.

A low value for β means that the kinetic pressure, p, is low compared to the magnetic pressure at the edge and most of the energy in the system is in the form of externally-supplied magnetic energy. The plasma pressure roughly indicates the fusion yield, the magnetic pressure provides a measure of input energy. The value of β thus indicates how efficient is the use of the magnetic field. Economic constraints impose a minimum β-value of 5 - 10% because of the cost of producing high-strength magnetic fields. Ignoring other factors, an economic fusion reactor would require a high beta.

12.4.1   The pinch effect

The magnetic field configuration within the plasma largely determines plasma behaviour. Flows of current and plasma may also induce changes in the field topology, since magnetic fields are produced by the motion of charged particles (Fig. 12.1(a)).

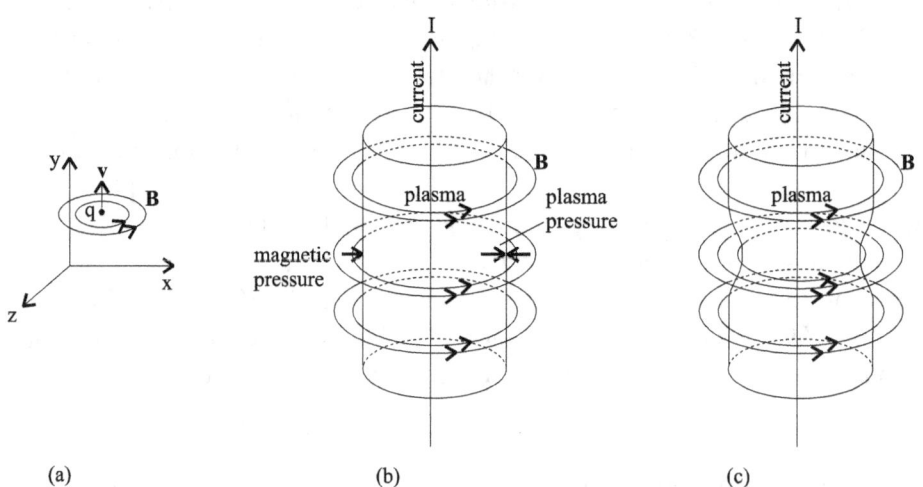

Fig. 12.1   Pinch effect in a column of plasma

An electric current flowing longitudinally through a cylindrical column of plasma produces a magnetic field perpendicular to it (Fig. 12.1(b)). The plasma contracts radially along its length under the influence of the magnetic field until the inward $J \times B$ (Lorentz) force is balanced by the outward pressure gradient force, ∇p, (Section 8.3). This self-constriction of the plasma is the *pinch effect*. The plasma is forced away from the discharge tube walls, forming a narrow column along the axis. If the current is pulsed, a radially-imploding shock wave results which can heat the plasma, but not enough to produce fusion. A current around the column has a similar effect, producing a longitudinal magnetic flux just inside the containing-cylinder wall which drives the plasma towards the centre, heating it up.

A plasma confined by the pinch effect is not in a stable equilibrium. Fig. 12.1(c) illustrates a *pinch point* — an anomaly in the magnetic field where the field strength is greater than elsewhere. The magnetic force, $qv \times B$, drives the charged particles towards the centre of the column. If the magnetic force is sufficient to overcome the kinetic pressure of the particles (a low-β plasma), the plasma column will contract radially. A small disturbance from cylindrical symmetry can therefore disrupt the pressure balance, causing the growth of instabilities such as were discussed in Chapter 8 (kink, sausage, etc.). These make the plasma break up and go to the walls of the containment vessel, losing energy and destroying the plasma column.

12.5   Magnetic confinement systems

Finding a magnetic field configuration which is able to confine sufficient fuel for a long enough time is one of the major problems associated with developing nuclear fusion as an energy source. The different methods of magnetic confinement of plasma form two groups, based on the topology of the magnetic field lines: open systems and closed systems.

12.5.1   Open systems

If some magnetic field lines leave the system, it is described as "open". A simple example is the magnetic "bottle" described in Section 4.9.1. A current-carrying coil is wound around a cylinder, with more turns at the ends than in the middle, producing a magnetic field in the cylinder which is stronger at the ends than in the middle, creating two magnetic

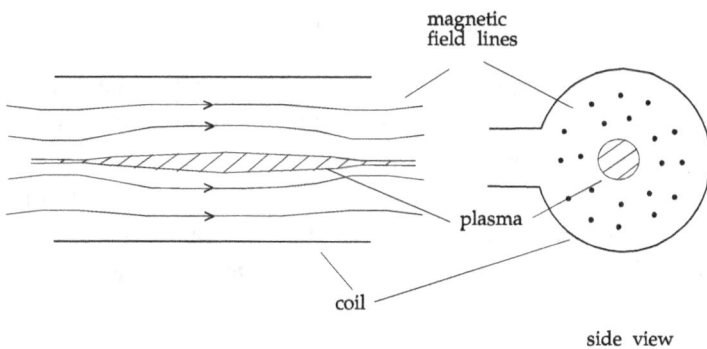

Fig. 12.2   Theta pinch

204

"mirrors". Charged particles tend to be "tied" to magnetic field lines while moving freely along them, so if field lines can leave the system, charged particles can also escape. A containment device such as the magnetic bottle requires a constant supply of plasma into the confinement region to replace the end losses.

The *theta pinch* (Fig. 12.2) is a pulsed system which simultaneously produces and heats the plasma. An electric current around the plasma column creates an axial magnetic field which confines the plasma. To overcome end losses, the machine is very long — the thetatron experiment at the UKAEA's fusion research centre in Culham in the late 1960s used a straight tube eight metres long. It is therefore not favoured as a prototype fusion reactor.

12.5.2 Closed systems

Open systems end losses can be avoided by forming the cylindrical column of plasma into a circle, like a ring doughnut — a shape known as a *torus*. The longitudinal magnetic field lines of the original cylinder close on themselves to form rings, which, to reduce particle loss, must not intercept the walls of the containment vessel. The open ends of the cylinder now form the *poloidal*[c] *cross-section* of the torus.

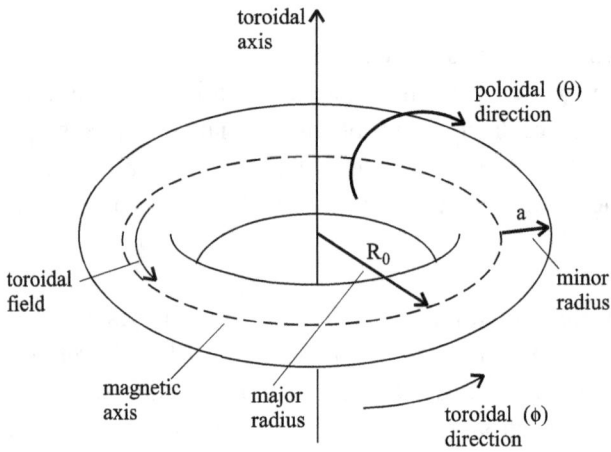

Fig. 12.3 Toroidal geometry

Fig. 12.3 shows the basic geometry of toroidal systems. The torus is axisymmetric: it is symmetric with respect to rotation about the major, or toroidal, axis which by convention is vertical. The magnetic axis is a single toroidal field line which generally marks the peak of the plasma density profile. The *major radius*, shown as R_0, measures the distance of the magnetic axis from the toroidal axis. The radius of the torus in the poloidal direction, known as the *minor radius*, is marked as a. The ratio of major to minor radius, R_0/a, is the toroidal

---

[c]   *from pol*ar and tor*oidal*

aspect ratio and gives an indication of how compact the torus is. Rotation about the toroidal axis is conventionally represented by ϕ, rotation in the poloidal direction by θ.

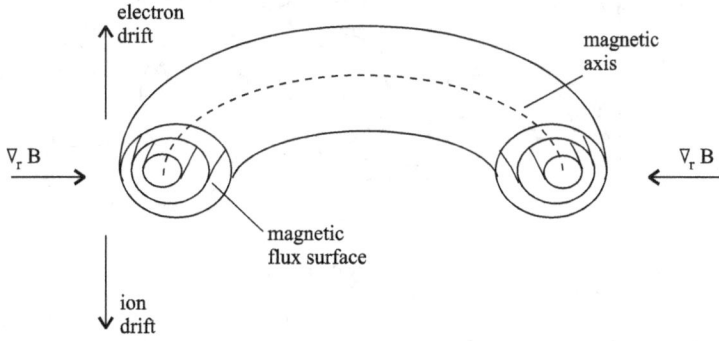

Fig. 12.4   Poloidal cross-section of torus

While a toroidal magnetic field eliminates end losses, the field curvature creates other problems. Radial magnetic field variations occur in the poloidal cross-section of the torus because the surfaces of constant magnetic pressure, although nested, are not concentric (Fig. 12.4). This outward shift of the inner surfaces is often referred to as the *Shafranov shift*.

A simple toroidal magnetic field, B_ϕ, around the major axis is insufficient to confine the plasma. Magnetic field variations across the minor cross-section (indicated by $\nabla_r B$ in Fig. 12.4) cause particles to drift to the outer edge of the vessel — electrons to the top of the torus, ions to the bottom — producing an unwanted charge separation. Centripetal acceleration produces a drift velocity (equation 4.36) whose magnitude is given by

$$v_d = \frac{1}{\omega_c} \frac{v_\parallel^2}{R_0} = \frac{m v_\parallel^2}{q B R_0} \tag{12.1}$$

where ω_c is the cyclotron frequency (equation 4.16). The drift velocity causes charge separation, through the influence of q. The vertical electric field created by these oppositely-directed drifts of electrons and positive ions interacts with the toroidal magnetic field to produce an outward $\boldsymbol{E} \times \boldsymbol{B}$ movement of the plasma, affecting its stability. An additional magnetic field around the minor axis (in the poloidal direction) is therefore needed to keep the plasma under control.

Most toroidal systems are built with large transformer coils passing through the central hole in the torus to produce the required fields (Fig. 12.5(a), overleaf). A set of toroidal field coils around the minor circumference produce a *toroidal magnetic field, B_ϕ,* around the major axis (Fig. 12.5(b)). As the current rises in the primary coils of the transformers, a voltage is induced around the torus establishing a ring current in the plasma and driving the plasma around the main axis of the torus. The ring current generates a *poloidal field, B_θ,* around the minor axis (Fig. 12.5(c)), resulting in a "sheared" or helical

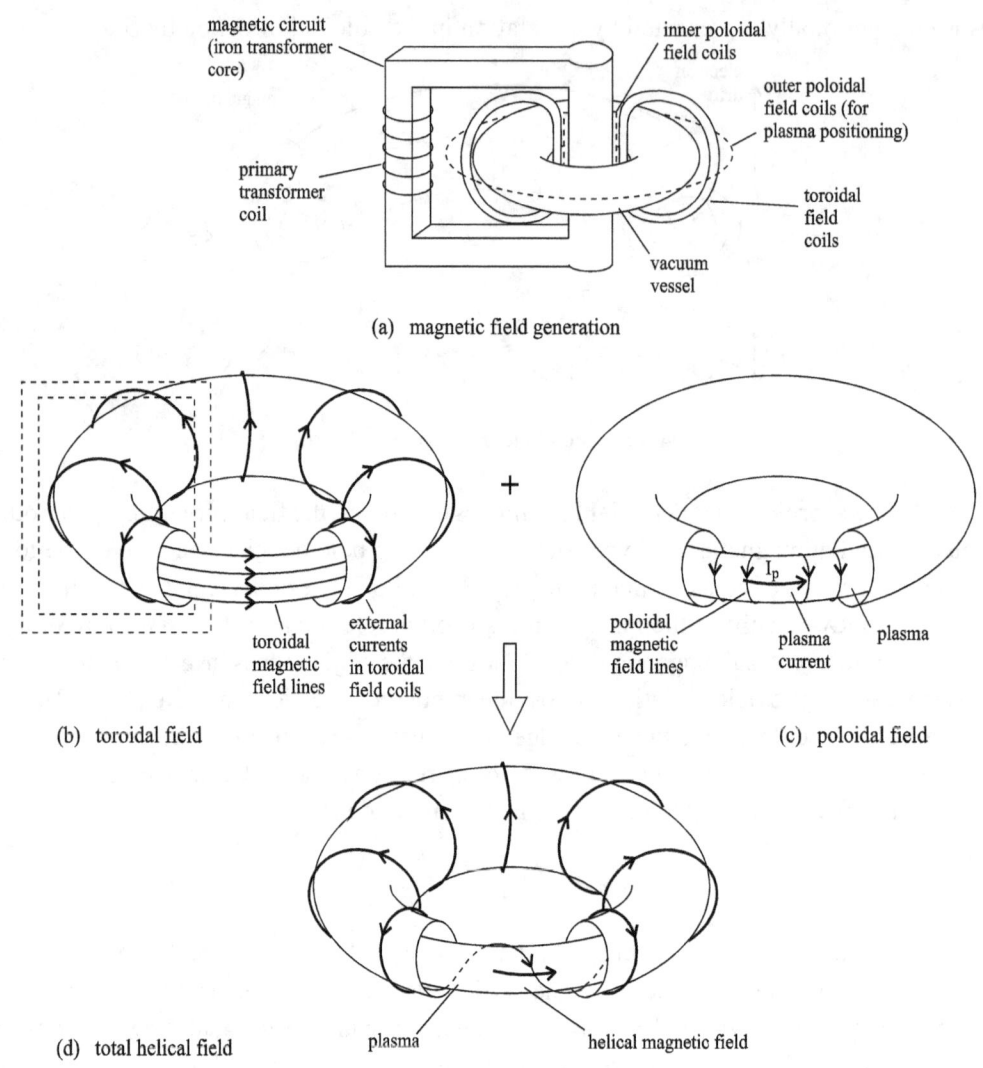

Fig. 12.5   Toroidal fields and currents

magnetic field in which the field lines twist as they encircle the torus (Fig. 12.5(d)). The poloidal field, by twisting the field lines, is the most important component of the magnetic field configuration as far as equilibrium, stability and transport processes are concerned. The helical field prevents the drift of particles to the walls. The degree of *magnetic helicity* indicates the structural complexity of a magnetic field configuration. A toroidal surface covered by magnetic field lines in this way is called a magnetic surface, or *flux surface*.

The magnetic topology is the same for all the major confinement systems: a system of toroidally-nested, closed magnetic surfaces produced by helical magnetic fields, shown in cross-section in Fig. 12.6. Equilibrium in the poloidal cross-section is a balance between the inward force due to the poloidal field, B_θ, the outward force due to plasma pressure, and compression of the toroidal field, B_ϕ. The plasma can only escape by crossing the lines of magnetic flux, either in collisions with other particles or through the development of instabilities which cause the plasma to move to the wall.

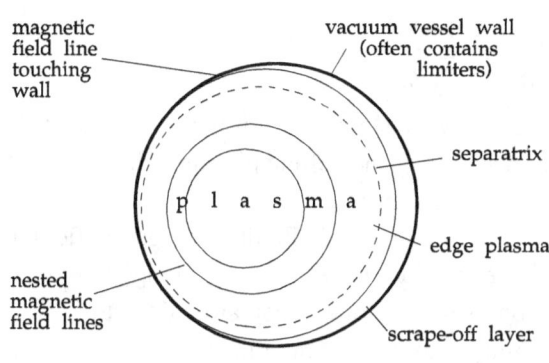

Fig. 12.6   Cross-section of torus

The *separatrix* (shown as a dotted line in Fig. 12.6) is the magnetic surface dividing the core plasma in the centre from the outer region, where field lines intercept the walls of the containment vessel. Plasma confinement away from the wall is only possible within this surface. Any plasma diffusing across the separatrix is "scraped off" onto the walls beyond, giving rise to the term *scrape-off layer*. The *edge plasma* or *radiating layer* is located just inside the separatrix. The effects of impurities and energy losses from radiation and particles hitting the walls of the vessel means that these two regions have a major influence on the confinement and transport properties of the bulk of the plasma.

The vacuum vessel containing the plasma is usually made of metal and must withstand a variety of forces, such as those due to atmospheric pressure and to changing magnetic fields which induce currents in it. To reduce contamination by impurities from the containment vessel and the energy losses which result, *limiters* are often built into the walls. These are bars of graphite which protrude from the wall and collect particles heading towards the walls. Since graphite has a smaller atomic number than most of the materials used in vacuum vessel walls, the limiters reduce the amount of pollution from heavy ions entering the plasma.

12.5.3   Safety factor, q

Two important confinement parameters are the pressure ratio, β, and the *safety factor*, or *winding number*, q.[d] The safety factor is a geometric property of the magnetic field: it

---

[d]   Not to be confused with the *confinement* or *"engineering"* Q, sometimes called the Q-value. This depends on the fusion triple product and is the ratio of fusion power produced to heating power supplied. If Q is large enough (between 10 and 20), the device is potentially a useful fusion reactor.

measures the "twistedness" of a magnetic field line. The slowly spiralling field lines each lie on nested surfaces of constant radius r (where the maximum value of r is the minor radius shown as a in Fig. 12.3). The rate of twisting is obtained from

$$\frac{r\,d\theta}{B_\theta} = \frac{R_0\,d\phi}{B_\phi}$$

where $r\,d\theta$ is rotation in the poloidal direction and $R_0\,d\phi$ is rotation in the toroidal direction. Thus

$$q = \frac{d\phi}{d\theta} = \frac{r B_\phi}{R_0 B_\theta} \tag{12.2}$$

B_ϕ is the toroidal field and B_θ the poloidal field. By substituting the minor radius, a, for r in equation (12.2) we obtain the value of q on the outer surface of the plasma. This is the ratio of toroidal to poloidal magnetic field divided by the aspect ratio.

In making one circuit of the magnetic axis (a poloidal circuit), a magnetic field line will make q circuits of the toroidal axis. Values of q are generally in the range 0.5 to 4. A value of $q = 1$ implies that a field line joins on itself after one transit around the torus; $q = 2$ requires two toroidal circuits. For MHD stability, q must be greater than 1. A small value of q (q very much less than 1) corresponds to a tightly-wound helix; infinite q describes a simple toroidal circuit. Each toroidal magnetic surface has a fixed value of q, but not all field lines have the same pitch, so q varies within the confining volume, usually increasing from the magnetic axis to the edge of the plasma. The radial variation of q from surface to surface, dq/dr, is called the *magnetic shear*. It measures the change of field line pitch from one surface to the next and is a significant factor in plasma stability.

Magnetic fluctuations which appear as discrete oscillatory motions or *modes* are often referred to by the numbers m and/or n (e.g. "n=1 mode") which define respectively the poloidal and toroidal mode number as shown in Fig. 12.7. The example shows the m=4 mode.

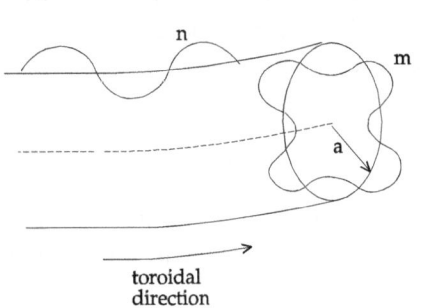

toroidal
direction

Fig. 12.7   Mode numbers

The safety factor can be expressed as the ratio of these two mode numbers: $q = m/n$. For low m and n mode numbers, the frequencies of the fluctuations are usually less than 50 kHz. In tokamaks, low-numbered modes may be involved in major disruptions while the higher modes may be responsible for particle and energy transport. If current and pressure conditions allow, MHD instabilities tend to form and grow on flux surfaces with low-order rational safety factors ($q = m/n$), for example: (m,n) equal to (1,1), (2,1), (3,2), etc. These surfaces tend to be the least stable and are a common source of problems in tokamak plasmas. Tearing modes usually occur at the q=2 surface and are mainly due to the

$m=2$, $n=1$ mode. A sheared magnetic field, whose direction changes as a function of position, can be used to localise instabilities in a plasma.

12.6 Confinement geometries

The most common magnetic confinement geometries have both toroidal and poloidal field components. They differ in the way they twist the magnetic field lines and can be divided into two groups: externally-generated and self-generated magnetic fields.

12.6.1 Externally-generated magnetic confinement

This group includes the tokamaks and most stellarators. Beta values are usually small and q is greater than 1, so the helical magnetic field twists very gradually around the torus.

a) The tokamak

The major radius, R_0, of most tokamaks is in the range 0.3 - 3 m. The aspect ratio (R_0/a) lies between 1.3 and 4 and is generally closer to 4.[e] JET has a vacuum vessel which is D-shaped in cross-section, a major radius of 3 m and a minor radius of 1.2 m, giving a rather tight aspect ratio of 2.5. The toroidal field is between 3 and 4 tesla. JT-60, which began operating at Tokai, Japan, in 1985, was built with a circular-cross-section vacuum vessel, a major radius of 3 m and minor radius of 0.95 m.

The relative strengths of the toroidal and poloidal fields are important for stability and plasma confinement. The poloidal field, B_θ, does most of the confinement. The toroidal field, B_ϕ, improves stability and is much stronger than B_θ, being generally between 1 and 5 T. At the edge, B_θ is generally less than B_ϕ by a factor of $2a/R_0$. As a result, q is greater than 1, and because of the large toroidal field, B_ϕ, the overall beta is small, usually only a few percent.

In the early tokamaks with a circular cross-section, beta values were limited to one or two percent of the confining magnetic pressure. The change to a more triangular or D-shaped vacuum vessel cross-section in many of the later machines produced beta values greater than 5%.

"Poloidal beta" is the ratio of plasma thermal energy to poloidal field energy and can be estimated from:

$$\beta_{pol} \approx \left(\frac{B_\phi}{B_\theta} \right)^2 \beta \qquad (12.3)$$

Values for β_{pol} can be very high compared to β. The outward shift of the plasma in a toroidal device is related to the value of poloidal β, with a maximum of $\beta_{pol} = R_0/a$. Beyond this value, outward displacement of the plasma is so large that the outer magnetic surfaces are no longer closed.

---

[e] Early tokamaks had larger aspect ratios. The Soviet machine T3, for example, had $R_0 = 1$ m and $a = 0.2$ m, giving an aspect ratio of 5.

210

In the 1980s, the high confinement mode of operation, known as "H-mode", was developed. Energy confinement times were improved by adjusting the heating conditions in the plasma and using divertors to control the plasma boundary layer.

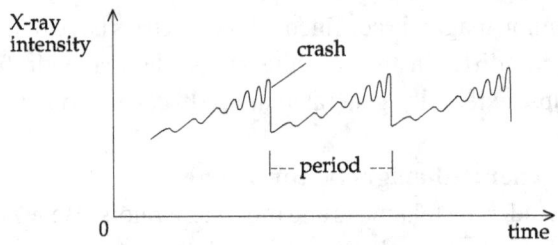

Fig. 12.8   Sawtooth oscillations

Tokamaks are particularly affected by *"sawtooth" oscillations* (Fig. 12.8). Over a period of between 10^{-3} and 10^{-1} seconds, the intensity of X-ray emissions from the central region of the plasma rises gradually, develops growing oscillations (which ultimately look like the teeth of a saw), and then falls abruptly ("crashes"). An increase in electron temperature and plasma pressure on the magnetic axis, where q is greater than 1, may cause current density to rise and the safety factor, q, to fall. As q becomes less than 1, an $m = 1$, $n = 1$ instability develops. The sawtooth crash occurs when the magnetic field lines reconnect and q again becomes greater than one. Tokamaks can, however, operate with axial values of q less than 1 without sawtooth oscillations developing.

b)   Stellarator

The stellarator obtains both toroidal and poloidal magnetic fields from currents in external coils wound helically around the torus. There is little or no induced plasma current. The currents in alternate coils flow in opposite directions to produce the required helical magnetic field (Fig. 12.9). As with tokamaks, the poloidal magnetic field, B_θ, is less than the toroidal field, B_ϕ. The stellarator has been shown to have a confinement potential at least as good as the tokamak. Its advantage over the tokamak is that it does not rely on the induction of currents which can decay due to the finite resistivity of the plasma. It can therefore be a steady-state device. Unfortunately, its field configuration is more complicated and beta values can be even lower than those of tokamaks and large machines have proved more difficult to build than large tokamaks because of the helical winding of the coils.

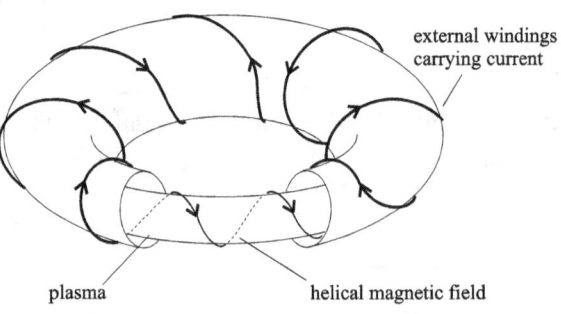

Fig. 12.9   Stellarator fields and currents

12.6.2   Self-generated magnetic confinement

Together with the stellarator, this group constitutes the "alternative lines" to tokamak fusion research. Machines with largely self-generated magnetic confinement generally have a relatively high beta value, around 10%, and q less than 1. The helical magnetic field is therefore more tightly-wound than in the tokamak and stellarator. These high-β machines rely more on internal currents for plasma confinement and less on external field coils, although some confinement must always be provided by external coils. Examples include the stabilized Z-pinch, the spheromak and various other machines with a more compact toroidal geometry.

a)   Reversed-field pinch (RFP)

The Z-pinch confines the plasma using an axial current and the poloidal magnetic field it generates. It often has an aspect ratio greater than 5. One of the most studied forms of the Z-pinch is the reversed-field pinch.

In 1963 the UK's toroidal pinch discharge, ZETA (major radius = 1.5 m; minor radius: 48 cm.), spontaneously produced a period of quiescence and improved confinement which was found to be a reversed-field pinch configuration. Investigation of the processes involved culminated in Brian Taylor's theory of relaxed states[9] in the 1970s, and ideal MHD theory was extended to include the resistive effects (Section 8.4.2) which play an important role in real plasmas.

In the RFP, external coils create the toroidal field and a toroidal current, I, is induced in the plasma. This current generates a poloidal field with a pinch-effect which compresses the plasma. Diamagnetic effects make the toroidal field change sign, or "self-reverse", in the outer region of the plasma with respect to its direction at the centre — the pitch of the helical field lines reverses (Fig. 12.10). The toroidal field, B_ϕ, is weaker than in the tokamak and is comparable in strength with the poloidal field, B_θ. As a result, the field lines spiral around the magnetic axis many times in one journey around the torus.

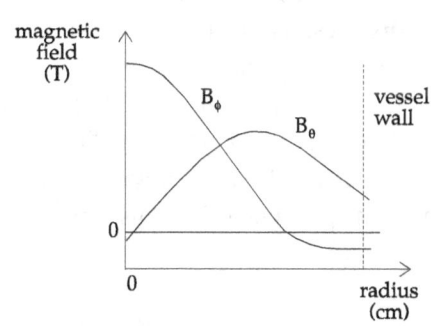

Fig. 12.10   RFP magnetic fields

The self-reversal process improves stability. Taylor explained it in terms of plasma relaxation to a state of near-minimum magnetic energy. After an initial unstable phase in which it moves, often violently, and dissipates energy, the plasma relaxes to a highly stable quiescent state which is limited by the decay of the reversed field. This quiet, "resting" state led to improved confinement.

RFP configurations are described by two parameters: the field-reversal ratio and the

212

pinch ratio. The field-reversal ratio, F, is the ratio of the toroidal field at the wall to the spatial average of the toroidal field:

$$F = \frac{B_{\phi \, wall}}{\langle B_\phi \rangle} \qquad (12.4)$$

F is therefore the normalised toroidal magnetic field. The pinch ratio, θ, is the ratio of the poloidal field at the wall to the spatial average of the toroidal field:

$$\theta = \frac{B_{\theta \, wall}}{\langle B_\phi \rangle} = \frac{\mu_0 I}{2\pi a \, \langle B_\phi \rangle} \qquad (12.5)$$

since, for a long straight conductor, $B = \mu_0 I / 2\pi r$. When the pinch ratio is greater than about 1.2 the relaxation of the plasma is accompanied by the generation of a reversed toroidal field in the outer regions of the plasma. When θ exceeds a second critical value of about 1.6 the final state is helically deformed. As θ increases, for example by increasing the plasma current, F must decrease. Values for F and θ, when plotted on an F-θ diagram (Fig. 12.11), show marked similarities between RFPs operating under widely differing conditions. The solid line shows the concentration of points, indicating that F is a function of θ, irrespective of the initial conditions. The dotted lines indicate the limit of scatter.

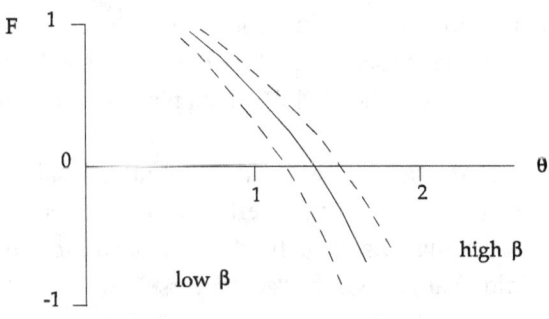

Fig. 12.11   Example of F-θ diagram

Reversed-field pinch machines were generally much smaller than tokamaks: ETA BETA II in Padua, Italy, which produced its first results in 1979, had a major radius of 65 cm and a minor radius of 12.5 cm; HBTX (High Beta Toroidal Experiment — $\beta \approx 10\%$), in operation throughout the 1980s at Culham in the UK, had major and minor radii of 80 cm and 26 cm respectively. By the late 1980s, when work on RFPs began to decline, electron temperatures of 0.8 - 1 keV with τ_E around 0.5 - 1 ms were being achieved.

b)   Spheromak

The spheromak is a spherically-shaped magnetic confinement system with an aspect ratio which is close to one. The magnetic field lines are closed as in the tokamak, but, unlike tokamaks and RFPs, the containment vessel is not ring-shaped — there is no hole in the middle for conductors or vacuum vessel walls to pass through. This feature could make the construction and maintenance of fusion reactors much simpler if problems such as confinement and raising plasma temperatures could be overcome. Experiments began on spheromaks in 1979 but because temperatures were consistently much lower than tokamaks and RFPs, interest waned in the 1990s.

In a typical spheromak, the plasma forms in a coaxial gun in line with the geometric axis (Fig. 12.12(a)). It is injected as a central column of plasma into the containment vessel, or flux conserver, by its own $J \times B$ force and carries with it the poloidal field provided by the gun solenoid (Fig. 12.12(b)). The plasma then relaxes, rather like a smoke ring, into the spheromak configuration: a ring or *annulus*, as shown in Fig. 12.12(c). This relaxation process amplifies the poloidal magnetic field and generates a toroidal field current.

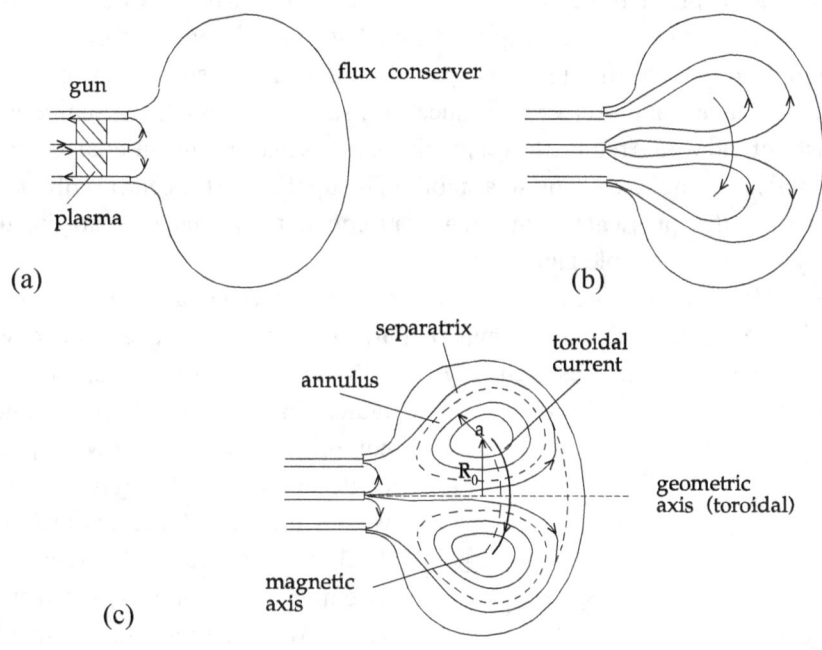

Fig. 12.12 gun-injected spheromak — geometry and formation (cross-section)

Unlike the tokamak, there are no toroidal field coils. The toroidal magnetic field is generated entirely by plasma currents and currents induced in the wall of the containment vessel. It therefore vanishes outside the plasma, ideally at the wall. Toroidal and poloidal magnetic fields are of similar magnitude. The magnetic field consists of nested toroidal surfaces in the annulus surrounding the magnetic axis. The separatrix divides the closed flux surfaces of the annulus from the open flux surfaces which remain attached to the gun. It occurs within the plasma volume in the spheromak, unlike in tokamaks and RFPs where it lies outside the plasma.

12.6.3   Spherical tokamaks — the latest design

Tokamaks built in the early 1960s had a very open ring-shape, rather like a car tyre, and aspect ratios between 4 and 5. Later versions were more compact, with a "D"-shaped poloidal cross-section. JET, for example, has an aspect ratio of 2.5. A spherical tokamak is the low aspect ratio limit of the tokamak. In 1987 the Heidelberg spheromak was fitted with a central current-carrying rod to produce tokamak-like plasmas — imagine a thin rod along the geometric axis of the spheromak in Fig. 12.12(c). A similar experiment was conducted on the UK's spheromak, SPHEX. Tokamak-like plasmas were obtained in both cases, but electron temperatures were only about 2.5 eV. The plasma remained cold.

Theoretical modelling indicated that tokamaks with aspect ratios close to 1 would have different plasma properties from conventional tokamaks. Higher values of β had been found on "D"-shaped machines. Theory predicted that, since β depended on plasma shape, the highly elongated and more triangular-shaped plasmas of a low aspect ratio tokamak would provide higher β values and greater resistance to instabilities. Such a machine would be easier to construct and would require lower magnetic fields than a conventional tokamak. The toroidal magnetic field, needed for plasma stability, is supplied by the central column and can be ten times less for the spherical tokamak than for a conventional machine carrying the same current, implying a considerable gain in efficiency.

The world's first high-temperature spherical tokamak began operating in 1991 at Culham, in the UK. Named START (Small Tight-Aspect Ratio Tokamak), the device was 2 metres in diameter and 2 metres high (Fig. 12.13). The plasma was induced at a major radius of about 40 cm, and then compressed into a low aspect ratio configuration. Aspect ratios as low as 1.2 were achieved although, with R_0 between 30 cm and 37 cm and a between 22 cm and 28 cm, R_0/a generally lay between 1.3 and 1.45. At these ratios, the plasma became separated from the central column. Electron temperatures around 500 eV and a world-record tokamak β of 40% were achieved.[f] The high beta value confirmed the increased efficiency while the low aspect ratio provided the predicted greater stability and better confinement.

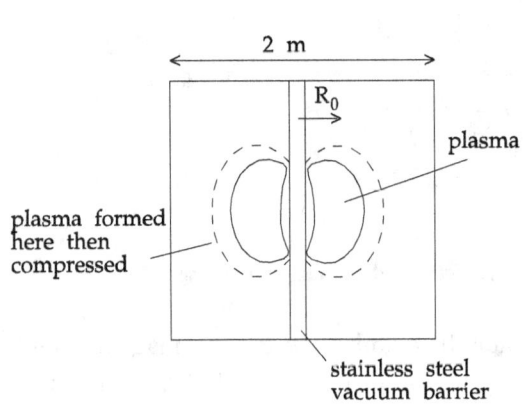

Fig. 12.13   START plasma (sketch)

START was built mainly from surplus and recycled equipment to test the theory. It shut down in March 1998 but had been so successful that a custom-made replacement, MAST (Mega-Ampère Spherical Tokamak — designed to produce a 2 MA plasma current) was built at Culham to continue the work. MAST is 4 metres in diameter and over 4 metres high, with

[f]   The previous record was 12.6% on DIII-D in San Diego, USA, with an aspect ratio of about 2.8.

a major radius of 85 cm and a minor radius of 65 cm. The first toroidal plasmas were achieved in December 1998 and by 2002, H-mode plasmas were regularly achieved under a range of plasma conditions.[10]

The results from START indicate that the spherical tokamak is the equal of conventional tokamaks. Its simpler design, greater efficiency and resistance to major disruptions, which in a conventional tokamak would lead to sudden termination of the plasma current, aroused international interest and similar devices are appearing around the world.

12.7   Heating the plasma

The simplest way to create and heat a plasma is to pass an electric current through a gas: the charged particles gain energy in response to the electric field and collide with other particles. The plasma current plays a dual role in experimental devices: it contributes to the confining magnetic field, such as the poloidal field in the tokamak, and it heats the plasma through ohmic or Joule heating.   Since plasma resistivity decreases with increasing temperature, ohmic heating cannot produce reactor temperatures. Additional heating is needed to raise the temperature above about 1 keV. Methods include:

1.   *Neutral beam heating.*   Neutral beam heating was first used in 1972 on the CLEO tokamak at Culham using a source developed at the Oak Ridge National Laboratory in Tennessee. Hydrogen or deuterium ions are accelerated to energies of between 30 keV and 100 keV before being passed through a gas. This neutralises the particles, allowing them to cross the magnetic field around the plasma. Once inside the plasma, they are re-ionised but retain the high energy acquired outside, transferring it to the plasma via collisions with the plasma particles. If the neutral beams are injected tangential to the plasma major radius, they can drive a net plasma current, reducing the need for an induced current. This method of supplying additional heating appears, from experiments on START, to be a useful method of heating spherical tokamaks.

2.   *Radio frequency (r.f.) heating.*   In 1952, Peter Thonemann, working at the Clarendon Laboratory in Oxford, suggested using electromagnetic waves and radio frequency power to produce circulating currents in a plasma. By 1964, it was known that large amounts of power could be coupled to an "ion cyclotron wave" from a r.f. generator.[11]   An ion cyclotron resonance heating system was installed on the C-Stellarator at Princeton in 1965, producing temperatures of about 50 eV in plasmas with densities in the region of $10^{18}\,m^{-3}$.

Electromagnetic waves from an antenna close to the containment vessel wall deliver radio frequency signals to the plasma. If the correct resonant frequencies are used, charged particles in the plasma acquire energy from the waves via processes such as Landau damping. Ions are heated by radio waves in the 30-200 MHz region (the ion cyclotron frequency or its harmonics). Waves in the 30 - 150 GHz range — the electron cyclotron frequency — heat electrons by resonating at their gyro-frequency and its harmonics. Other useful frequencies include Alfvén waves in the low MHz range and the ion plasma frequency between 1 and

4GHz. The accelerated particles transfer their energy to the plasma through collisions with other particles, heating the plasma in the process. If the radio waves are made to travel in a toroidal direction, momentum is transferred to the electrons, producing a toroidal plasma current. Since radio frequency generators can be operated continuously, a steady-state tokamak operation becomes feasible.

3.   In toroidal vessels, *magnetic compression* across the minor radius can be achieved by increasing the toroidal field on a time scale comparable to the energy confinement time. A special case is the pinch effect where the magnetic field is self-induced by a current in the plasma. Particle energies increase because of the compression, as with compression of an ordinary gas. If a plasma column is compressed slowly with no loss of particles, its temperature and volume V are related by:

$$TV^{\gamma-1} = constant \qquad (12.6)$$

The symbol γ is the ratio of specific heats and is determined by the number of degrees of freedom N via the condition $\gamma = (2 + N)/N$. In three dimensions, $\gamma = 5/3$ and equation (12.6) becomes $TV^{2/3}$ = constant. As V increases, T decreases. Such compression is quite successful at heating the plasma, but increasing the toroidal field requires a considerable input of power.

12.8   Turbulence

Collisions between particles are unpredictable and produce random fluctuations in plasma behaviour. Instabilities can amplify these random variations, producing large-scale unpredictable motion, or turbulence, within the plasma. Plasma turbulence is similar in many respects to ordinary fluid turbulence. It occurs in both astrophysical plasmas and magnetically-confined laboratory plasmas, where it may be the underlying cause of many instabilities and oscillations which limit confinement.

Charged particles in a turbulent plasma will behave differently from those in a static, stable plasma. The particles can be accelerated by the turbulence and both particles and energy may be transported across the magnetic field. Tokamak experiments in the 1980s showed that thermal energy and momentum transport across the magnetic field is many times faster than was predicted by the theory, which only considered the effects of Coulomb collisions. Although the mechanisms involved are not fully understood and there is, as yet, no generally-accepted fundamental theory of turbulence, the theoretical models of plasma behaviour now take the effects of plasma turbulence into account.

12.9   The prospects for fusion

Whatever system is eventually chosen for a fusion reactor (assuming it is a magnetically-confined system), a decision must be made about whether to let the fuel burn out, then reload and reheat the reactor after each pulse, or to try to continuously refuel the plasma and remove the waste products. These waste materials include the main product from the reaction, helium, and the build-up of radioactive isotopes, such as tritium.

12.9.1   ITER

In the late 1980s, international discussions began on a successor to machines such as JET and TFTR which would establish the technological feasibility of fusion and study the problems of tritium production in the lithium blanket and radiation damage to the vacuum vessel walls. The size of reactor believed to be necessary and the costs involved in building and maintaining it mean that no single country can afford to take on such a project. It requires long-term worldwide commitment and collaboration to prove that a magnetic confinement device will be successful as a nuclear fusion reactor.

Talks began in January 1992 between the four main fusion groups — USA, Russia, Japan and Europe — to design a fusion experiment which would operate under reactor-relevant conditions: the International Thermonuclear Experimental Reactor (ITER). It would be twice as big as JET and would take around 12 years to complete, at an expected cost (then) of £3 billion. It would be the first fusion device to produce thermal energy at the level of an electricity-producing power station. ITER, it was hoped, would demonstrate the scientific feasibility of magnetic fusion by achieving ignition, in that fusion energy released in the plasma would be sufficient to maintain its temperature. Ignition occurs when enough fusion reactions take place for the process to become self-sustaining. No additional external heat is needed. While it is not essential in a fusion power reactor, operating close to ignition is desirable to minimise external power requirements.

Discussions continued through the 1990s, culminating in the presentation in July 1998 of a mature design and estimate of the costs involved. These had doubled. The US Congress refused to continue supporting the project and the American team pulled out, reserving the option to return at a later date. Those remaining set about developing a new design for a minimum cost machine in which a burning plasma can be achieved. The result is ITER-FEAT (ITER Fusion Energy Advanced Tokamak), with a major radius of about 6 m and an aspect ratio in the range 2.5 to 3.5. Experiments on JET since 1997 have contributed to the design and development of the ITER project and used D-T plasmas in an ITER-like configuration. The new spherical tokamaks have also provided input to the programme.[12]

Exploratory discussions on implementation began in 2000 between the remaining partners: Europe, Japan and Russia, together with Canada, to select the construction site and decide how the costs would be shared. These discussions culminated in the announcement on 28 June 2005 that the partners, now including China, South Korea and the USA, had agreed to build ITER at Cadarache in south eastern France, with the aim of being operational by 2016. The main research facility and a later prototype commercial reactor will be built in Japan. The treaty launching the project was signed in Paris in November 2006.

12.10   Inertial confinement

In magnetic confinement, electromagnetic fields attempt to contain a fairly low-density plasma for long enough to enable sufficient fusions to take place, as expressed by

Lawson's criterion. The confinement time is limited by diffusion of particles and energy. Inertial confinement has no external means of containing the plasma. Instead, high-intensity lasers compress and heat fuel pellets of deuterium and tritium, creating a high-density plasma, with particle densities in the region of 10^{30} m^{-3}, which then expands under its own internal pressure. Lawson's criterion, $n\tau_E > 10^{20}$ s m^{-3}, still applies. The confinement time is limited by the inertial properties of the pellet — how long it can resist the outward forces produced by compression and heating and remain in its original form — hence the name "inertial confinement".

The fuel pellet is a sphere less than 1 mm in diameter and consists of a thin shell made of glass or similar material, containing the deuterium and tritium (Fig. 12.14(a)). It is injected into a vacuum chamber and the outer surface of the shell is irradiated uniformly by high-power lasers. Dust on any of the optics would cause the laser to damage the equipment, so the air in the laboratory must be kept scrupulously clean. When laser energy is focused onto matter, the material heats up rapidly and is ionised to form a plasma. In the early years of inertial fusion research, infra-red lasers were used, but by the late 1970s it had been found that less than half of the laser energy was absorbed by the fuel pellet. Short wavelengths transfer energy more efficiently than long wavelengths.

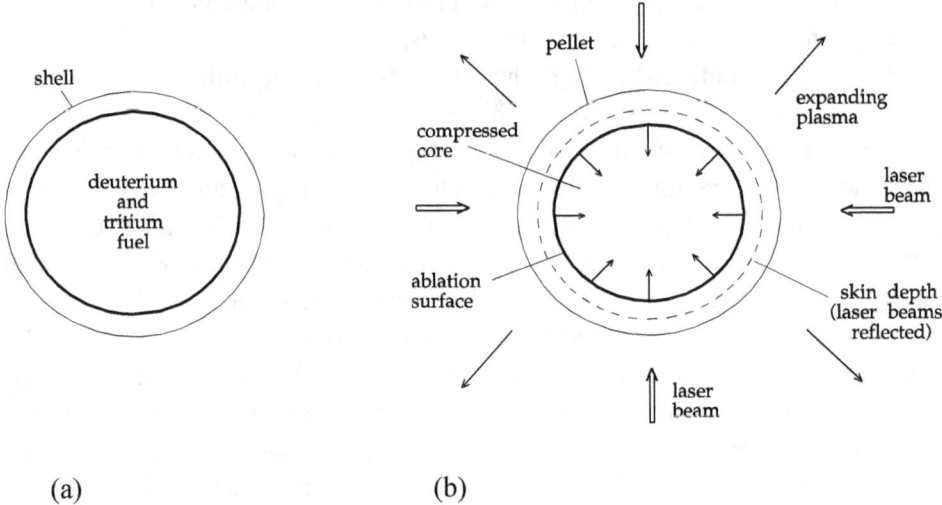

(a)   (b)

Fig. 12.14 Inertial confinement

The material forming the outer shell absorbs some of the laser energy and is heated rapidly to around 10^8 K (approximately 10 keV), increasing the pressure on the fuel inside (Fig. 12.14(b)). The outer surface of the shell evaporates (*ablates*) and expands as a plasma into the surrounding vacuum with a velocity in the region of 10^6 m s^{-1}. The laser beams can only penetrate a certain distance into the pellet: the skin depth, given by equation (6.20). Compared to the earlier (long wavelength) infra-red lasers, modern short-wavelength lasers can penetrate further into the shell, where electron densities are higher. Energy transfer from beam to plasma via electron and ion collisions is therefore more efficient. At the point where the angular frequency of the laser beam, ω, equals the electron plasma frequency, ω_{pe}, the laser beam is reflected. Electron density at this point is that given by equation (6.19). The energy which has been absorbed by the electrons at this "critical surface" is transported inwards to the ablation surface and outwards into the expanding plasma from the outer shell. The ablation surface marks the layer at which the plasma from the outer shell is created. Much of the energy deposited by the laser is converted into the outwardly-directed kinetic energy of this plasma.

Just as a rocket moves forward by expelling hot gases backwards, so the exhaust from the expansion of the outer shell accelerates the inner surface of the shell inwards, with equal and opposite force. As the shell implodes, it compresses and heats the fuel inside. During this time, some 10^{-10} s, the contents of the pellet are held together by inertia. Once the ignition temperature of about 10 keV is reached, the nuclei in the fuel pellet acquire sufficient energy to penetrate the Coulomb potential barrier. Nuclear reactions begin (equation 11.4) and the fuel pellet expands under the internal pressure generated by the fusion reactions, forming a hot plasma. In the short time it takes for the plasma to expand, the nuclear reactions produce more energy than is required to create and heat the plasma. It is this excess energy which could be used to generate electricity. The neutrons released in the D-T reaction pass through the surrounding plasma and can be collected by a lithium "blanket" surrounding the chamber or by fluid circulating in the reactor chamber. Here, the 14.1 MeV of energy each carries can be deposited and turned into electricity by conventional means. The remaining 3.5 MeV of energy carried by the helium nuclei (α-particles) is released in collisions with the surrounding fuel particles. When the energy returned to the fuel by the helium nuclei is greater than that needed to compress the fuel, the pellet is said to have ignited.

The confinement time, τ_E, is determined by the time it takes for the plasma to expand. This is inversely proportional to the ion sound speed, v_s, (equation 6.24), which is around 10^6 m s^{-1} at 10 keV. By the time the radius, r, of the plasma has expanded by one quarter, the reaction rate has decreased by a half and so the confinement time, τ_E, is considered to be approximately $0.25 r v_s^{-1}$. Once the reaction is completed, a new pellet is introduced into the vacuum chamber and the process is repeated.

Ideally, the pellet is a perfect sphere, since any departure from spherical symmetry lessens the effect of the implosion. Shock waves may develop as the shell implodes, greatly reducing compression. Alternatively, the shell may break up before compression is achieved. Asymmetries in the shell can produce plasma instabilities, in particular the Rayleigh-Taylor instability (Section 8.4.1), which amplify the effect of laser beam or pellet imperfections and disrupt the expansion of the plasma. For uniform implosion of the shell, laser intensity around the sphere must be uniform, but high-power laser beams often have variations in intensity across the beam. This can produce differences in acceleration over the surface of the shell, leading to imperfect implosion and compression. Inertial confinement is thus prone to rapidly-growing instabilities which result from our inability to produce perfect pellets and completely uniform laser beams.[13]

12.11   Summary

1.   Lawson's criterion and the ignition temperature together determine the feasibility of a fusion reactor.

2.   Magnetic confinement aims to contain a plasma within a vacuum vessel for long enough to enable fusion to take place in large quantities. Containment systems vary depending on the topology of the magnetic field lines. Closed systems, in particular the tokamak, are currently favoured.

3.   Two important confinement parameters are the pressure ratio, β, and the safety factor, q. The pressure ratio measures the stability of the plasma: for long-term stability, plasma pressure must be less than magnetic field pressure at the plasma surface. The safety factor indicates the helicity of the field lines: how many toroidal circuits a field line makes before one poloidal circuit is completed. The variation from one magnetic flux surface to another (the magnetic shear) is important in maintaining stability.

4.   Tokamaks have a very strong toroidal magnetic field parallel to the plasma current which is used to control MHD instabilities. Pinches have much weaker toroidal magnetic fields, comparable in strength with the poloidal field. Field lines therefore spiral many times in one circuit of the torus.

5.   As an alternative to magnetic confinement, inertial confinement uses lasers to implode high-density pellets of deuterium and tritium. Imperfections in both pellets and laser beams are a major source of plasma instability.

References

1. Ware, A.A.: *Phil. Trans.*, **243**, p.197-220 (1951)
    Cousins, S.W., & A.A. Ware: *Proc. Phys. Soc. B.*, **64**, p.159-66 (1951)

2. *Plas. Phys. & Cont. Fusion*, **30**, (14) p.1991-2102 (1988)

3. Artsimovich, L.A.: *Nucl. Fusion (Internat.)*, Suppl. pt.I, p.15-20 (1962)

4. Peacock, N.J., *et al.*: *Nature*, **224**, p.488-90 (31 Oct. 1969)

5. Pease, R.: *Nature*, **354**, p.95 (14 Nov. 1991)
    Physics Today: **47**, p.17 (Jan. 1994); p.55 (May 1997); **52**, p.45-6 (July 1999)

6. Raizer, Yu. P.: *Sov. Phys. Uspekhi*, **8**, p.650-73 (Mar/Apr. 1966)
    Dawson, J.M.: *Phys. Fluids*, **7** (7), p.981-7 (1964)
    Basov, N.G., et al.: *Sov. Phys. JETP*, **24** (4), p.659-66 (1967); *Appl. Optics*, **6** (11), p.1814-7 (1967)

7. Lawson, J.D.: *Proc. Phys. Soc. B*, **70**, p.6-10 (1957)

8. *Physics Today*, **47**, p.17 (Jan. 1994)

9. Taylor, J.B.: *Phys. Rev. Lett.*, **33**, p.1139 (1974)

10. Field, A.R., *et al.*: *Plas. Phys. & Cont. Fusn.*, **44**, p.A113-21 (May 2002)

11. Thonemann, P.C., W.T. Cowhig & P.A. Davenport: *Nature*, **169**, p.34-5 (1952)
    Hooke, W.M., & M.A. Rothman: *Nucl. Fusn. (internat.)*, **4**, p.33-47 (1964)

12. *Plas. Phys. & Cont. Fusn.*, **42** (supp. 12B), p.B385-96 (Dec. 2000)

13. Nuckolls, J., *et al.*: *Nature*, **239**, p.139-42 (15.9.1972)
    Craxton, R.S., R.L. McCrory & J.M. Soures: *Sci. Am.*, **255** (2), p.60-71 (Aug. 1986)
    Town, R.P.J., & A.R. Bell: *Phys. Rev. Lett.*, **67**, p.1863-6 (30.9.1991)

222

Chapter 13

Aurorae and sunspots

Until the early 1900s, when observation equipment improved and plasma theory developed, the composition of stars such as the Sun, and the origin of the strange lights of the aurora mystified observers. Our present understanding is that (ignoring dark matter) a significant part of the universe consists of largely-collisionless plasma. Solid matter occurs in planets, cometary heads, meteorites, interstellar dust and dark nebulae, while the gaseous state is found in planetary atmospheres and non-ionised interstellar gases. Of the rest, the stars, large parts of planetary atmospheres, and much of the interplanetary and interstellar gases are in a state of partial or even total ionisation.

References to the aurora appear in the ancient writings of both East and West, but because their cause was unknown, the lights in the sky were regarded with fear and superstition, believed by many to be the spirits of the dead. Galileo Galilei and the French mathematician and astronomer Pierre Gassendi are variously credited with first using the term *aurora borealis* (Latin for "dawn of the north") around 1620. Gassendi suggested that whatever caused the auroral displays must occur at great height since widely-separated observers reported the same effects. One popular theory was that reflections from ice crystals in the air at high latitudes were responsible.

Sunspots, the magnetised cool spots in the solar surface, are also mentioned in early Chinese records. The first recorded observations using telescopes were made during the winter of 1610-11, when they were thought to be dark bodies moving above the surface of the Sun. From the mid-1640s to about 1715 reduced solar activity prevented further investigation as both auroral displays and the number of sunspots declined. This period is now known as the Maunder Minimum — named after the British astronomer E.W. Maunder who noticed, in 1890 while examining old records, that virtually no sunspots had been observed between 1645 and 1715. This coincided with freak weather conditions in England in the 1680s when "frost fairs" were held in London in winter on the frozen River Thames.

In 1716, English astronomer Edmund Halley (1656-1742) observed his first auroral display.[1] He proposed that the cause was magnetic particles issuing from the Earth's magnetic poles: the lights had never been seen near the equator, and appeared more frequently in Iceland and Greenland than in Norway — the northern magnetic pole, at that time, was situated close to Greenland. French philosopher J.J. de Mairan, writing in 1731, disputed Halley's theory and poured scorn on the then popular idea that the aurora was a reflection of polar snow and ice. He believed that the aurora was linked to the solar atmosphere, having noticed a connection between auroral displays and the return of sunspots.[2] Nevertheless, Halley's idea that the aurora was caused by geomagnetic effects became widely accepted, and was confirmed on 5 April 1741 by O. P. Hjorter, supported by George Graham, a London instrument maker, who had observed strong geomagnetic activity in London on that day. Some thirty years later, J.C. Wilcke noticed that auroral rays extend upward in the direction of the magnetic field.

Benjamin Franklin's work on lightning in the early 1750s led to increased interest in atmospheric effects. Tiberius Cavallo announced to the Royal Society of London in December 1776 that there was always some electricity in the atmosphere and it was always positive.[3] Louis Guillaume Lemonnier, a French physician, reported that the atmosphere is always electrically charged, even when there are no clouds.

In 1790, English chemist and physicist Henry Cavendish (1731-1810) calculated the height of a luminous arch of the aurora to be between 52 and 71 miles [83 - 114 km]. John Dalton, the founder of atomic theory, took reports of auroral observations in northern England and Scotland in March 1826 and used triangulation to estimate the height of a luminous arch as being almost 100 miles [160 km] above the Earth's surface.[4] We now know that the heights of aurorae vary between a lower limit at about 98 km and an upper one at 300 km, with maximum activity occurring around 110 km.

The Sun was also under investigation. William Herschel reported in 1801 that he had observed sunspots with clouds moving across them. He identified a "planetary atmosphere extending to a great height and of great density, being subject to agitations such as with us are occasioned by winds. The gases of the solar atmosphere are transparent," he wrote.[5]

In 1826, Heinrich Schwabe (1789-1875), a pharmacist in Dessau near Leipzig in eastern Germany, began recording sunspot numbers. After collecting data for 25 years, he reported in 1851 that the intervals between sunspot maxima varied between 7 and 13 years, with an average period of about 10 years. Having heard of Schwabe's work, Dr. Rudolph Wolf began a similar series of observations in Zurich in 1849 which revealed a minimum in the number of sunspots in 1856 and 1867, an interval of 11 years. English amateur astronomer Richard Christopher Carrington (1826-75) reported in 1858 that the latitude of sunspots changed with the solar cycle. Throughout the two years before the solar minimum in February 1856, the spots were confined to an equatorial belt between 20° latitude north and south of the solar equator. Shortly after the minimum, that belt disappeared and two (apparently) new belts were seen between latitudes 20° and 40° in both hemispheres.[6]

During the 1850s, Edward Sabine published a series of papers[7] on the large-scale magnetic disturbances commonly known as "magnetic storms". He had predicted in 1843 that, by studying their effects on the local magnetic direction and force over a long period of time, these magnetic storms would be shown to be periodic and could therefore be classified as a particular magnetic phenomenon. By 1852, using hourly observations of the magnetic declination — the angle on the Earth's surface between the directions of the geographic and magnetic north pole — in various British colonies, Sabine had found regular daily and annual fluctuations, together with a periodic variation of about ten years. This corresponded, both in duration and timing of maximum and minimum variation, with the fluctuations in frequency and magnitude of solar spots recently announced by Schwabe. Whether this was mere coincidence or an indication of a causal connection he could not say.

Solar eclipses provided an opportunity to observe features of the Sun not normally visible, and the eclipses of the mid-1800s in particular were soon followed by several announcements. French astronomer Dominique-Francois-Jean Arago (1786-1853) made an extensive study of the 1842 eclipse and declared that the Sun is wholly gaseous, describing what later became known as the chromosphere as "clouds at the base of the corona". He suggested that prominences were simply clouds floating in the Sun's atmosphere and that these clouds contained the sunspots.[8]

In 1852, Warren de la Rue, a pioneer of astrophotography, obtained the first good photographs of the Moon and, five years later, of the Sun. His photographs of the total eclipse of 1860 in Spain, together with the observations of the Italian Jesuit astronomer Father Angelo Secchi (1818-78), finally proved that prominences do in fact belong to the Sun and not to the Moon. These red streamers of glowing gas were first described in detail in 1733 by the Swedish observer Birger Vassenius, having witnessed the solar eclipse in Gothenburg. He, like many later observers, thought they were lunar phenomena, but de la Rue, in a lecture to the Royal Society in 1862, declared that the progressive covering of the prominences on the east, in the direction of the Moon's motion across the Sun, and the gradual uncovering of fresh prominences on the west, indicated that they belonged to the Sun.[9]

While working at his observatory in Surrey on 1st September 1859, Richard Carrington noticed a great flare of white light on the Sun — the first recorded observation of a solar flare. At the time of the flare, a disturbance in the Earth's magnetic field was detected at the Kew Observatory in London. Eighteen hours later, there began one of the strongest magnetic storms ever recorded and aurorae were seen as far south as Puerto Rico. (To have arrived so quickly, the cause of the disturbance must have travelled from the Sun at more than 2300 km s^{-1}, a high velocity even for a disturbed solar wind, the average velocity of which is $250 - 800$ km s^{-1}.) Scottish physicist Balfour Stewart (1828-87) described auroral displays of "almost unprecedented magnificence" accompanied by "excessive disturbances of the magnetic needle" being seen throughout Europe, America and Australia. Telegraph communication was also disrupted by the current produced in the wires.[10] A large sunspot was observed and was thought to be linked to the auroral displays and the magnetic storm.

Stewart suggested that the lower levels of the atmosphere acted as an insulating layer between the Earth, with its magnetic field, and the upper layers of the atmosphere. This region, being much less dense, became electrically conducting, forming a "secondary coil" around the Earth's soft iron core. Currents from the Sun, maintaining their general direction during the period of the storm, would, he suggested, act on the Earth's "magnetic matter". He had also observed pulsations in the Earth's magnetic field, superimposed on the main disturbance, with periods of a few minutes. We now know that the magnetosphere pulsates at a variety of intervals.

In 1861, de la Rue obtained a stereoscopic view of a sunspot, which seemed to confirm the suggestion made in 1774 by Alexander Wilson (1714-86) of Glasgow University, that sunspots are depressions in the solar atmosphere.[11] Wilson had reached this conclusion after five years of observation of the way the umbra changed around the nucleus as the spot moved across the Sun. By 1865, de la Rue, Stewart and Benjamin Loewy, working at the Kew Observatory, had concluded that sunspots exist below the level of the photosphere, which they presumed to be a gas or cloud. The relative brightness of faculae, large bright areas of the photosphere seen most easily near sunspots, indicated that "faculous matter" was ejected from the Sun, leaving behind the sunspot as a cavity. Since the central part of the spot is less luminous than the photosphere, it must, they decided, be of a lower temperature.[12]

In 1868, British astronomer Sir Joseph Norman Lockyer (1836-1920) and Pierre Jules César Janssen (1824-1907) discovered helium in the Sun's spectrum some 30 years before it was found on Earth. Lockyer declared that observations of the spectrum of solar prominences indicated that they are merely "local aggregations of a gaseous medium which entirely envelopes the Sun". He suggested the name *chromosphere* for this layer, whose thickness he had determined to be about 5000 miles [8000 km].[13]

Following the widely observed aurorae of early September 1859, Professor Elias Loomis of Yale University analysed data from several observation stations in the northern hemisphere and found that the region of maximum auroral occurrence lay in an oval around the north pole.[14] We now know that the auroral zone forms an oval band around the magnetic pole, about 20-25° from it. At that time it was widely believed that the geographical distribution of auroras was linked with that of thunderstorms, but Loomis pointed out that thunderstorms are more frequent in equatorial than in polar regions.

In 1870, after further investigation, Loomis reported that auroral displays had the same approximately ten-yearly period as sunspots. Within the zone of greatest frequency, aurorae occurred almost daily with no noticeable periodicity in number, but perhaps with periodic variations in intensity. Further away from the Pole, where the annual number of aurorae was around 20 to 25, the ten-yearly period of sunspots was distinctly visible. Auroral displays in the middle latitudes of America were generally accompanied by an unusual disturbance of the Sun's surface on the same day as the aurora. Large disturbances of the Earth's magnetism (magnetic storms) were also accompanied by unusual disturbances of the Sun's surface (sunspots) on the day of the storm. Both effects, he believed, were due to "some influence which emanates immediately from the Sun".[15] What this "influence" was would remain a mystery for many years.

This period of the 19th century was one of fascination with the effects of electrical discharges through different rarefied gases. Several researchers, beginning with William Watson in 1752, had noticed similarities between the visual effects in the discharge tube and the display produced by the aurora. In Geneva in the 1850s, Auguste Arthur de la Rive (1801-73), was investigating the effect of magnetism on a voltaic arc and passed a current of electricity through a rarefied gas, producing a luminous ring which rotated around the pole of the magnet, not unlike the aurora.[16] By then, many believed that the aurora had its source above the terrestrial atmosphere, with perhaps a cosmic origin. G.B. Donati, working in Florence in 1872, suggested that an exchange of electricity between the Sun and the planets altered the electrical state of the Earth, producing the effect.[17]

Warren de la Rue and Hugo Müller tried to assess the height of the aurora using their work on discharge tubes. In April 1880 they reported that the height of maximum brilliancy was 37.67 miles [60 km]. At a height of 124.15 miles [198.6 km] the pressure would be too low for an electrical discharge to occur. "The greatest exhaust we have produced, 0.00055mm of mercury, corresponds to a height of 81.47 miles and we have failed to produce a discharge in hydrogen at this low pressure," they declared. The colour of the discharge varied with the density of the air at the same potential: at a pressure of 62 mm the discharge was a bright red colour, like that so often seen in the aurora. This, they calculated, corresponded to an altitude of 12.4 miles. At a pressure of 1.5 mm, corresponding to 30.86 miles, the discharge was salmon-coloured. In air, the discharge at the cathode always had a violet hue and this tint in the aurora, they believed, indicated closeness to the "negative source". The following year, Eugen Goldstein suggested that solar particles might cause the aurora.[18]

In 1880, Balfour Stewart proposed that daily fluctuations in the Earth's magnetic field may be due to electric disturbances or an electrically-conducting layer of gas in the atmosphere — a similar suggestion is attributed to Gauss, in 1839. Stewart envisaged convection currents, produced by solar heating of the upper atmosphere, moving from the equator to the poles in the upper reaches of the Earth's atmosphere. These currents, he said, could be regarded as conductors moving across lines of magnetic force. Daily variations in the convection currents thus caused the daily variations in intensity of the terrestrial magnetic field. Whenever there was a small but abrupt change in the Earth's magnetism — a magnetic storm — violent earth currents, alternately positive and negative, were observed, accompanied by auroral displays. Stewart concluded that earth currents and aurorae are secondary discharges caused by sudden changes in the Earth's magnetism, suggesting that movement of convection currents in the upper atmosphere, produced by the Sun, might cause electrical phenomena which could explain the changes in terrestrial magnetism. His ideas, which Sydney Chapman later described as a bold piece of speculation, were very close to the modern atmospheric-dynamo theory, but it was left to Arthur Schuster to define the theory quantitatively. He showed, in 1889, that the daily variations in terrestrial magnetism originated outside the Earth's surface, concluding that they were caused by electric currents circulating in the upper regions of the atmosphere.[19]

So, by the end of the 19th century, it was clear that auroral displays and magnetic storms were in some way linked to disturbances in the Sun, as evidenced by sunspots; while similarities with electrical discharges suggested that aurorae were also electrical phenomena. The possibility of an electrical connection between the Sun and Earth had also been put forward, but the underlying mechanisms were still the subject of much discussion. In November 1892 at a meeting of the Liverpool Physical Society, Oliver Lodge (1851 - 1940) observed that a period of several hours elapsed between the appearance of a sunspot and the magnetic storm which resulted from it. If the sunspot was the direct cause of the storm, he reasoned, the effect should reach Earth at the speed of light; that is, within a few minutes of the sunspot's occurrence. This was not the case. Prof. George Francis FitzGerald (1851 - 1901) of Trinity College, Dublin, had, said Lodge, explained the storm by supposing that a cloud of electrified particles was projected from the Sun at the time the spot occurs. These particles, if they passed near the Earth, would create a magnetic storm. FitzGerald expanded on this idea a few months later, suggesting that if magnetic storms were due to a solar electromagnetic effect it must be confined to "some ray projected from the Sun and which happens to cross the Earth. It is difficult to see how such a ray can be produced unless it is a projection of electrical ions in some of the outlying streamers of the corona."[20]

In the late 1890s, the Norwegian physicist Kristian Olaf Birkeland (1867-1917) began a series of expeditions to northern Norway to study the aurora. The data on the associated magnetic disturbances which he collected during his 1902-3 expedition led him to the view that large electric currents flowed along magnetic field lines during auroral displays. Borrowing from the ideas of Crookes and Thomson regarding cathode rays, he tried to verify his theories in the laboratory using a magnetic dipole inside a model of the Earth, an instrument which he called a *terrella*. When it was placed in a vacuum tube and exposed to cathode rays, he found that electrons directed towards the terrella produced luminous spirals and other patterns similar to the auroral effects seen during magnetic storms. He proposed that charged particles from the Sun, in the form of cathode rays, were responsible for the auroral displays.[21] His fellow countryman, Carl Størmer, developed a mathematical analysis of Birkeland's ideas and observations, producing models (including a magnetic "bottle") for the theoretical paths of the diverted rays using the differential equations of cathode rays in a magnetic field.[22] Both men's ideas were way ahead of their time. It would be well into the 1950s before Størmer's contributions were fully appreciated, and the late 1960s before Birkeland's theories were understood and accepted.

Meanwhile, discussion continued about the cause of the daily and periodic variations in terrestrial magnetism. In 1892, Lord Kelvin decided that magnetic storms could not be due to direct solar action because of the enormous energy which the Sun would have to supply,[23] a view which was widely supported for several years. Stewart and Schuster's theory that electrical currents in the upper atmosphere caused the daily variations in terrestrial magnetism was also becoming generally accepted. In 1902, the year after Marconi's transmission of radio signals across the Atlantic Ocean, Sir Oliver Heaviside (1850-1925) and Arthur Edwin Kennelly (1861-1939) drew on this idea when they (separately) predicted the presence of an electrically-charged layer in the upper atmosphere. This layer would, they suggested, enable electromagnetic waves to be reflected around the curved surface of the Earth and beyond the horizon. By late 1910, Schuster had concluded that, if Birkeland's idea was correct — that the connection between solar outbursts and magnetic storms on the Earth was due to an ejection of charged particles by the Sun — it confirmed his long-held belief that the particles increase the ionisation of the outer regions of the atmosphere. This, he said, allowed the ever-present electromotive forces due to the Earth's rotation to locally increase the intensity of the electric circulation. He was, however, far from convinced that magnetic storms were caused by the direct magnetic action of swarms of electrified particles travelling as cathode rays from the Sun.[24] In Catania in Sicily, Riccò, who had been studying magnetic storms since 1882, calculated that the supposed rays emitted from the Sun travelled to Earth with a velocity of 900 - 1000 km s^{-1}.[25] This is now considered to be the maximum speed of what has become known as the solar wind.

American professor George E. Hale, working at the Mount Wilson Solar Observatory in California in 1908, discovered the magnetic fields associated with sunspots. Photographs of the Sun's surface, using the hydrogen line H_α, indicated that all sunspots are vortices and revealed the existence of currents in the solar atmosphere. A gyratory motion in sunspots and a possible magnetic effect had been suggested by several earlier observers. Hale proposed that segregation of positively- or negatively-charged particles caught in the stream of a solar vortex would produce magnetic lines of force at right angles to the plane of the vortex. Laboratory experiments indicated that such a field would split the spectral lines into circularly and oppositely polarised components — the Zeeman effect. Both Hale and Zeeman had detected this doubling and sometimes tripling of the lines in the Sun's spectrum.[26]

Sidney Chapman, at Trinity College, Cambridge, outlined his first theory of magnetic storms in 1918. He suggested that, since all great magnetic storms begin almost simultaneously over the whole Earth, their cause might be a system of electric currents flowing horizontally in the upper atmosphere and inducing currents within the Earth. During the first brief phase of the storm the magnetic field strength increases. This is followed by a decrease of much greater amplitude which lasts for several hours, and then a period of recovery lasting several days. Chapman proposed that directional changes in the atmospheric currents might be the cause of the fluctuations. The storm itself, he postulated, was generated by a singly-charged beam of solar particles entering the Earth's atmosphere, an idea previously rejected by Schuster. Frederick Lindemann pointed out that mutual electrostatic repulsion would destroy such a beam and suggested instead equal numbers of positively- and negatively-charged particles — what we now call plasma.[27]

In 1923, E.A. Milne suggested that particles might be emitted from stars such as the Sun not as a beam but radially, from the entire surface. Six years later and now at Imperial

College, London, Chapman and his student Vicenzo Ferraro built on this idea when they proposed that an expanding cloud of charged particles from the Sun was the primary cause of magnetic storms and auroral displays. In discussing the electrical state of this flow of particles, Chapman and Ferraro employed the same equations that Langmuir had recently used to define the basis of plasma physics, but made no reference to either Langmuir or plasma.[28]

By 1930 the two men felt able to outline the main sequence of events in a magnetic storm, although a mathematical analysis still had to be developed. In a series of papers over the next few years they laid the foundations of our understanding of the processes involved.[29] They suggested that magnetic storms were primarily caused by the approach of an electrostatically-neutral stream of ionised solar particles (plasma) towards the Earth. Electric currents flowing parallel to the Earth's surface would be induced in the outer regions of the geomagnetic field as the stream entered and compressed the Earth's magnetic field, the compression being detected at ground level as a sharp increase in the terrestrial magnetic field. The stream itself would carry neither a significant current nor an appreciable magnetic field. On reaching the Earth, Chapman and Ferraro estimated the ion density of the stream would be between 2×10^7 and $2 \times 10^{15} \, \mathrm{m}^{-3}$ with a speed of at least $10^6 \, \mathrm{m\,s}^{-1}$. It seemed likely that a westerly current was set up around the Earth, the gradual dissipation of this ring current corresponding to the final (recovery) phase of the storm.

Ferraro continued to ponder on how the charged particles were transported from Sun to Earth. In April 1937, in a paper on the Sun's magnetic field, he suggested that if the field was not as severely constrained as was then thought, it would not vanish at the solar surface. Surface charges would rapidly disperse into space under the mutual repulsion of their like sign. It was probable that the density of these charged particles would be small and Ferraro imagined them "hooking" themselves on to the magnetic lines of force. The flow of charged particles from Sun to Earth would then be almost entirely along the magnetic field.[30] It was from this idea that the image developed of magnetic lines of force "frozen into" a plasma (Section 8.4.1).

While Chapman was developing his theory of magnetic storms throughout the 1920s, others were investigating the (then theoretical) conducting layer in the upper atmosphere, believed to be responsible for the daily variations in terrestrial magnetism. In Austria in 1911, Victor Franz Hess (1883-1964) made several ascents in a balloon, reaching heights in excess of 5 km to collect ionisation data. Although nowhere near the conducting layer, his results indicated that ionisation increases markedly with altitude above about 150 metres. Sir Joseph Larmor had shown that the long free path of the electrons in the upper atmosphere would produce scattering and refraction of radio waves, while a fairly sparse distribution of ions — densities in the region of $10^{10} \, \mathrm{m}^{-3}$ were suggested — would be sufficient to bend the rays round the Earth.[31]

In 1925, Smith-Rose and Barfield pointed out that the Kennelly-Heaviside layer theory could explain long-distance radio transmission and daily variations in signal intensity and direction but more evidence was needed of the layer's existence. Appleton and Barnett announced later that year that the variation in strength of short wave signals was due to interference between ground waves and waves reflected through large angles by the upper atmosphere.[32] Appleton had persuaded the BBC to vary the frequency of the Bournemouth transmitter while he recorded the strength of the received signal in Cambridge. By identifying the stronger signal produced by constructive interference between ground waves travelling direct from transmitter to receiver (*T - R* in Fig. 13.1) and waves reflected off the electrically-charged layer proposed by Kennelly and Heaviside (*TABCR* in Fig. 13.1), Appleton used

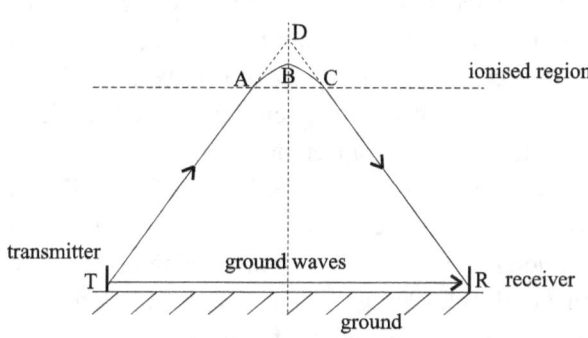

Fig. 13.1   Radio wave tracks

triangulation to estimate the night-time height of the reflecting region to be about 80 km for 400 metre length waves. This could only be an approximation, since there is no actual plane of reflection (point *D* in Fig. 13.1) — the course of the wave is "bent" progressively — and some of the time delay is due to signal-propagation time.

Eighteen months later, Russian-born Gregory Breit and American Merle Anthony Tuve confirmed the existence of the conducting layer. By sending short pulses of radio waves vertically and timing the arrival of a reflected signal, they calculated that the effective height of the reflecting layer was between 80 and 208 km, with seasonal and possibly daily variations: higher in autumn than in summer, and in the afternoon than in the morning.[33] Hafstad and Tuve later reported a marked increase in the height of the layer plus changes in echo-patterns during a magnetic storm. At night, signals were weaker, with greater scattering, similar to the effect normally observed just before sunrise.[34]

Chapman interpreted night-time reflection of wavelengths around 200 m as due to free electrons. The upper atmosphere was, he said, an ionised layer "of unknown thickness" whose lower level was 40 - 50 km above ground by day and about 90 km at night. Night-time electron density at altitudes of around 90 km was at least 10^{11} m^{-3}. The greater thickness of the daytime layer was obviously due to solar action, probably UV radiation, while progressive recombination of lower level ions after sunset caused the night-time rise of the undersurface. Auroral height measurements indicated that the streams of charged particles from the Sun (which, according to Birkeland and Størmer's theory, produced the aurora) penetrated to about 90 km above the ground in polar regions. Since aurorae were observed up to a height of several hundred kilometres, said Chapman, the zone of ionisation must be very deep.[35]

Sir Edward Appleton, meanwhile, had noticed the simultaneous reception of two or more reflected rays from one signal shortly before dawn which he attributed to a second, higher reflecting region (later named the Appleton layer). By late 1929 he had found that the height of the reflecting zones increased slowly after sunset, continuing through the night and reaching a maximum of around 126 km just before dawn. It then fell rapidly about 30 minutes before sunrise until the lower daytime value was reached. The Kennelly-Heaviside layer (which Appleton labelled the E-region, having used the letter E for the electric vector of the reflected wave) normally deflected 400 metre waves but became penetrable in the few hours before dawn. For a short time, two sets of atmospheric waves, one from the E region and the other from a higher (F) region, were simultaneously received, with the waves from the F region becoming the main one. Measurements indicated that the average height of region F was about 230 km. About 30 minutes before sunrise at ground level, the E region waves were recorded once more, as sunrise occurred in the upper atmosphere.

All of this indicated that ionisation in the upper atmosphere by solar UV radiation caused the variations in the E and F regions. After sunset this radiation ceased and recombination of ions and electrons produced the slow increase in height of the reflecting layer suggested by Chapman. The reduced intensity of the reflected waves, observed during daylight hours, was perhaps due to absorption by an ionised layer (labelled the D region by Appleton) which formed below the normal deflecting E region. By 1931 double reflections from the F region were being detected, indicating the possibility of two layers, F_1 and F_2, in the upper region (Fig. 13.2). Four years later, Appleton and Naismith reported that, unlike regions E and F_1, the seasonal variation of F_2 could not be satisfactorily explained by ionisation due to solar UV radiation. In F_2, they suggested, the cause was seasonal variations in molecular temperature, which in summer they estimated to be at least 1200 K.[36]

Earlier discussions on radio wave propagation through the upper atmosphere had ignored the effect of the Earth's magnetic field. Appleton had pointed out in 1924 that, if electrons rather than positive ions were responsible for the upper atmosphere conductivity, radio wave propagation would be greatly affected by the Earth's magnetic field, since this would alter the reflection frequency (via equation (4.16), the cyclotron frequency). In 1931, D.R. Hartree introduced magnetic fields into the general theory, developing what became known as the magneto-ionic theory of the ionosphere.[37] The term "ionosphere" was suggested by the Scottish physicist Robert Alexander Watson-Watt (1892-1973) in the early 1930s to denote the entire ionised region of the upper atmosphere.

Radio exploration of the upper atmosphere continued throughout the decade. All the information obtained prior to the 1940s came from ground-based observations of terrestrial magnetism, aurorae and meteorites, and radio-wave exploration. After World War II a new era of direct exploration began with the development of the sounding rocket — an unmanned rocket-powered vehicle which followed a sub-orbital trajectory and was used for research purposes. In 1947, Sir Edward Appleton was awarded the Nobel Prize for Physics for his work on the physical properties of the upper atmosphere, and in particular for his discovery of the F region — the Appleton layer.

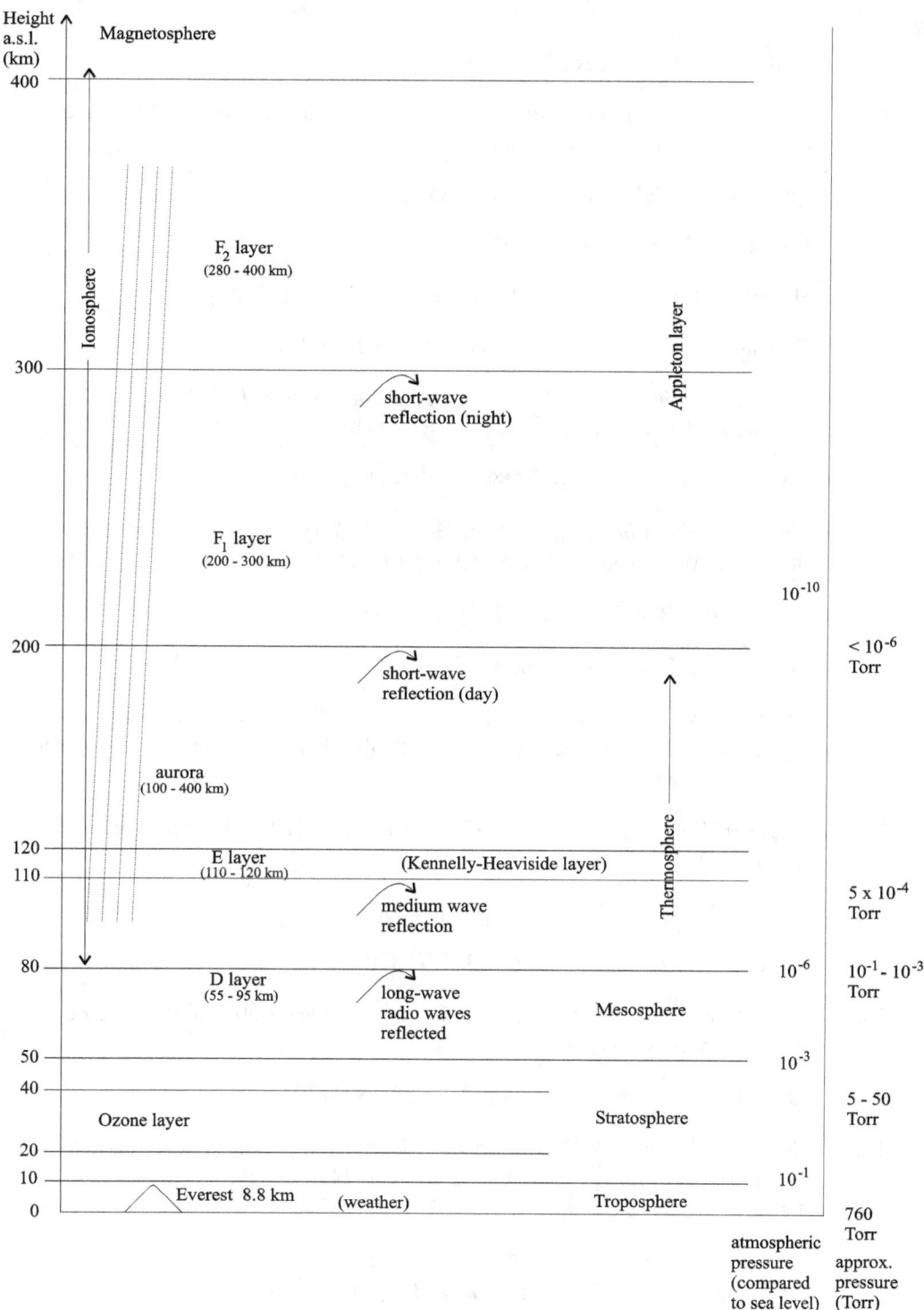

Fig. 13.2   Atmospheric zones

References

1. Halley, E.: *Phil. Trans.*, **29**, p.406-28 (1714-6)

2. Mairan, J.J.: *Traité physique et historique de l'aurora boréale*. Mémoires de l'Académie Royale des Sciences (1731)

3. Cavallo, T.: *Phil. Trans.*, **67**, p.48-55 (1777)

4. Dalton, J.: *Phil. Trans.*, (1828) p.291-302

5. Herschel, W.: *Phil. Trans.*, **91**, pp.265-318, 354-62 (1801)

6. Carrington, R.C.: *Monthly Notices, Roy. Astron. Soc.*, **19**, p.1-3 (1858)

7. Sabine, E.: *Phil. Trans.*, **141**, p.123-9, 635-41 (1851); *Phil. Trans.*, **142**, p.103-24 (1852); *Proc. Roy. Soc.*, **8**, p.40 (1856-7) [abstr.]

8. Lockyer, J.N.: *Phil. Trans.* **159** I, p.425-44 (1869)

9. Vassenius, B.: *Phil. Trans.*, **38**, p.134-5 (1733-4)
   de la Rue, W.: *Proc. Roy. Soc.*, **12**, p.58-64 (1862-3)

10. Stewart, B.: *Roy. Soc. Proc.*, **11**, p.407-10 (1860-2)

11. Wilson, A.: *Phil. Trans.*, **64**, p.1-30 (1774)

12. de la Rue, W., B. Stewart and B. Loewy: *Roy. Soc. Proc.*, **14**, p.37-9 (1865); *Phil. Mag.* 4th ser., **29**, pp.237-8, 390-4 (1865); *Roy. Soc. Proc.*, **16**, p.447 (1867-8)

13. Lockyer, J.N.: *Roy. Soc. Proc.*, **17**, pp. 91, 131-2 (1868-9); *Phil. Trans.*, **159** I, p.425-44 (1869)

14. Loomis, E.: *Am. J. Sci.*, **30**, p.89-94 (1860)

15. Loomis, E.: *Am. J. Sci.*, **50**, p.153-71 (1870)

16. de la Rive, A.A.: *Proc. Roy. Soc.*, **5**, p.659-61 (1843-50); obit. *Proc. Roy. Soc.*, **24**, p.xxxvii-xl (1875-6)

17. Donati, G.B.: *Comptes Rendus*, **74**, p.884-5 (1872)

18. de la Rue, W., and W.H. Müller: *Roy. Soc. Proc.*, **30**, p.332-4 (1879-80)
    Goldstein, E.: *Ann. d. Phys. & Chem.*, **12**, p.249-79 (1881)
    Ferraro, V.C.A.: *Adv. in Phys.*, **2**, p.265-320 (1953)

19. Stewart, B: *Nature*, **22**, p. 146-7, 202-3 (June, July 1880)
    de la Rue, W., & H. Müller: *Nature*, **22**, p. 169 (1880)
    Schuster, A.: *Proc. Roy. Soc.*, **45**, p.481-6 (1888-9); *ibid.*, **80**, p. 80-2

(1907-8); *Phil. Trans.*, **180**, p.467-518 (1889); *ibid.*, **208**, p.163-204 (1907)
Chapman, S.: *Proc. Phys. Soc.*, **37**, p.38D-45D (1925)

20.   Rowlands, P.: *Oliver Lodge and the Liverpool Physical Society.* p.83 Liverpool University Press (1990)
FitzGerald, G.F. *Electrician*, **31**, p.389-90 (1893)

21.   Birkeland, K.: *Comptes Rendus*, **148**, p.30-3 (1909); *ibid.*, p.1556-9 (1909); *ibid.*, **150**, p.246-8 (1910); *Nature*, **87**, p.483-4 (1911)

22.   Størmer, C.: *Comptes Rendus*, **142**, p.1580-3, (1906); *ibid.*, **143**, p.140-2 (1906); *ibid.*, **147**, p.733-5 (1908)

23.   Kelvin: (in President's Address) *Proc. Roy. Soc.*, **52**, p.307 (1892)

24.   Schuster, A.: *Proc. Roy. Soc.*, **85**, p.44-50 (1911)

25.   Riccò, A.: *Nature*, **82**, p.8 (1909-10)

26.   Hale, G.E.: *Nature*, **78**, p.368-9 (1908); *Astrophys. J.*, **28**, pp. 100-16, 315-43 (1908)
Zeeman, P.: *Nature*, **78**, p.369-70 (1908)

27.   Chapman, S.: *Proc. Roy. Soc.*, **95**, p.61-85 (1918)
Lindemann, F.: *Phil. Mag.*, **38**, p.669-84 (1919)

28.   Milne, E.A.: *Trans. Camb. Phil. Soc.*, **22**, p.483-517 (1923); *Monthly Notices, Roy. Astron. Soc.*, **86**, p.459-73 (1926)
Chapman, S.: *Mon. Not. Roy. Astron. Soc.*, **89**, p.456-70 (1929)
Chapman, S., & V.C.A. Ferraro: *Mon. Not. Roy. Astron. Soc.*, **89**, p.470-9 (1929)

29.   Chapman, S. & V.C.A. Ferraro: *Nature*, **126**, p. 129 (1930); *Terr. Mag.*, **36**, p. 77-97, 171-86 (1931); *ibid.*, **37**, p.147-56 (1932)
Chapman, S.: *Terr. Mag.*, **40**, p. 349-70 (1935)

30.   Ferraro, V.C.A.: *Mon. Not. Roy. Astron. Soc.*, **97**, p.458-72 (1937)

31.   Discussion on the electrical state of the upper atmosphere. 4 March 1926 *Proc. Roy. Soc.*, **111**, p.1-13 (1926)

32.   Smith-Rose, R.L., & R.H. Barfield: *Proc. Roy. Soc.*, **107**, p.587-601 (1925)
Appleton, E.V., & M.A.F. Barnett: *Proc. Roy. Soc.*, **109**, p.621-41 (1925)

33.   Breit, G., & M.A. Tuve: *Phys. Rev.*, **28**, p.554-75 (1926)

34.   Hafstad, L.R., & M.A. Tuve: *Terr. Mag.*, **34**, p.39-44 (1929)

35.   Chapman, S.: *Roy. Met. Soc. J.*, **52**, p.225-36 (1926); *Proc. Phys. Soc.*, **37**, p.38D-45D (1924-5)

36.   Appleton, E.V.: *Proc. Roy. Soc.*, **126**, p.542-69 (1930); *Proc. Phys. Soc.*, **42**, p.321-39 (1930)
    Appleton, E.V., & G. Builder: *Proc. Phys. Soc.*, **44**, p.76-87 (1932); *Proc. Phys. Soc.*, **45**, p.208-220 (1933)
    Appleton, E.V., & R. Naismith: *Proc. Roy. Soc.*, **150**, p.685-708 (1935)

37.   Hartree, D.R.: *Proc. Camb. Phil. Soc.*, **27**, p.143-62 (Jan. 1931)
    Appleton, E.V.: *Rep. Prog. Phys.*, **2**, p.129-65 (1935)

Chapter 14

Astrophysical, space and atmospheric plasmas

14.1   Introduction

Astrophysical plasmas include both stellar and interstellar plasmas; space plasmas occur outside the Earth's ionosphere, but within the solar system. Natural plasmas within the Earth's atmosphere include lightning and *St. Elmo's fire* — a visible bluish flame-like glow discharge from isolated points above the ground such as ships' masts, church spires and aircraft.[1]

Direct exploration of the Earth's plasma environment began in 1946 using V-2 rockets, but the real impetus came in the late 1950s with the development of satellites. Since then, ground-based experiments and simulations combined with satellite exploration have enabled the broad structures of the solar-system plasmas to be worked out.

Particle density and temperature vary widely throughout the universe. Intergalactic space may consist of low-density plasma with less than ten particles per cubic metre compared to perhaps 10^{30} m $^{-3}$ and above in stellar interiors. Magnetic fields, of various sizes and strengths, usually occur in conjunction with these plasmas. They can become important, even at low field strength, because collision frequencies are often very low. Space plasmas, although modelled as homogeneous, often have a complicated structure in reality, with no Maxwellian velocity distribution, making non-Maxwellian effects important. In an overview such as this, it is only possible to look briefly at a few examples.

14.2   The interstellar medium

Early investigations had underestimated the importance of interstellar matter, but studies of absorption lines and variations in stellar visibility in the 1930s revealed a variety of dust particles and diffuse gases in interstellar space. The presence of dust and gas was confirmed by early satellite observations which also provided the first indications of a general galactic magnetic field.[2]

Much of the interstellar medium is a tenuous plasma of ionised hydrogen with an electron density of about 10^5 m $^{-3}$. It exists as a plasma simply because of its low density — the probability of an ion recombining with an electron is very low. Conditions are a long way from thermodynamic equilibrium and there is no uniquely defined "temperature". Nevertheless, electrons will often have a Maxwellian distribution, with a (kinetic) temperature of between 10^2 K and 10^4 K (0.01 - 1 eV). Cosmic rays — highly energetic relativistic electrons, protons and alpha particles emitted from distant galaxies — can have individual energies of the order of 10^{10} eV and sometimes up to 10^{20} eV even in our own quiet area of the universe, and may influence local temperatures.

Dusty plasmas are perhaps the most common. The dust particles acquire electrostatic charge from the plasma or are photo-ionised by ultraviolet radiation. If their density is high enough, they will interact with each other, producing collective behaviour similar to that of the ions and electrons in a plasma. If the average distance between the charged grains is greater than the plasma Debye length the system is described as "dust in plasma". A true dusty plasma has an inter-grain spacing much less than λ_D. These charged dust particles can affect the screening process, altering the plasma's composition, density and energy distribution and affecting the dispersion relations of waves propagating through the plasma.

Large regions of dust, gas and plasma (*nebulae*) are often associated with powerful sources of spectrum-wide electromagnetic emission. Some of this is strongly polarised, suggesting that it may be cyclotron radiation from relativistic electrons spiralling in an extensive magnetic field. An example is the radiation from the diffuse part of the Crab Nebula.

We saw in Chapter 6 that electromagnetic waves can propagate at frequencies below the plasma frequency when plasma is immersed in a magnetic field. The frequency and sense of polarisation of the electromagnetic wave are particularly important when the wave frequency, ω; the plasma frequency, ω_p; and the electron cyclotron frequency, ω_{ce}, are similar in magnitude. For example, when ω is approximately equal to ω_{ce}, electrons in the plasma can gain energy from the wave's electric field, accelerating to relativistic levels when the electric field is particularly strong. This process may contribute to the origin of cosmic rays and relativistic electrons in radio sources.

In plasma, electromagnetic waves of different frequencies propagate at different group velocities; with lower frequencies taking longer to travel the same distance due to dispersion. A signal's travelling time can therefore provide information about the medium through which it has passed. With astrophysical plasmas, we do not know when the signal was first sent, so we compare the relative arrival times of the various frequency components of a signal from a distant object.

Plasma frequency ω_p is proportional to charged particle density (equation 6.14), so density fluctuations in the interstellar medium cause local variations in both velocity and direction of electromagnetic waves travelling through it. The dispersion occurs because the phase velocity and group velocity of electromagnetic waves in a plasma are frequency-dependent. Phase velocity is the velocity of an individual frequency component of a signal and increases with electron density. Group velocity is the velocity of the wave packet forming the signal and containing all the frequency components. Since these components travel at different velocities, the effective path length of signals from a distant radio source, such as a pulsar, to an Earth-bound observer is altered by variations in density, broadening the shape of the signal. The lower the frequency, the more noticeable the effect. The change in the group velocity from that in a vacuum is extremely small, but because the path length is so large the arrival time of pulses at different radio frequencies is measurably different.

If the wave frequency, ω, is very much greater than the plasma frequency, ω_p, as is the case with radio waves passing through the interstellar medium, we can estimate the distance of the source from the Earth and obtain information about the interstellar medium from the dispersion of the signal. For example, if a pulsar signal is received with two separate frequencies, f_1, and f_2 (f_1 being the higher frequency and therefore the first to arrive), and the interval between arrival times is Δt, then we can use the formula

$$\Delta t = \alpha \left(\frac{1}{f_1^2} - \frac{1}{f_2^2} \right)$$ (14.1)

to estimate the distance the signal has travelled. A plot of Δt for each adjacent pair of frequencies in the signal is a straight line passing through the origin. The slope of the line indicates the distance to the pulsar; the value of α provides a measure of the dispersion. The plane of polarisation of the two frequencies may be measurably different, due to Faraday rotation, and when used with the time delay, can provide information about the magnetic fields of the interstellar medium.

14.3   Stellar plasmas — the Sun

Apart from the outermost surface layers, conditions inside stars are usually close to local thermodynamic equilibrium, so there is a well-defined temperature (usually in the range 10^7 K to 10^8 K (between 1 - 10 keV)) governing all processes. Since the outer layers are not in thermodynamic equilibrium, different methods of measuring and calculating temperature can yield different values.

Stars like the Sun produce energy via fusion reactions in their interior which transform hydrogen into helium. They remain in equilibrium because the gravitational force confining the plasma balances the plasma pressure from particle motion and fusion reactions within the star. The ideal MHD equation of motion (equation 8.7) must therefore incorporate the effects of gravity rather than electric fields:

$$\rho_m \frac{\partial v}{\partial t} = \rho_m g + J \times B - \nabla p$$ (14.2)

For magnetic equilibrium, $\rho_m \partial v/\partial t$ must be zero. The gravitation term, $\rho_m g$, and the pressure gradient, ∇p, are the dominant terms. As the nuclear fuel is used up, the outward plasma pressure decreases and gravitational forces cause the star to collapse.

In stars of similar mass to the Sun, the central temperature is less than 1.6×10^7 K (about 1.4 keV) — comparable with many terrestrial magnetic confinement devices. The Sun consists of about 74% hydrogen and 25% helium, with other elements making up the remainder. In 1929, English physicist Robert d'Escourt Atkinson, working in Berlin with

Fritz Houtermans, suggested that solar energy results from thermonuclear fusion reactions of hydrogen nuclei, the impacts between protons producing heavier elements with the emission of radiation.[3] This proton-proton chain was only accepted as the main source of solar energy when the Sun's composition was finally established in the 1950s. It has two main routes:-

1. Two protons ($_1^1$H) combine to form a deuteron, one of the protons turning into a neutron, releasing a positron (e^+) and a neutrino (ν):

$$_1^1\text{H} + _1^1\text{H} \rightarrow _1^2\text{D} + e^+ + \nu + 1.44 \text{ MeV}$$

A third proton combines with the deuteron to produce a low-mass isotope of helium ($_2^3$He) and releases energy in the form of a gamma-ray photon (γ):

$$_1^1\text{H} + _1^2\text{D} \rightarrow _2^3\text{He} + \gamma + 5.49 \text{ MeV}$$

Finally, two $_2^3$He nuclei combine to produce an alpha particle ($_2^4$He) with the release of two protons:

$$_2^3\text{He} + _2^3\text{He} \rightarrow _2^4\text{He} + _1^1\text{H} + _1^1\text{H} + 12.86 \text{ MeV}$$

85% of the Sun's energy is produced this way.

2. The second route provides about 15% of the Sun's energy. Beryllium (Be) and lithium (Li) are temporarily produced, together with gamma rays and neutrinos:

$$_2^3\text{He} + _2^4\text{He} \rightarrow _4^7\text{Be} + \gamma + 1.59 \text{ MeV}$$

The beryllium nucleus captures an electron and becomes $_3^7$Li:

$$_4^7\text{Be} + e \rightarrow _3^7\text{Li} + \nu + 1.37 \text{ MeV}$$

Finally, the lithium nucleus captures a proton and breaks apart into two helium nuclei:

$$_3^7\text{Li} + _1^1\text{H} \rightarrow 2_2^4\text{He} + \gamma + 17.35 \text{ MeV}$$

Stars with a greater mass than the Sun have a central temperature higher than 1.6 x 10^7K. In these stars, energy is produced mainly through a series of reactions called the *CNO* or *carbon cycle*, first described by Hans Albrechte Bethe in 1938.[4]

14.3.1   Structure of the Sun

Fig. 14.1 shows the principal zones of the solar atmosphere and interior. The fusion reactions outlined above occur in the hot central core which has a central temperature of about 10^7K and a radius approximately one quarter of the Sun's total radius. The energy from the reactions is released chiefly as gamma ray photons and is carried outwards by a process known as radiative transfer through the *radiation zone* into a region where convection becomes important. The radiation zone extends from the core to about three-quarters of the solar radius. From here to the "surface" is the convection zone. The high-energy photons can take up to 10 million years to reach the surface via numerous absorptions and re-emissions in the radiative zone. During this time, radiation energy is transferred to other particles and less-energetic photons and the gamma ray photons eventually emerge at the surface as visible light.

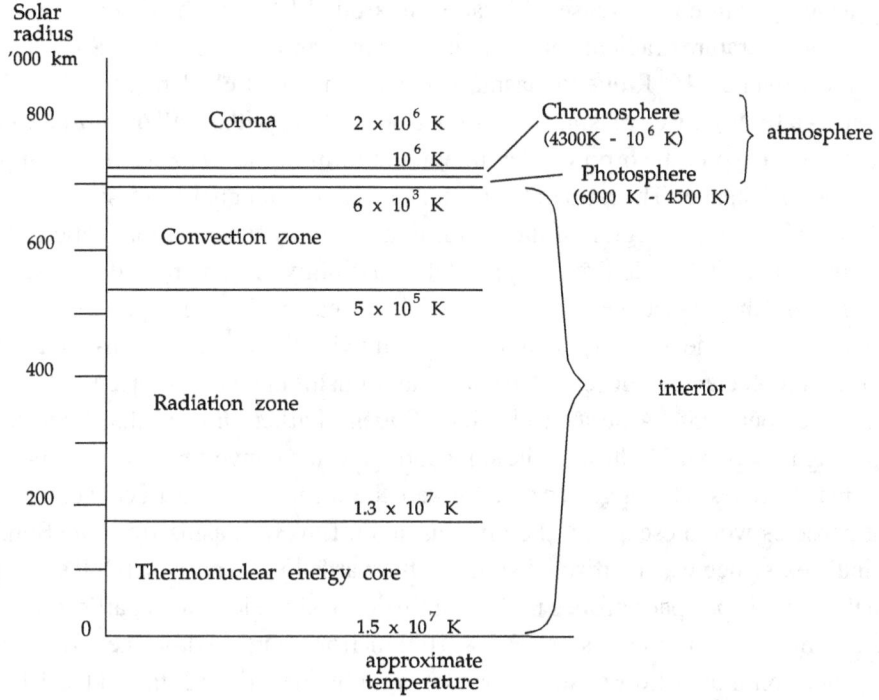

Fig. 14.1 Simplified solar structure (solar radius = 6.96 x 10⁸ m)

Hot plasma rises through the *convection zone* in cells (Fig. 14.2), spreading out and cooling at the photosphere before sinking back into the convection zone. The convection cells appear at the surface as bright, rapidly-evolving *granules* approximately 10^3 km in diameter,

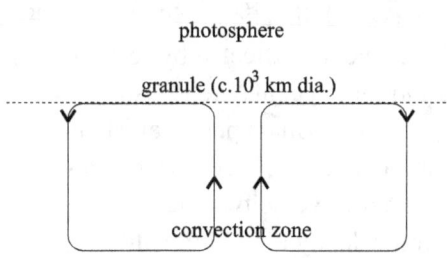

Fig. 14.2   Solar convection cells

against the darker photosphere, giving the Sun a mottled appearance. These granules last for between five and ten minutes. Giant cells, or super-granules, some $10^4 - 10^5$ km in diameter and lasting for about 24 hours may also occur.

The solar atmosphere consists of three layers: the photosphere, the chromosphere and the corona, with transition regions between them. The *photosphere* is the visible surface of the Sun and forms a layer of weakly-ionised (about 0.1%) hydrogen some 500 km thick which emits light and heat. The temperature drops from about 6000 K near the convection zone to about 4500 K at the outer limit. Particle density is about 10^{23} m^{-3}.

The *chromosphere* is a tenuous layer some 2000 km thick with an electron density of about 10^{18} m^{-3}. Plasma waves, possibly Alfvén waves, emerge from the turbulent convection region and are damped. The resulting continuous release of energy heats the chromosphere, causing the temperature to increase with radius to around 10^6 K at the outer limit.

The temperature gradient continues into the outermost region of the Sun, the *corona*, reaching more than 2 x 10^6 K due to plasma-wave heating. Particle density is about 10^{12} m^{-3}. The corona is a fully-ionised plasma of hydrogen and helium with small quantities of heavier ions. It extends beyond the orbit of Neptune in the form of the *solar wind*, its temperature decreasing with distance. The ratio of the plasma pressure to magnetic pressure in the corona is small ($\beta \approx 10^{-2}$), so this region of the Sun is dominated by the solar magnetic field.

Lodge and FitzGerald first suggested the possibility of a solar wind in 1892. Nearly forty years later, Chapman and Ferraro proposed that streams of low energy particles emerged in all directions from the Sun. German physicist Ludwig Biermann used this idea in 1951 to explain the high acceleration observed in comet tails, but his theory was rejected. Seven years later, in November 1958, American physicist Eugene Parker showed that hydromagnetic waves propagating outwards through the solar corona would convert most of their energy into suprathermal particles. He suggested that this kept the corona in a state of constant expansion and that particles would escape all the time, as an outflow of plasma from the Sun.[5] The solar wind's existence was confirmed shortly afterwards by the early satellites. Using ion traps on the early *Luna* space probes in 1960, a flux of positive ions with particle densities of around 10^{12} m^{-2} s^{-1} was found some 2.5 x 10^5 km from Earth, while the strongest direct evidence for a continuous flow of solar particles was obtained towards the end of 1962 by the positive-ion spectrometer on board *Mariner 2*.[6]

A continuous, supersonic stream of plasma, the solar wind emerges from the Sun as the outer region of the expanding solar corona. It results from the pressure difference between the corona and interstellar space which overcomes the effect of solar gravity, and extends as far as the *heliopause*, the edge of the solar system, some 100 - 150 AU[a] from the Sun. The fastest streams come from holes in the corona that lie over areas where no surface activity occurs and where the solar magnetic field is mainly in a radial direction, with the field lines stretching out into interplanetary space. Solar wind behaviour is affected by solar flare and sunspot activity. Gusts can enhance auroral displays and cause geomagnetic disturbances on Earth. It has been estimated that the Sun loses about 10^9 kg of its mass per second to the solar wind. Particle density is about 10^7 m^{-3}, with temperatures of about 10^4 K, electron temperature being slightly higher at approximately 5 x 10^4 K. Particle drift velocity varies between 300 - 600 km s^{-1} with gusts of up to 1000 km s^{-1}. The electron velocity distribution differs greatly from that of the ions, electron velocities being much greater than the drift velocity.

14.3.2   The solar magnetic field

The solar magnetic flux density at the photosphere is about 10^{-4} T. This is a highly turbulent region, with much of the activity being associated with magnetic fields. Dark

---

[a]   AU = Astronomical Unit — the distance from the Sun to Earth, some 1.5 x 10^8 km

sunspots (cooler by some 1500 K) with diameters up to 10^4 km and strong magnetic fields of about 0.1 T, appear on the photosphere. Their average number varies over a period of about 11 years. They tend to cluster in groups, the larger groups lasting for several weeks. Their magnetic effects reach into the corona where they cause distinct emission lines in the spectra of highly ionised atoms, and because this activity takes place in the solar atmosphere at great distances from the actual surface of the Sun, Earth-based astronomers can study the solar cycle in the otherwise hidden polar regions. The magnetic effect of a new sunspot cycle begins in the polar regions at latitudes of 75° north and south of the solar equator several years before the sunspots from the previous cycle have faded away. These sites of magnetic disturbance gradually move towards the equator, so that, at sunspot maximum in the middle of the cycle, most sunspots occur in the region between 10° and 15° north and south of the equator. At the same time, the north and south magnetic poles exchange their positions over an 11 year period from one sunspot maximum to the next, producing a *solar cycle* with a period of 22 years.

H.W. Babcock, working in California in 1961, established the modern sunspot theory. He assumed that the solar magnetic field is located in the convection zone and that the lines of force run (initially) from one pole to the other, as with the Earth's magnetic field. The period of solar rotation is about 25 days near the equator and some five or six days longer in the polar regions. This differential rotation distorts the field lines in the equatorial region, slowly twisting the magnetic field into a series of rings on either side of the equator. Rotation rates also vary with depth because the Sun consists entirely of gas. Convection currents (in the convection zone) tangle the field still further. Babcock suggested that differential rotation led to distortion, and ultimately to severing and reconnection of the lines of force in the corona together with expulsion of flux loops, producing pairs of sunspots with opposite polarities. After sunspot maximum the circular fields diffuse polewards, ultimately forming a dipolar field of reversed polarity. The process repeats itself, reproducing the initial conditions after a complete 22 year magnetic cycle.[7]

Above the photosphere, huge arching columns of cool dense gas known as *prominences* can extend from the chromosphere up to heights of 2×10^8 m, similar in size to the radius of the Sun. These prominences often appear above sunspots and are a dramatic example of how magnetic fields can affect plasmas. Electrons spiralling in the looped magnetic fields above the sunspots produce cyclotron radiation at radio frequencies. *Quiescent prominences* may hang suspended for days and even months above the solar surface, being supported against gravity by the Sun's magnetic field. Their temperatures are around 5×10^3 K with particle densities of about 10^{16} m^{-3}, compared to 10^6 K and 10^{12} m^{-3} in the surrounding corona. *Eruptive prominences* blast cold dense plasma with particle densities of around 10^{17} m^{-3} outward from the Sun at speeds of around 10^3 km s^{-1}. It is thought that these effects are produced by the Lorentz force, from the local currents and magnetic fields, balancing the Sun's gravitational force: $\boldsymbol{J} \times \boldsymbol{B} = \rho_m \boldsymbol{g}$.

Violent localised eruptions, known as *solar flares*, occur above sunspots in the upper chromosphere and inner corona. Their frequency is closely related to that of sunspots. The temperature in a small region may increase dramatically by some 10^3 K, rising to 5 x 10^6 K, as the flare reaches maximum brightness within minutes, before fading away in about an hour. One of the most powerful solar flares on record was that observed by Carrington in September 1859 when the Sun's brightness doubled for several minutes. Large amounts of particles and radiation, much of it in the X-ray and extreme UV regions of the spectrum, may be thrown into space at speeds sometimes exceeding 500 km s^{-1} in a *coronal mass ejection* (CME). The massive cloud of particles erupts from the solar atmosphere like a huge bubble, spreading outwards as it travels into space. In April 1997, an ejected cloud was some 5 x 10^{10} m in diameter when it reached Earth three days later. The path of the cloud can be tracked by the twinkling effect it has on external radio sources. If the Earth is in its path, the cloud can cause major disruption as its magnetic field interacts with that of the Earth — "when the Sun sneezes, the Earth catches a cold" is a popular description. Satellites, navigational instruments, communications systems and power supplies are all at risk of damage. Similar events are believed to occur elsewhere in the universe: huge flares have been detected as an increase in intensity on other stars.

Solar flares produce intense radiation in the visible, ultra-violet and X-ray regions of the spectrum and increase the production of low energy cosmic rays by the Sun. They are probably caused by *magnetic reconnection* (Section 8.4.2), and may involve sudden changes in regions where the Sun's magnetic field is strongly concentrated. In the corona, magnetic pressure dominates both gravitational forces and plasma pressure and, in equilibrium, the equation of motion (equation (14.2)) reduces to $J \times B = 0$. A magnetic field which satisfies this equation is a *"force-free" field*, since the Lorentz force vanishes. Its structure is complicated, with no neatly-nested field lines. Tangled magnetic fields which break and reconnect generate powerful electric fields that accelerate charged particles to very high speeds. Such a field contains more energy than a conventional magnetic field, and the excess may be released by the flare.

The Sun's magnetic field is carried into space by the solar wind — it is, in effect, "frozen into" the solar wind plasma and forms the *interplanetary magnetic field* (IMF), discovered by *Pioneer V* in 1960.[8] The direction and magnitude of the IMF can be determined by measuring the Faraday rotation of circularly-polarised radiation. Its polarity and strength vary with solar activity and changes in the Sun's magnetic field. Near Earth it is generally about 5 x 10^{-9} T.

14.4   The Earth's magnetosphere

A planetary magnetic field deflects the charged particles of the solar wind, creating an elongated cavity known as the *magnetosphere* within the solar wind plasma (Fig. 14.3). Exploration of the Earth's magnetosphere began in the early 1900s when Birkeland attempted to construct the current system during magnetic disturbances and realised that it must extend

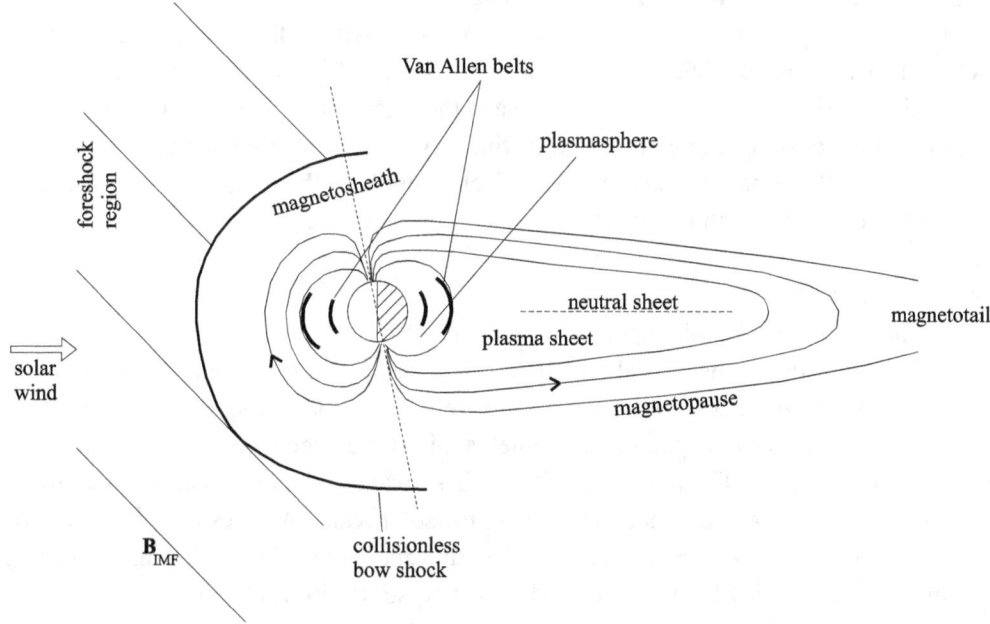

Fig. 14.3   The Earth's Magnetosphere (sketch)

beyond the upper atmosphere.   Various theories emerged in the early 1930s but nothing definite was known about the magnetosphere until 1958 when satellites and space probes enabled direct measurement of magnetic fields in space.   Early observations revealed that the geomagnetic field was confined and shaped by the solar wind.   Thomas Gold (1920-2004) suggested the name "magnetosphere" for the region of ionised gas in the outer atmosphere of the Earth, above the ionosphere.   Observations by Van Allen and others indicated that this zone extended out to between five and ten Earth radii, depending on the level of magnetic disturbance.[9]

The Earth's magnetosphere is now loosely defined as the region in which plasma behaviour is controlled by the geomagnetic field rather than by the IMF.   It starts about 300 - 500 km above the Earth's surface, overlapping the upper reaches of the atmosphere and extending outwards roughly 70,000 km from the Earth[b] in the sunward direction, where the supersonic flow of charged particles in the solar wind compresses the geomagnetic field. On the side away from the Sun (the night-side), the solar wind draws the magnetosphere far beyond the orbit of the Moon in a tail more than a million kilometres long.   The magnetospheric plasmas contain charged particles from both the solar wind and the ionosphere. Plasma waves may bring in ionospheric ions by heating the upper levels of the

---

[b]   About 11 Earth radii, measured from the centre of the Earth: $R_E \sim 6380$ km

ionosphere. The heated ions travel along magnetic field lines, becoming trapped in the magnetosphere. Various electric fields are generated within the magnetosphere by the movement of charged particles through the Earth's magnetic field and by the interaction between the rotating Earth, its magnetic field and the solar wind. Spacecraft observations of the Earth's magnetosphere have revealed distinct regions with different physical properties separated by well-defined boundaries. The electric fields in the magnetosphere create and maintain these structures and connect one region with another.

14.4.1 Bow shock

The solar wind is still supersonic, with speeds in excess of 2.5×10^5 m s^{-1} when it reaches Earth. Particle density is about 10^7 m^{-3}, with temperatures around 10^4 K (between 1 and 10 eV). Magnetic flux density is approximately 10^{-8} Tesla. From equations (6.31) and (6.34) this gives an Alfvén speed and magnetosonic wave speed for the solar wind in the region of 7×10^4 m s^{-1}. The phase velocity of these waves is approximately equal to their actual velocity, and all are less than the speed of the solar wind. As a result, a standing MHD shock wave, or *bow shock*, forms when the solar wind encounters the Earth's magnetosphere. It is similar to the shock which develops when a supersonic aircraft breaks the sound barrier in the Earth's atmosphere.

Shocks form wherever flows of plasma and field energy undergo a sudden change in speed, field strength, temperature or density. Space plasmas are generally collisionless, the mean free path between collisions being greater than the size of the plasma, and the shocks which form in these plasmas are *collisionless shocks*: the scale-length of the shock is much less than the mean free path, making collisional effects insignificant. The solar wind plasma, with a collisional mean free path of the order of 1 AU, is essentially collisionless, so the Earth's bow shock is a collisionless shock wave.

The bow shock forms a distance of a few Earth radii in front of the magnetosphere, around 14 R_E from the centre of the Earth — its actual position varies in response to the solar wind. It marks the boundary where the supersonic solar wind is abruptly slowed to subsonic speeds and is heated as it passes through the shock layer, some 5×10^5 m thick. Much of the kinetic energy of the particles is converted into thermal energy.

Upstream of the bow shock is the *foreshock* region, a zone of intense plasma particle and wave activity. The IMF lies at an angle of approximately 45° to the Earth-Sun line. The first field line that touches the Earth's bow shock forms a tangent to it, as shown in Fig. 14.3. The foreshock region lies downstream of this line. It is a region peculiar to collisionless plasma shocks and contains high-energy particles which have been accelerated by the shock, and which, because of their energy, produce waves and turbulence as they are scattered by irregularities in the magnetic field. The foreshock has two main regions: the electron foreshock and the ion foreshock; the highest energy particles (the electrons) being found closest to the tangent field line.

14.4.2   Magnetopause

By convention, outer boundaries of atmospheric regions are termed "pauses". The outer boundary of the magnetosphere, the *magnetopause*, separates the solar wind and IMF from the planetary magnetic field and plasmas. The boundary of the geomagnetic field was predicted by Chapman and Ferraro in 1931 and detected experimentally in 1960.[10] It lies about 10 R_E from the centre of the Earth and marks the "surface" where the outward magnetic pressure of the Earth's field exactly balances the pressure of the solar wind. The size of any planetary magnetosphere is determined by this pressure balance. No magnetic field lines cross the magnetopause. A boundary layer, or sheath, between five and ten ion Larmor radii in thickness has been observed at the magnetopause.

Like most plasma boundaries, the magnetopause carries a thin sheet of electric current. In the Earth's case it is about 10^5 m thick and known as the Chapman-Ferraro current. A *current sheet* is a thin surface across which magnetic field direction and/or strength change significantly. Such variations produce electric fields and currents. In the magnetopause, these currents close on themselves, providing the necessary $J \times B$ force (equation (8.7a)) to balance the rate of change of solar wind momentum and deflect the solar-wind plasma. Strong plasma waves have been observed in the magnetopause.

The solar wind particles form a highly-conducting plasma, seemingly frozen into the interplanetary magnetic field. The IMF cannot penetrate the Earth's magnetic field, and the particles, too, have difficulty. Most are deflected around the Earth, flowing subsonically in the *magnetosheath* — a turbulent region outside the magnetopause, between it and the shock wave — eventually rejoining the solar wind beyond the tail of the magnetosphere. The magnetosheath plasma has a higher temperature and greater density than the solar wind plasma because of the effect of the bow shock and turbulence.

When the IMF points southward, it connects with the northward-directed geomagnetic field through the dayside magnetopause. Magnetic flux is stripped off the dayside magnetopause and carried downstream in the magnetosheath to be deposited in the magnetotail along with flux from the IMF. The dayside magnetopause contracts while the tail section expands. James Dungey put forward this idea of a magnetosphere open to magnetic reconnection in the 1960s, but it was not until 1978 that observations produced clear evidence of reconnection and accelerated plasma flows at the magnetopause and in the tail.[11]

14.4.3   Magnetotail

The night-side geomagnetic field extends in a tail almost parallel to the flow direction of the solar wind. Oppositely-directed field lines are separated by a layer of hot plasma: the *plasma-sheet*. This lies in the equatorial plane of the Earth's magnetosphere and contains particles with energies of between 1 and 10 keV and density in the region of 10^6 m^{-3}. Ion temperatures are generally six or seven times greater than electron temperatures. The plasma flows rapidly in various directions, chiefly earthwards. A diamagnetic current flows across the plasma-sheet from east to west (dawn to dusk) maintaining the tail geometry and connecting with currents flowing in the magnetopause. Embedded in the plasma-sheet is another current sheet of almost zero magnetic field strength, the *neutral sheet*.

To the north and south of the plasma-sheet lie the north and south lobes of the magnetotail. In the southern lobe the magnetic field lines extend from the south polar region of the Earth, away from the Sun towards the distant tail-end. In the northern lobe they come sunward from the distant tail, terminating at the north polar cap. The magnetic field strength in the tail slowly increases with distance from the neutral sheet. In the lobes, the magnetic field is stronger than in the plasma-sheet and both plasma density and temperature are lower, around $10^4 \, m^{-3}$ and 100 eV respectively.

14.4.4   Magnetic storms

Solar flares enhance the solar wind, causing disturbances in the magnetosphere some 30 hours later (occasionally, the time lag is much less) and producing magnetic storms. The sudden strong variation in the Earth's magnetic field is detected simultaneously all over the Earth. Magnetic storms can produce gradients and instabilities in the plasma of the magnetosphere which lead to bulk plasma motion.

The initial phase of a magnetic storm, shown as (1) in Fig. 14.4, can last from tens of minutes to hours. Compression of the Earth's magnetic field by the increased solar wind pressure resulting from the solar flare causes the geomagnetic field strength to increase rapidly in the space of a few minutes. It remains at this increased level throughout the initial phase. During the main phase (2), the magnetic field strength at the surface of the Earth falls to below its normal value as energetic plasma penetrates deep inside the magnetosphere, forming a ring current around the equatorial zone. The

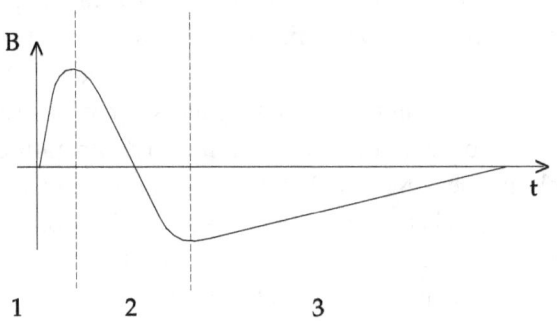

Fig. 14.4 Magnetic field variation during magnetic storm

decrease in strength is greatest at the geomagnetic equator. This phase can last between six and 48 hours. In the decay phase (3), which lasts for several days, the field strength gradually returns to normal as particles are lost from the Van Allen radiation belts.

14.4.5   Van Allen belts

Low energy particles in the solar wind, including most of the protons, cannot enter the Earth's magnetic field and are deflected around the Earth. Highly-energetic particles, in particular electrons, cross the magnetopause in large numbers, especially near the poles, and help to populate a permanent zone of radiation around the Earth: the *Van Allen radiation belts*. Discovered by the first American satellite, *Explorer I* in 1958 and named after James Van Allen who led the team analysing the data,[12] the radiation belts lie within the inner magnetosphere in the region often called the *plasmasphere*. This is a region of cold plasma with a temperature of around 1 eV and density of $10^8\,m^{-3}$ which extends from the edge of the ionosphere, at an altitude of about 1000 km, to about 4½ R_E.

The Van Allen belts contain charged particles trapped in the magnetic "bottle" that is the Earth's non-uniform magnetic field (Fig. 14.5). Originally thought to be two separate "belts" of radiation, they in fact form a continuous distribution of charged particles with two zones of maximum energy: the inner and outer belts. The distribution is not symmetric around the Earth because of the effects of solar radiation. Density varies along the field lines, being greatest in the equatorial plane. It falls at low altitudes, through loss of particles to the neutral atmosphere below. The outer radiation belt is bounded on the sunward side by the magnetopause. On the night-side its outer limit lies much closer to Earth.

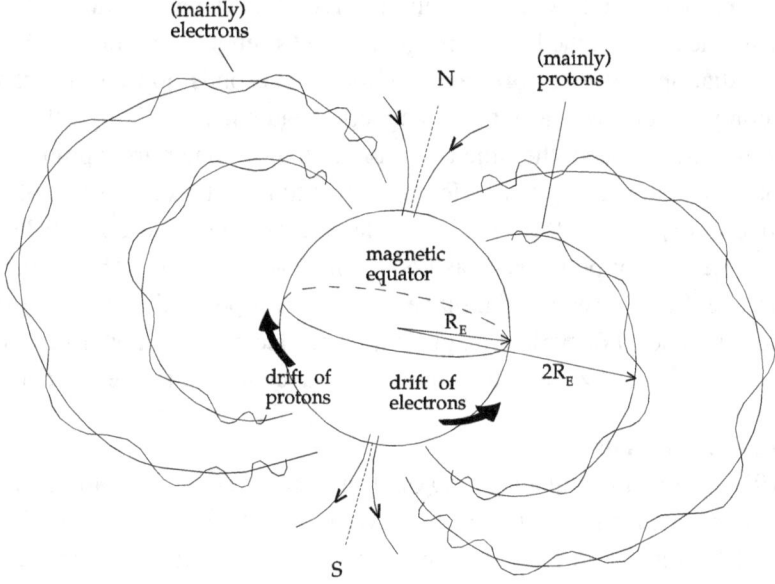

Fig. 14.5   Van Allen radiation belts (not to scale)

The inner Van Allen belt lies between the ionosphere and about 2 R_E, extending from about 2000 km to 5000 km above the Earth's surface at the equator. Protons with kinetic energies greater that 30 MeV collect here. Electrons with energies greater than 1.6 MeV are concentrated in the outer belt, about 8000 km thick and extending outwards from an altitude of about 16,000 km above the equator (situated between about 3 and 4½ R_E). Particles with lower energies, of 40 keV and above, cannot be assigned to distinct belts. They spread over the whole region, extending to about 50,000 km (approximately 9 R_E) from the Earth.

Carl Størmer's earlier mathematical treatment of Birkeland's theories had suggested a zone to which charged particles from the Sun would have no access. Within this region, charged particles spiralling around the Earth's magnetic field lines would bounce back and forth, reflected by the converging magnetic field. With the discovery of the radiation belts, his ideas were finally accepted. A charged particle, spiralling around a field line, encounters an increasingly strong magnetic field as it approaches one of the poles. Eventually the particle reaches the mirror point of the magnetic "bottle", approximately 100 km above the Earth's surface, and is reflected towards the other pole where the process is repeated. Any particles penetrating below 100 km are lost to the atmosphere, producing auroral displays. The trapped particles bounce back and forth between mirror points until a collision enables them to escape. At very low altitudes, particles are lost to the Earth's atmosphere through Coulomb collisions, and at high altitudes through wave-particle scattering.

The Larmor period, $\tau = 2\pi/\omega_c = 2\pi m/|q|B$, shows that the time taken for a particle to complete a single twist in its helical path is independent of the radius of the helix and the velocity of the particle and depends only on the strength of the magnetic field. At 2000 km above the magnetic equator, the Earth's magnetic field strength is about 5×10^{-5} T. A proton in the inner radiation belt therefore takes about 10^{-3} seconds and an electron about 10^{-6} seconds to complete an orbit around a magnetic field line. Although the orbit time is independent of particle speed, the time taken to move from one mirror point to the other is not. A proton with kinetic energy of 1 MeV at this altitude will take approximately 2 seconds to travel from pole to pole; an electron with similar energy would take about 0.1 seconds. The oscillation between mirror points is measured by the *bounce period*. This is the time it takes a particle to travel from the equatorial plane to one mirror point, then to the other, and back to the equatorial plane. For typical magnetospheric electrons with energies of 10 - 50 eV travelling from one auroral zone to the other, the bounce period is a few seconds.

14.4.6   The ring current

In 1911, Størmer postulated a ring of electric charge revolving around the equatorial region of the Earth, with a radius large compared to that of the Earth. Ideas about magnetic storms in the 1930s included a current ring a few Earth radii in radius in the plane of the magnetic equator, "should it be possible for such a current to exist".[13]

As well as orbiting field lines and bouncing between mirror points, particles trapped in the Earth's magnetic field undergo a transverse drift which makes the whole helical trajectory gradually rotate about the Earth's axis. Particles closest to the Earth are in a stronger magnetic field than those furthest away. The combination of magnetic field gradient and curvature (Section 4.9.1) makes positive ions drift slowly westward and electrons eastward, ultimately drifting all the way round the Earth. The 1 MeV proton discussed earlier would take about 30 minutes to drift around the world. An electron with similar energy and altitude will take about 50 minutes. The movement of the ions and electrons produces an electric current, the ring current, which flows in a clockwise sense from east to west. It extends outwards from the ionosphere, where collisions inhibit particle drift, to altitudes of about 65,000 km (11 R_E) where magnetic field distortions prevent particle drift. Satellite observations have shown that most of the particles responsible are about 20,000 - 30,000 km above the Earth's surface (4 - 5½ R_E) at the equator. Particle energy in the ring current is in the region of 85 keV.

Current strength varies with solar activity. The "quiet time" ring current is intensified during times of solar and auroral activity by the injection of particles from the solar wind and the ionosphere. The current measures some 10^6 A during a moderate aurora, rising to several million amperes during a magnetic storm. In severe storms the ring current may be just 6000 km above the Earth's surface and can produce major deviations in the surface magnetic field.

14.4.7 Aurorae

Any acceptable theory of the aurorae must explain not only their geographical distribution, concentrated around the polar regions, but also factors such as height, form, daily variations and spectrum, together with the close relationship between aurorae, magnetic storms and solar disturbances. Much observational work was done in the 19th century, but little progress was made in establishing the underlying mechanisms until the late 1940s.

In August 1947, scientists at Jodrell Bank demonstrated the plasma characteristics of aurorae by bouncing radio signals off a luminescent cloud near the end of an auroral streamer. Electron density was estimated to be about 5 x 10^{13} m $^{-3}$, some 100 times greater than the normal night time ionisation density in the upper levels of the ionosphere. Ground-based observations of auroral arcs indicated that most of the light came from energetic electrons entering the atmosphere from higher altitudes. The first measurements of intense electron fluxes in the region of an auroral arc were made in 1960 by rocket-borne detectors. Positively- and negatively-charged particles were separated using magnetic fields and it was found that electrons with energies less than 10 keV produced much of the auroral light.[14]

The solar wind is an excellent electrical conductor. As it sweeps past the Earth's magnetic field it makes the magnetosphere behave like an enormous natural generator, producing very large electrical currents. These currents flow along magnetic field lines down into the upper atmosphere surrounding the north and south magnetic poles, accelerating electrons and ions to produce highly-structured streams of energetic charged particles with kinetic energies between 100 and 10^5 eV. Electrons from the plasma-sheet are also drawn in along the magnetic field lines. As the high-speed charged particles collide with gases in the upper atmosphere, the gases fluoresce producing the electrical display of the aurora which is often accompanied by plasma waves and turbulence in the magnetosphere.

The Aurora Borealis and Aurora Australis occur most frequently at latitudes of about 70° north and south of the equator, within the *auroral oval*, an asymmetrically displaced "ring" around each geomagnetic pole. The displays take place at heights of 100 km to 400 km above the Earth's surface, and may be seen in temperate zones when major disturbances in the Sun affect the solar wind, causing the auroral ovals to broaden and expand. Particularly active aurorae occur when solar activity disturbs the solar wind and when the interplanetary magnetic field has a southward-pointing component. During moderate solar activity, a typical aurora dissipates some 10^{14} joules of energy into the Earth's atmosphere in just a few hours.

Distinct structures such as arcs, spirals and rays are visible close to the polar regions. The equatorward edge of the aurora is usually diffuse and lacking in form. Nitrogen produces the weaker blue and violet features in low-altitude aurorae below 100 km. The frequently-seen green colour is produced by oxygen atoms radiating at 557.7 nm at altitudes between 100 and about 250 km. Sometimes the higher-altitude (above 250 km) red emissions of atomic oxygen at 630 nm are visible.

14.5   The Earth's ionosphere

The Earth's ionosphere is a weakly-ionised transition zone between the fully-ionised plasma of the magnetosphere above it and the neutral atmosphere below. It is a collision-dominated region of diffuse hydrogen and helium extending into the magnetosphere from an altitude of 80 km to about 600 km. Although ionisation increases with altitude, it is only about 0.1% even at the highest levels. The two main sources of charged particles are photo-ionisation on the sunward side by ultraviolet solar radiation at wavelengths between 10 nm and 100 nm, and collisional ionisation by particles with kinetic energies greater than 1 keV which enter the atmosphere from the magnetosphere at high latitudes in the auroral zone. Electrical conductivity results from a mixture of electromagnetic fields, the interaction of the Earth and the solar wind, plus atmospheric heating caused by solar ultraviolet radiation. Electric currents in the ionosphere are produced by charge separation of positive ions and electrons due to collisions and gravitational drift (Section 4.7.1).

Table 14.1   Parameters of the Earth's magnetosphere

region	electron density (m^{-3})	temperature (eV)	magnetic field strength (Tesla)
solar wind	10^7	1 - 10	10^{-8}
magnetosheath	10^7	10	5×10^{-8}
magnetotail	5×10^4	85	3×10^{-8}
plasma-sheet	5×10^6	10^3	5×10^{-9}
magnetosphere	10^7	5×10^3	5×10^{-8}
plasmasphere	10^8	0.4 - 1	10^{-5}
Van Allen Belts	10^6	4×10^3	5×10^{-7}
ionosphere	10^{11}	0.01	10^{-5}

Table 14.1 illustrates how electron density, temperature and magnetic field strength vary throughout the Earth's plasma environment. Electron density is always a balance between ionisation and recombination, together with diffusion of neutral and ionised particles from one area to another. In the ionosphere, it varies with altitude, latitude, time of day, the seasons and the 11 year sunspot cycle, since the initial ionisation is produced by solar radiation. Fig. 14.6 shows variations with altitude and time of day. Electron density increases

with altitude up to about 300 km, then decreases slowly. There is a general reduction at night. Near the outer limit of the atmosphere there is much ultraviolet radiation but few atoms and molecules, so electron density is low. Closer to Earth, UV radiation is increasingly absorbed so, although there are more atoms and molecules, there is less ionisation and electron density is again low. The optimum lies between the two.

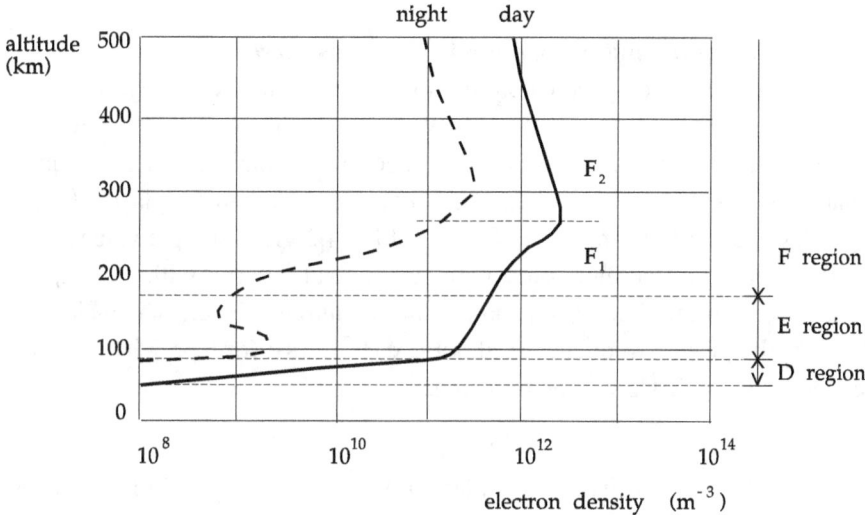

Fig. 14.6 Electron density in the ionosphere

The ionosphere can be divided into four "layers" produced by changes with altitude in both the chemical composition of the atmosphere and the solar spectrum. The layers are not distinct bands, as was originally thought, but simply variations in ionisation levels.[15] The lowest "layer", at altitudes between 60 and 90 km, is more accurately described as an ionised gas. Called the D layer, it is a region of low electron density close to the Earth's surface, where the temperature is approximately 200 K. Electron densities rise to about 10^9 m^{-3} at midday. Collision damping causes rapid recombination of ions and electrons and the layer disappears rapidly after sunset as photo-ionisation ceases.

The E layer begins at about 90 to 100 km. Atmospheric neutral-atom density is around 10^{20} m^{-3} and electron density is about 10^{11} m^{-3}, so even here the degree of ionisation is low. The layer is formed by the absorption of longer-wavelength ultraviolet radiation (about 90 nm) which can permeate the less dense upper regions of the ionosphere. Charged particle motion in the E layer is affected by solar and lunar tidal forces plus winds and convection systems in the atmosphere below. These move the particles across the Earth's magnetic field, producing charge separation and electric currents parallel to the Earth at altitudes of 100 to 150 km. Solar and auroral activity increase the intensity of these currents which in turn affect the Earth's magnetic field. The E layer is thus often referred to as the *dynamo layer*.

254

In the F layer, ionisation is generally much higher. The F_1 layer is a day-time feature produced by ultraviolet radiation at wavelengths between 20 and 80 nm; it therefore decreases at night. The highest region, the F_2 layer, is the more important of the two and contains the most dense plasma in the Earth's environment. Electron densities can reach $10^{12}\,m^{-3}$ and the temperature, at around 2000 K, is ten times that of the D layer. Ionospheric storms, possibly caused by solar particles, increase ionisation in the D region and decrease it in the F region.

14.5.1 Wave reflection and absorption in the ionosphere

A radio wave can only penetrate the ionosphere if its frequency is higher than the plasma frequency; otherwise it is reflected. We saw in Chapter 6 that a wave is absorbed when its angular frequency, ω, equals the ion or electron cyclotron frequency (equation 4.16) — a resonance — and is reflected when ω is equal to or less than the plasma frequency, ω_p. Since ω_p varies with altitude and time of day (Table 14.2 gives sample values), the point at which the ionosphere reflects electromagnetic radiation also varies with the frequency of the radiation, altitude and the time of day. For example, in the E layer, where electron density is $5 \times 10^{11}\,m^{-3}$, the plasma frequency is about $4 \times 10^7\,rad\,s^{-1}$ (or 6.35 MHz). Waves with a frequency less than 6.4 MHz will be reflected.

Table 14.2

layer	electron density (m^{-3})	plasma frequency $(rad\,s^{-1})$	radio frequencies reflected below:
D	day: 10^9	1.8×10^6	285 kHz
E	day: 5×10^{11}	4×10^7	6.4 MHz
	night: 10^9	1.8×10^6	285 kHz
F_1	day: 10^{12}	5.6×10^7	9 MHz
	night: 5×10^{10}	1.3×10^7	2 MHz
F_2	day: 5×10^{12}	1.3×10^8	20 MHz
	night: 5×10^{11}	4×10^7	6.4 MHz

Electron density varies continuously with altitude throughout the ionosphere. There are no sudden changes or definite boundaries. Since ω_p is density-dependent, both plasma frequency and refractive index also vary continuously with altitude. Electromagnetic radiation is gradually bent until it is almost parallel with the ground, before returning to Earth by a similar path (Fig. 14.7).

We therefore cannot use Snell's law in the form $N_1 \sin \theta_1 = N_2 \sin \theta_2$ to plot the path of terrestrial radio waves and find their point of reflection in the ionosphere. Since the refractive index, N, changes only slightly over one wavelength, a simplified model of the form

$$N \sin \theta = \sin \theta_0 \qquad (14.3)$$

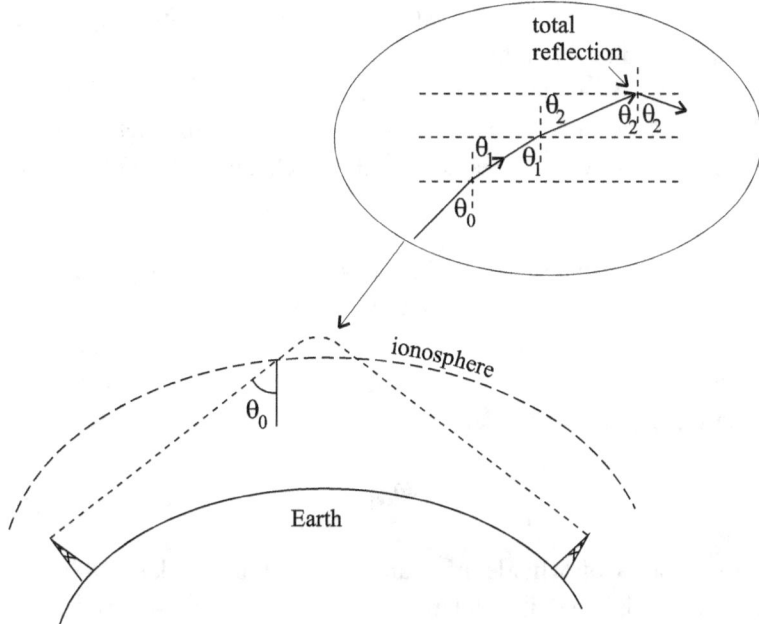

Fig. 14.7   Radio wave reflection by the ionosphere

is used instead, where N is a continuously varying function of altitude and θ_0 is the angle at which the wave encounters the first layer of the ionosphere. Total reflection occurs when $\theta = 90°$; that is, when $N = \sin \theta_0$. Substituting equation (6.12) for N:

$$N^2 = 1 - \omega_p^2/\omega^2 = \sin^2 \theta_0$$

and using $\sin^2 \theta + \cos^2 \theta = 1$, a wave of frequency ω is reflected by the ionosphere when

$$1 - \sin^2 \theta = \cos^2 \theta_0 = \omega_p^2/\omega^2$$

so: $$\omega = \omega_p/\cos \theta_0 \qquad (14.4)$$

which is at the lowest altitude at which a particular electron density is encountered. A wave entering obliquely is less penetrating since it is reflected at lower values of N and n_e than for normal incidence. A wave will pass through the plasma if n_e is below the critical value for the frequency of the wave.

In the D layer, free electron density is low, so ω_p is low (less than 300 kHz) and only low frequency waves are reflected, unless θ_0 is large. This region is characterised by day-time absorption of low-frequency radio waves from ground level. When the D layer disappears at night, absorption falls. Long wave radio reception is strongly affected by these electron density variations: distant radio stations which can be received at night, and which may interfere with local radio stations, are unobtainable during the day. The E region reflects radio waves at frequencies up to a few MHz during the day, while the F region reflects even higher frequencies.

In theory, the method of equation (14.3) can be used to investigate electron density distribution as a function of height in the ionosphere, by measuring the time difference between a signal being transmitted and received. In practice, the presence of the Earth's magnetic field introduces an additional complication: electromagnetic waves are split into ordinary and extraordinary waves (Section 6.6.3) which have different velocities in plasma.

14.5.2   Whistlers

An example of plasma waves in the ionosphere is the propagation mode known as a Whistler. These are transverse waves, with frequencies well below the electron cyclotron frequency (the R-waves of Section 6.6.1), which propagate along the Earth's magnetic field in the magnetosphere; the wave vector, k being parallel to B. When k is small, the angular frequency of the wave is approximately

$$\omega_w = \omega_{ce} \left(\frac{ck}{\omega_{pe}} \right)^2 \tag{14.5}$$

The first reports of whistlers began to appear in the late 1880s. W.H. Preece described a disturbance in the British telephone system: "twangs were heard, as if a stretched wire had been struck, and a kind of whistling sound". The earliest scientific account was published in 1919 by Heinrich Barkhausen. His theory of their origin, published in 1930, correctly identified whistlers as having been launched at the Earth's surface by lightning bolts, but he thought that their descending tone was due to multiple reflections from the ionosphere. T.L. Eckersley corrected this five years later. He suggested that whistlers originate in atmospherics, the change in frequency being caused by dispersion, and showed that at frequencies below the gyro-frequency, radio waves could be propagated freely through the ionosphere in the extraordinary mode.[16]

In the early 1950s, L.R.O. Storey defined whistlers as atmospherics at frequencies below 15 kHz, giving a characteristic whistling tone of descending pitch. They are produced, he said, by dispersion in the ionosphere of the impulsive atmospheric "clicks" or pulses of electromagnetic radiation emitted by lightning discharges.[17] The resulting waves travel along magnetic field lines from one hemisphere of the Earth to the other, often in "ducts" where electron density is greater. The pulse contains a range of frequencies, covering a band from perhaps a few hundred to maybe 10^4 Hz. Since phase and group velocities are functions of frequency, the components of the pulse travel with different speeds, the higher frequency waves having higher group and phase velocities. The length of time a wave takes to reach a distant receiver is determined by the group velocity. Since v_g is frequency-dependent, the waves become dispersed in the ionosphere and magnetosphere. The high-frequency waves travel faster and arrive at a distant receiver before the lower-frequency waves, producing ionospheric radio noises with descending tones of decreasing pitch, in the audible frequency range between 1 and 10 kHz. The usual sound is a whistle beginning at high frequency and descending to lower frequencies over a period of a few seconds, the amount of dispersion increasing with the length of the path travelled. The rate at which the frequency decreases is indicative of plasma density along the path from source to receiver. Whistler rates are highest at mid- to high-latitudes in regions where thunderstorms are most common. Similar waves have also been detected on Jupiter and Neptune.

14.5.3 Lightning

Lightning has been studied since 1752 when Benjamin Franklin first showed that it was an electrical phenomenon. With the development of spectroscopy in the 1860s, attempts were made to observe the lightning spectrum. In 1868, Lt. John Herschel made several observations during the Indian monsoon. He reported seeing a "more or less continuous spectrum, crossed by numerous bright lines", one of which he believed to be nitrogen. C.T.R. Wilson, at the Solar Physics Laboratory in Cambridge, showed in 1916 that net changes in the Earth's field due to lightning flashes were mostly positive. Ten years later, Appleton took measurements further from the discharge channel and found the opposite to be the case, concluding that a thundercloud is bipolar, with the positive charge uppermost. This was later confirmed by Schonland and Craib's observations of South African thunderstorms. In 1940, Simpson and Robinson reported that many storm clouds had an additional collection of positive charge at the base of the cloud. The electric field on the ground, below the main body of the thundercloud, was found to be more often negative than positive, with predominantly positive fields as the cloud approached and receded.[18]

A fully-developed thundercloud forms an electric dipole with an excess of positive charge towards the top and negative charge at the base of the cloud, separated by perhaps 5km or more of cloud (Fig. 14.8). An electric field is also produced below the cloud, at ground level. Sometimes, as Simpson and Robinson found, there is a small additional pocket of positive charge at the base of the cloud. The charge separation is believed to result from the motion of water droplets within the cloud: falling raindrops and ice crystals may become electrically polarised as they fall through the atmosphere's natural electric field.

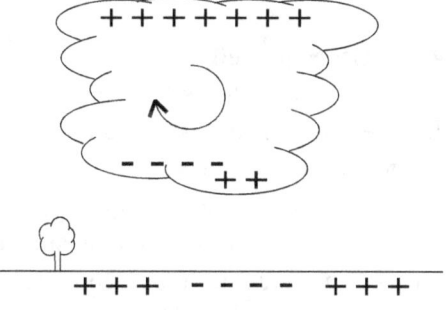

Fig.14.8 Charge distribution in thundercloud

Cloud electrification appears to need the strong up-draughts found in large thunderclouds which carry water droplets upwards, super-cooling them to around -20°C. The developing hailstones are tossed about, causing collisions with other ice crystals. The smaller particles lose negative ions, becoming positively-charged and are carried upwards on the up-draughts to collect in the top of the cloud. The larger particles gain negative ions and become negatively charged, falling towards the base of the cloud because of their weight and building up an excess of negative charge in the lower regions of the cloud. As the thundercloud moves over the Earth's surface, an equal but opposite charge is induced on the ground, which follows the movement of the cloud.

As the field strength increases between the regions of charge separation, free electrons begin to move under the influence of the external electric field. Channels of ionised air begin to form in the negatively-charged region of the thundercloud. These negatively-charged "stepped leaders" proceed downward in a series of jumps, up to 50m long, branching as they go. This initial, almost invisible phase takes perhaps a tenth of a second and involves comparatively small currents (tens or hundreds of amperes). Schonland first reported in 1935 that lightning develops as a series of steps, with several leader branches and lateral forks, creating a channel through the insulating layer of the cloud or the air below.[19] As the stepped leader approaches the ground, positive charge on the Earth's surface enhances the electric field. This is particularly so on trees, tall buildings and other "point" objects which naturally concentrate the charge. A positive streamer can develop at these locations.

Eventually, the positive streamer connects with the descending stepped leader from the cloud; the electric current greatly increases and travels back up the stepped leader into the cloud. The narrow channel thus formed is much more strongly illuminated by this return stroke. The air or cloud vapour in its path is ionised and becomes conducting — a plasma. As it does so, it is heated and expands. The large current flowing through the channel during the (visible) return stroke creates a shock wave heard as thunder. Currents of 10^4 A and temperatures in excess of 10^4 K are common. The voltage carried by the lightning bolt depends on the length of the bolt: most are about a mile (1.6km) long. If the discharge occurs within the cloud, the electrons at the base can flow up through the cloud in a diffuse flash of sheet lightning — diffuse because the actual path of the discharge is hidden by the cloud. If the air beneath the cloud breaks down, the lightning appears forked.

14.6   Other planets

Magnetospheres have been observed around many planets in the solar system and identified around pulsars and radio galaxies. Lightning and auroral displays have also been detected on other planets.

The inner regions of Jupiter's magnetosphere contain a hot plasma of electrons and protons with some ions of helium, sulphur and oxygen, many of which may originate in the volcanoes of the Jovian moon, Io. These charged particles are caught up in Jupiter's rapidly rotating magnetic field and are accelerated to extremely high speeds. The plasma temperature is around 30 keV, making it the hottest in the solar system (the centre of the Sun is only about 2 keV), but it is very thin, with an average density of about 10^4 m^{-3} in the hottest regions. Its high temperature means that Jupiter's plasma contains a large amount of energy and can resist the solar wind, although the pressure balance between the two is precarious: a gust in the solar wind can blow away some of the plasma, making the magnetosphere deflate rapidly, sometimes to half its original size. The lost plasma is quickly replaced with fresh electrons and ions accelerated by Jupiter's rotating magnetic field and the magnetosphere expands again.

Jupiter's rapid rotation spreads the charged particles trapped in its magnetosphere into a huge electrically-charged current sheet which lies in the plane of the magnetic equator. In the Jovian equivalent of the Earth's Van Allen belts, MeV electrons continuously lose energy emitting cyclotron radiation at radio wavelengths. Observations of polarised radio noise from Jupiter in 1960 were interpreted as indicating magnetically-trapped radiation of greater intensity and higher electron energy than occurs in the Earth's plasma environment. Evidence of auroral emission from Venus suggested similar phenomena there too.[20]

The magnetospheres of the outer planets of the solar system may be less affected by the interplanetary magnetic field because its field strength decreases with distance from the Sun. These planets have a number of moons which, unlike our Moon, orbit deep inside the planetary magnetosphere and will affect the dynamics of the magnetospheric plasmas.

14.7 Summary

1. The plasma frequency, ω_p, and the interactions of electric and magnetic fields with the charged particles of astrophysical and atmospheric plasmas have a fundamental impact on their behaviour and characteristics.

2. Radiation provides a useful diagnostic tool; for example, observation of pulsars provides information about the interstellar medium through which the pulses have passed.

3. Solar activity — sunspots, flares and prominences — affect the behaviour of the solar wind, which in turn affects the planetary magnetospheres, producing magnetic storms, turbulence and intensified auroral displays.

4. Rotation of the Earth's radiation belts produces a ring current flowing in an east-west direction above the ionosphere, which depresses the geomagnetic field.

5. The Earth's ionosphere is a weakly-ionised collision-dominated region of the upper atmosphere whose varying electron density affects plasma frequency. Radio waves are reflected at different altitudes depending on their frequency and the time of day.

References

1.   Schuster, A.: *Nature*, **53**, p.207-12 (1896)

2.   Eddington, A.S.: *Proc. Roy. Soc.*, **111**, p.424-56 (1926)
   Dieter, N.H., & W.M. Goss: *Rev. Mod. Phys.*, **38**, p.256-97 (1966)

3.   Atkinson, R.d'E., & F.G. Houtermans: *Nature*, **123**, p.567-8 (1929)

4.   Bethe, H.A.: *Phys. Rev.*, **55**, p.434-56 (1939)

5.   Biermann, L.: *Z. Astrophys.*, **29**, p.274-86 (1951)
   Parker, E.N.: *Astrophys. J.*, **128**, p.677-85 (1958); *Sci. Am.*, **210**, p.66-76 (Ap. 1964)

6.   Snyder, C., M. Neugebauer & U.R. Rao: *J. Geophys. Res.*, **68**, p.6361-70 (1963)

7.   Babcock, H.W.: *Astrophys. J. (USA)*, **133**, p.572-87 (1961)

8.   Coleman, P.J., et al., *Phys. Rev. Lett.*, **5**, p.43-5 (1960)

9.   Gold, T.: *J. Geophys. Res.*, **64** (9), p.1219-24 (1959)

10.   Chapman, S., & V.C.A. Ferraro: *Terr. Mag.*, **36**, p.77-97, 171-86 (1931)
    Cahill, L.J., & V.L. Patel: *Planetary & Space Sci.*, **15**, p.997-1033 (1967)

11.   Dungey, J.W.: *Phys. Rev. Lett.*, **6** (2), p.47-8 (1961); *J. Phys. Soc. Japan*, **17** Suppl. A-II, p.15-9 (1962)
    Russell, C.T., & R.C. Elphic: *Space Sci. Rev.* (Neth.), **22** (6), p.681-715 (Dec. 1978)

12.   Van Allen, J.A., & L.A. Frank: *Nature*, **183**, p.430-4 (Feb. 1959)
    Gold, T.: *Nature*, **183**, p.355-8 (Feb. 1959)

13.   Vestine, E.H., & S. Chapman: *Terr. Mag.*, **43**, p.351-82 (1938)

14.   Lovell, A.C.B., J.A. Clegg & C.D. Ellyett: *Nature*, **160**, p.372 (Sept. 1947)
    McIlwain, C.E.: *J. Geophys. Res.*, **65**, p.2727-47 (1960)

15.   Jackson, J.E., & J.C. Seddon: *J. Geophys. Res.*, **63**, p.197-208 (1958)

16.   Preece, W.H.: *Nature*, **49**, p.554 (12 April 1894)
    Barkhausen, H.: *Phys. Zeits.*, **20**, p.401-3 (1919); *Proc. Inst. Rad. Engrs. (NY)*, **18**, p.1155-9 (1930)
    Eckersley, T.L.: *Nature*, **135**, p.104-5 (1935)

17.   Storey, L.R.O.: *Phil. Trans.* A, **246**, p.113-41 (1953)

18.   Herschel, J.: *Proc. Roy. Soc.*, **17**, p.61-2 (1868-9)
Wilson, C.T.R.: *Proc. Roy. Soc.*, **92**, p.555-74 (1916); *Phil. Trans.* A, **221**, p.73-115 (1920)
Appleton, E.V., R.A. Watson Watt & J.F. Herd: *Proc. Roy. Soc.*, **111**, p.615-54 (1926)
Schonland, B.F.J., & J. Craib: *Proc. Roy. Soc.*, **114**, p.229-43 (1927)
Simpson, G., & F.J. Scrase: *Proc. Roy. Soc.*, **161**, p.309-52 (1937)
Simpson, G., & G.D. Robinson: *Proc. Roy. Soc.*, **177**, p.281-329 (1941)

19.   Schonland, B.F.J., D.J. Malan & H. Collens: *Proc. Roy. Soc.*, **152**, p.595-625 (1935)

20.   Van Allan, J.: *Radiation Research*, **14** (5) p.540-50 (1961)

Chapter 15

Plasma classification

The impetus for the growth of plasma as a branch of physics came largely from the search for an abundant and cheap source of power which began in earnest in the 1950s. Investigations of plasma properties and behaviour have been undertaken in laboratories around the world to achieve that end, and much has been learnt as a result. Unrelated technologies have found uses for the gas discharges which, even in the 1970s, were considered much less exciting than fusion research. While the search for an answer to the world's energy needs continues, environmental and industrial applications of gas discharges are now being developed on an increasing scale. The constant demand from the computer industry for smaller and faster processors, together with concerns about the effects of hydrocarbons and greenhouse gases on global warming, and the need to dispose of toxic waste products safely and swiftly have opened up important new avenues of research. As more is known about how and why plasmas behave as they do, more commercial applications will be found. It may be that much of the funding for research in the future will be focused on environmental and industrial uses of plasma rather than the more expensive long-term fusion projects.

We have established that a plasma is a quasi-neutral mixture of charged particles characterised by collective behaviour, but the transition from weakly-ionised gas to plasma is often hard to define. There is a grey area in the middle where one person's ionised gas is another's weakly-ionised plasma — it is often a question of interpretation. In an ionised gas, the density of charged particles is so low that all interactions between charges can be ignored. Their effect on the behaviour of the gas is minimal. In a plasma, charged-particle densities are large compared to neutral particles, and interactions between the charged particles determine the behaviour of the plasma.

The movement of ions and electrons in a plasma both produces and results from electric and magnetic fields within the plasma. A charged particle does not generally move in a straight line but tends to both follow electric fields and be influenced by any magnetic fields present. Local concentrations of electric charge create long-range Coulomb forces which influence the motion of the charged particles a long way from the centres of concentration, producing the characteristic collective behaviour that distinguishes a plasma from a neutral gas. It is this interaction between particles and fields that is responsible for the difference in properties and behaviour of ionised gases and plasmas.

In Chapter 5 we established three criteria for an ionised gas to be a plasma. A fourth was added in Chapter 6 (section 6.4.1):-

1. Charged particle density must be large enough for the Debye length, λ_D, to be much less than the physical dimensions of the system. This enables collective shielding to prevent charge imbalance within the body of the plasma and ensure quasi-neutrality.

2. For effective screening and collective behaviour, the number of charged particles in a Debye sphere, N_D, must be much greater than 1. Each charged particle only interacts collectively with the charges within its Debye sphere.

3. Local concentrations of charge are confined to volumes the size of the Debye sphere. Outside these small volumes, the charge density of ions is equal to the electron density and the plasma is electrically neutral.

4. The collision frequency, υ_{en}, between electrons and neutral atoms must be very much smaller than the plasma frequency, ω_p, to keep collision damping to a minimum: $\upsilon_{en} \ll \omega_p$. An alternative expression which is sometimes used is: $\upsilon_{en}\,\tau > 1$ where τ is the mean time between the collisions. If this condition is not met, the electrons will not act independently of the neutral particles but will be forced into equilibrium with them, and the plasma becomes an ionised gas.

Six major parameters can be identified which characterise the behaviour of plasma. Three, linked to electron motion, represent high-frequency activity. These are: electron plasma frequency, ω_{pe}; electron cyclotron frequency, ω_{ce}; and the Debye length, λ_D. The other three, representing low-frequency behaviour, are linked to the motion of the heavier ions: ion plasma frequency, ω_{pi}; ion cyclotron frequency, ω_{ci}; and the Alfvén speed, v_A. The last named is characteristic of very low frequency fluid-like plasma behaviour. Table 15.1 gives approximate values for these parameters, together with indications of temperatures, density and magnetic fields in a variety of plasmas.

Table 15.1 - Approximate magnitudes

Type of plasma	n (m^{-3})	T (K)	B (Tesla)	λ_D (m)	N_D	ω_{pe} $(rad\ s^{-1})$	ω_{ce} $(rad\ s^{-1})$	ω_{pi} $(rad\ s^{-1})$ $[Z=1]$	ω_{ci} $(rad\ s^{-1})$ $[Z=1]$	v_A $(m\ s^{-1})$
interstellar medium	$10^5 - 10^7$	$10^2 - 10^4$	$10^{-9} - 10^{-11}$	2.2	4.4×10^7	5.6×10^4	17.6	1.3×10^3	9.6×10^{-3}	2.2×10^3
solar atmosphere (chromosphere)	10^{18}	10^4	10^{-4}	6.9×10^{-6}	1.4×10^3	5.6×10^{10}	1.76×10^7	1.3×10^9	9.6×10^3	2.2×10^3
solar corona	$10^{12} - 10^{13}$	10^6	$10^{-3} - 10^{-4}$	3.1×10^{-2}	6.2×10^8	1.3×10^8	8.8×10^7	2.9×10^6	4.8×10^4	4.9×10^6
solar wind	$10^6 - 10^7$	10^4	$10^{-8} - 10^{-9}$	3.1	6.2×10^8	1.3×10^5	8.8×10^2	2.9×10^3	0.48	4.9×10^4
stellar interior	$10^{30} - 10^{33}$	$10^6 - 10^7$	10^{-4}	2.2×10^{-11}	2.2	4×10^{17}	1.76×10^7	9.3×10^{15}	9.6×10^3	3.1×10^4
Earth's ionosphere	$10^9 - 10^{11}$	10^3	$10^{-4} - 10^{-5}$	2.2×10^{-2}	4.4×10^5	5.6×10^6	8.8×10^6	1.3×10^5	4.8×10^3	1.1×10^7
Earth's magnetosphere	$10^6 - 10^7$	10^7	10^{-5}	97.6	1.9×10^{13}	1.3×10^5	1.76×10^6	2.9×10^3	957.7	9.75×10^7
tokamak	10^{19}	10^8	$1 - 5$	2.2×10^{-4}	4.4×10^8	1.8×10^{11}	5×10^{11}	4.2×10^9	3×10^8	2×10^7
glow discharge	$10^{15} - 10^{17}$	10^4	---	6.9×10^{-5}	1.4×10^4	5.6×10^9	---	1.3×10^8	---	---

Fig. 15.1 shows a more detailed classification of plasmas by temperature and density than that in Chapter 5. Plasmas range from the tenuous plasmas of the solar wind to the ultra dense plasmas of stellar interiors and white dwarfs, from the cool chemical plasmas used in industrial processes, to the thermonuclear plasmas in magnetic-confinement fusion devices. In the solar wind and interstellar medium, particle density is about $10^7 \, \text{m}^{-3}$. In present-day tokamaks, particle density is in the region of $10^{20} \, \text{m}^{-3}$, while in the quantum plasmas of white dwarfs particle densities can be greater than $10^{35} \, \text{m}^{-3}$. Temperatures in laboratory plasmas range from about 0.1 eV ($10^3 \, \text{K}$) in gas discharges, to 20 keV ($\sim 10^8 \, \text{K}$) in magnetic fusion devices. Scale lengths — the physical size of the plasma — also vary from a few millimetres in some laboratory plasmas to light years in the case of galaxies.

The various straight lines ($N_D = 1$, etc.) represent approximate "boundaries". At the top, the line labelled $k_B T = m_e c^2$ (at $T \approx 10^9 \, \text{K}$) separates the *relativistic plasmas* of pulsar magnetospheres from the rest. Pulsars are thought to have very strong magnetic fields which may alter the structure of their atmosphere. They also radiate large amounts of energy at various frequencies from radio through to X-rays and gamma rays. At certain frequencies, the charged particles trapped in the magnetic field will be accelerated to relativistic levels by this radiation. Relativistic effects become important at temperatures above $10^9 \, \text{K}$, because thermal energies become greater than the electron rest mass energy, and so $k_B T$ is greater than $m_e c^2$.

Below the line $k_B T = m_e c^2$ is the large group of so-called *classical plasmas*, with boundary conditions: $1 \, \text{eV} \le k_B T \le m_e c^2$; $N_D > 1$; and $k_B T < E_F$, where E_F is the Fermi energy, given by

$$E_F = \frac{h^2}{8 m_e} \left(\frac{3n}{\pi} \right)^{\frac{2}{3}} \qquad (15.1)$$

Classical plasmas include the man-made and naturally-occurring plasmas discussed in previous chapters. The lower line, marked $k_B T = 1 \, \text{eV}$, indicates the lower limit for adequate amounts of ionisation. At temperatures of $10^4 \, \text{K}$ and less, recombination of electrons and ions becomes significant and plasmas such as those found in the ionosphere and some gas discharges are only partially ionised.

Non-neutral plasmas consist of a collection of charged particles with no overall charge neutrality. Important amongst these are the single-component plasmas (pure ion or pure electron) which have applications in atomic clocks and mass spectrometry. They reveal a range of plasma behaviour, such as plasma waves, screening and instabilities and are frequently easier to confine than neutral plasmas in which inter-particle collisions lead to diffusion across the magnetic field. A confined single-component plasma in thermal equilibrium is in a state of minimum free energy and is therefore stable. Non-neutral plasmas are often characterised by strong self-generated electric fields, which can be used to accelerate the charged particles, in particular ions, for plasma heating and for microwave generation.

The diagonal line labelled $N_D = 1$ marks the boundary between the weakly-coupled

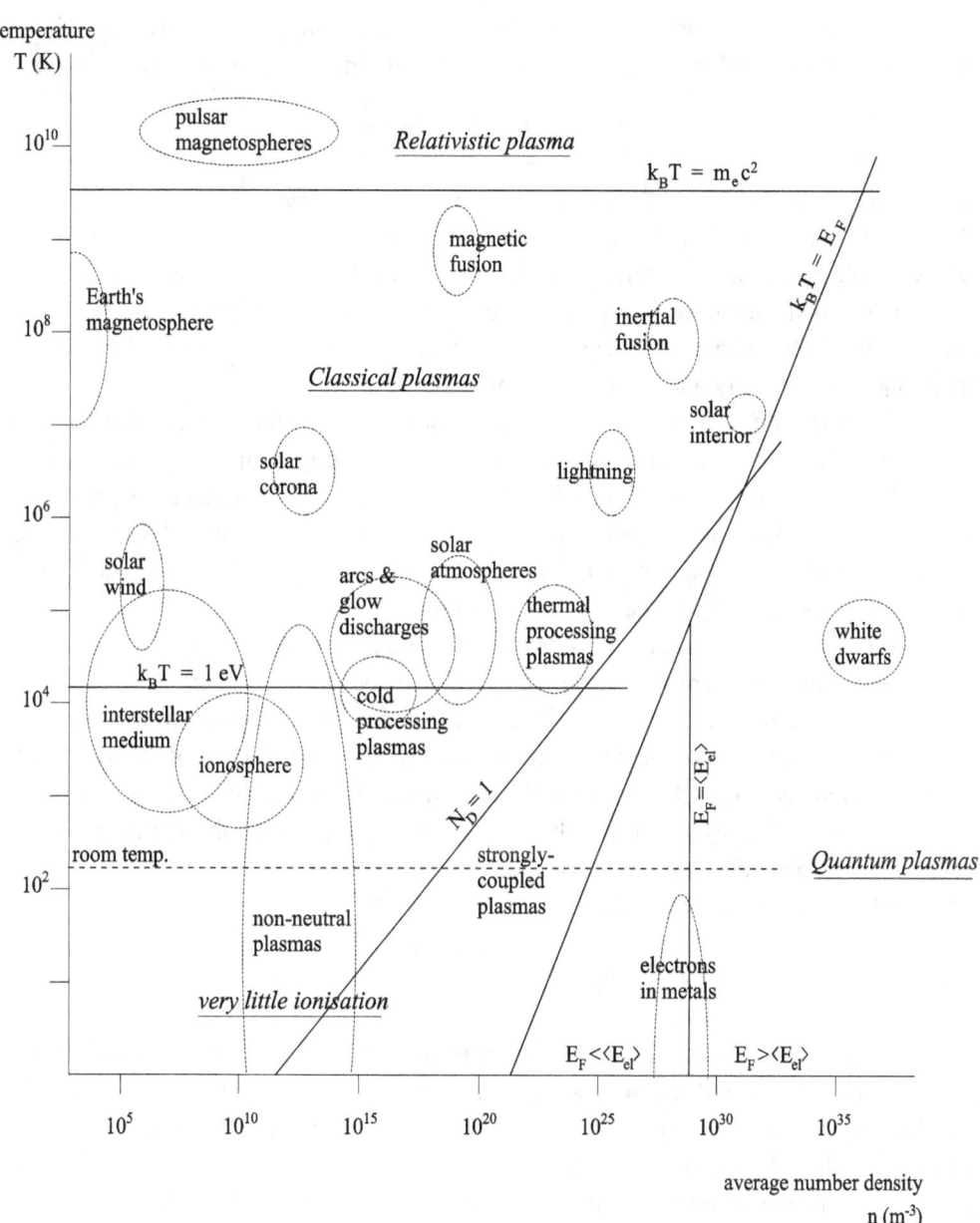

Fig. 15.1 Classification of plasmas by temperature and density

plasmas, to the left of the line, and *strongly-coupled plasmas*, to the right. In strongly-coupled plasmas, where $N_D = \frac{4}{3} \pi n \lambda_D{}^3$ is less than 1, the Coulomb interactions dominate thermal effects. These plasmas can occur in a variety of situations such as stellar interiors, laser-compressed plasmas and some gas discharges. The energy of the average Coulomb interaction between neighbouring particles — the Coulomb potential energy — is given by

$$\langle E_{el} \rangle \approx \frac{1}{4\pi\varepsilon_0} e^2 n^{\frac{1}{3}} \tag{15.2}$$

This is Coulomb's Law, where the potential energy of a charge q in the electric field of a collection of charges is $q^2/4\pi\varepsilon_0 r$. Here r is the average distance ($n^{-1/3}$) between the charges. In classical plasmas, this Coulomb potential energy is always much less than the average kinetic energy, $\frac{3}{2}k_B T$, of the particles. In strongly-coupled plasmas, high particle density make the ratio of the Coulomb potential energy to the kinetic energy quite large. Most laboratory plasmas have low density and are weakly-coupled.

The *plasma parameter* (equation 5.4) is sometimes called the coupling parameter because it provides a useful measure of how strongly- or weakly-coupled a plasma is. A value of $g = 1/(n\lambda_D{}^3)$ very much less than 1 indicates a weakly-coupled plasma in which $\langle E_{el} \rangle$ is much less than the kinetic energy of the plasma. As g becomes greater than 1, the plasma becomes more strongly-coupled and at temperatures and densities for strong coupling, the plasma properties become similar to those of fluids.

At very high densities the plasma no longer behaves like an ideal gas because of the effects of electrostatic interactions (Debye shielding) and, in the case of *quantum plasmas*, degeneracy.[a] The line labelled $k_B T = E_F$ marks the transition to quantum plasmas. These behave in ways similar to classical plasmas but their physical conditions of low temperatures or very high density add to their complexity and make them a specialised field of study.

As plasma density increases, the average distance between the particles decreases. Quantum effects become significant when the inter-particle distance, $n^{-1/3}$, is comparable with the thermal de Broglie wavelength of an electron:

$$\lambda_{DB} = \frac{h}{m_e} \left(\frac{2k_B T_e}{m_e} \right)^{-\frac{1}{2}} \tag{15.3}$$

For a plasma with an electron temperature of 1 eV, λ_{DB} is approximately 10^{-9} m. Thus, quantum effects will become significant when n is in the region of 10^{27} m^{-3}. The characteristic scale of electron kinetic energy in such a quantum plasma is the Fermi energy, given by equation (15.1). If $n = 10^{27}$ m^{-3}, $E_F = 5.8 \times 10^{-20}$ J. When $k_B T = E_F$, both λ_D and λ_{DB} are very close in size to the inter-particle distance $n^{-1/3}$. From this point, as particle density increases and E_F becomes greater than the thermal energy, $k_B T$, quantum effects begin to

---

[a]   Two or more quantum states have the same energy.

dominate. At densities of around $10^{29}\,\text{m}^{-3}$ (marked by the vertical line in Fig.15.1), when E_F is greater than the Coulomb potential energy $\langle E_{el} \rangle$ (equation 15.2), quantum effects dominate those from the Coulomb interaction, producing nearly ideal quantum plasmas.

White dwarfs provide an example of quantum plasmas. These are small hot stars in the final stage of their life. Of similar mass to the Sun, they have collapsed to about the diameter of the Earth and consist of a degenerate gas of extremely tightly-packed charged particles. These stars have stabilised with the gas pressure balancing their gravitational force, but their density means that a Maxwellian distribution is inappropriate.

Quantum effects are significant in solid-state plasmas, *electrons in metals* being an example. Crystalline solids usually have a definite melting temperature and are characterised by a recurring pattern in the positions of the atoms making up the solid, which can be represented in three dimensions as an array of points — a lattice. Metals have a crystalline structure in which the outermost electron in each atom becomes detached and is free to move through the crystal forming an "electron gas" around the lattice of fixed positive ions. A metal can thus be thought of as a contained plasma of positive ions and free electrons whose attraction to the ions holds the whole structure together. Under certain circumstances, the particles can show a wave-like behaviour and plasma oscillations may be seen. The density of metals is very high ($n \sim 10^{29}\,\text{m}^{-3}$) so the plasma frequency, ω_p, will also be very high — around $10^{16}\,\text{rad}\,\text{s}^{-1}$. When electrons with energies of around 1 keV are fired through a thin metal foil they lose energy in discrete amounts which are increments of $\omega_p h/2\pi$ as they pass through the foil.

The future

At some point, nuclear fusion on a commercial scale will be achieved, although whether it will be via current lines of research is hard to say. The history of electrical discharges, as outlined in Chapter 1, reveals that, for some 50 years, research was focused on the positive end of the discharge tube, and on the nature of striations. The cathode end of the tube was largely ignored. Only when interest turned to the cathode rays (and to electrons) in the 1890s did the science of the discharge tube (and, ultimately, plasma theory) begin to emerge. It may turn out that we have spent the last 50 years of the 20th century "looking at the wrong end of the tube" in our quest for a magnetically-confined toroidal fusion reactor. This has considerably extended our understanding of plasma, but it may not be what we end up with as the energy provider of the future.

The increasing size and construction costs of the envisaged commercial reactors give cause for concern. Most individual countries will, quite simply, be unable to afford their own electricity-generation plant. As we enter a period of declining fossil fuel reserves, such a situation has the potential to be economically (and politically) destabilising. There is also an argument which says that more than one reactor will be required by each nation (or region) to cover for times of maintenance and to avoid vulnerability to sabotage. We must, therefore, continue to pursue alternative ideas with equal vigour.

In the meantime, plasma physics must extend its applications in environmental and industrial sectors. Much work remains to be done also in atmospheric and space plasmas. The behaviour of the Earth's magnetosphere, plasma-sheet and aurora; the physics of magnetic storms; and the interaction of the solar wind with the Earth's plasma environment as a whole are all areas of active and necessary research. We need to know more about the behaviour of the star on which we depend — the Sun — in order to predict and shield ourselves from its periodic outbursts. Coronal mass ejections have affected orbiting satellites, communications and power supplies in the past. As we become increasingly dependent on hardware in orbit around the Earth we must find ways of anticipating problems and protecting our investment.

Suggested Further Reading

Baumjohann, W., & Treumann, R.A., *Basic Space Plasma Physics*, (Imperial Coll. Press, 1996) — geophysical and space plasmas

Bittencourt, J.A., *Fundamentals of Plasma Physics*, (Pergamon Press, 1986) — a mathematical treatment of plasma theory

Dendy, R.O. (ed.), *Plasma Physics: an introductory course*, (Cambridge University Press, 1993) — plasma theory; space, industrial and fusion plasmas

Goldston, R.J., & Rutherford, P.H., *Introduction to Plasma Physics*, (I.O.P. Publishing, 1995) — Plasma theory

Grill, A., *Cold Plasma in Materials Fabrication*, (I.E.E.E. Press, 1994) — plasma chemistry, reactors, diagnostics, deposition of films, etching

Howatson, A.M., *An Introduction to Gas Discharges*, (Pergamon Press, 1976 (2nd. ed.))

Hutchinson, I.H., *Principles of Plasma Diagnostics*, (Cambridge University Press, 1987) — magnetic field, density and temperature measurement; radiation processes (classical and quantum mechanical treatment); ion processes

Kitchin, C.R., *Optical Astronomical Spectroscopy*, (IOP Publishing, 1995) — introduction to spectroscopy

Manheimer, W., Sugiyama, L.E. & Stix, T.H. (eds.), *Plasma Science and the Environment*, (AIP Press, New York, 1997) — global and large-scale environmental processes; energy use and conservation; waste treatment using plasma processing.

Parks, G.L., *Physics of Space Plasmas, an Introduction*, (Addison-Wesley, New York, 1991) — geophysical and space plasmas.

See also Internet sites such as:
www.fusion.org.uk/ (EURATOM/UKAEA site: has a list of fusion sites world-wide)

Appendix A

A mathematical diversion on ω_p

The simplest model for an oscillating system is a particle of mass m moving in one dimension and acted on by a force:

$$F(x) = b - ax \qquad (A.1)$$

where a and b are constants with $a>0$. The equation of motion for this system is of the form

$$m\frac{d^2x}{dt^2} + ax = b \qquad (A.2)$$

Let us assume that the electrons in our plasma are forced to move in one dimension only, the x-direction. The initial situation is shown in Fig. A.1(a), where the shaded area δx contains n_0 electrons. (This will also be the density of the ions because the undisturbed plasma is electrically neutral.)

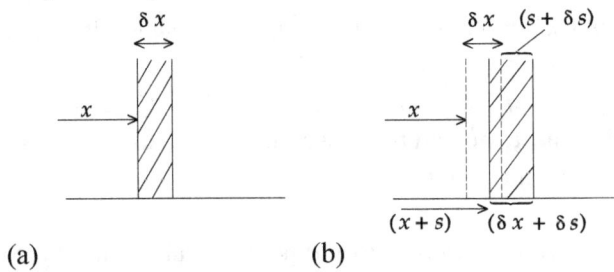

Fig. A.1   Displacement of electrons

A one-dimensional disturbance occurs which causes the electrons in the shaded area δx to be displaced in the x direction by a small distance $s(x,t)$, the displacement being a function of both distance x and time t. This is the situation represented in Fig. A.1(b). The electrons originally contained in δx are now in the shaded region ($\delta x + \delta s$). The new particle density, n, is given by

$$n = n_0\frac{\delta x}{(\delta x + \delta s)} = \frac{n_0}{(1 + \dfrac{\delta s}{\delta x})}$$

The change in density $(n - n_0)$ produced by the displacement will be small, enabling us to use the binomial expansion: $(1 + x)^r = 1 + rx + ...$ to write

$$n = n_0(1 + \frac{\delta s}{\delta x})^{-1}$$

as

$$n = n_0(1 - \frac{\delta s}{\delta x})$$

The change in density is therefore

$$(n - n_0) = -n_0 \frac{\delta s}{\delta x} \tag{A.3}$$

The charge density, ρ, at any point is given by

$$\rho = -n_0 q \frac{\partial s}{\partial x} = n_0 e \frac{\partial s}{\partial x} \tag{A.4}$$

for electrons. (Partial derivatives are used when a function of several variables is differentiated with respect to one variable only, the others being treated as constant.)

The charge density is related to the electric field produced by the charge separation, via Gauss's law

$$\nabla \cdot E = \frac{\rho}{\varepsilon_0} \tag{A.5}$$

Since we are dealing with electron movement in the x direction only, the electric field can be represented by E_x. Substituting equation (A.4) into equation (A.5) we obtain

$$\frac{\partial E_x}{\partial x} = \frac{n_0 e}{\varepsilon_0} \frac{\partial s}{\partial x} \tag{A.6}$$

which, after integration, produces:

$$E_x = \frac{n_0 e}{\varepsilon_0} s + C \tag{A.7}$$

where C is a constant. In equilibrium at time $t = 0$, before the electrons were displaced, $s(x,t) = 0$ and $E_x = 0$, therefore $C = 0$. From equation (4.1) the force on a displaced electron is

$$|F|(s) = -eE = -\frac{n_0 e^2}{\varepsilon_0} s \tag{A.8}$$

Comparing this with the general form $F(x) = b - ax$ given in equation (A.1) we find that

$$a = \frac{ne^2}{\varepsilon_0} \quad \text{and} \quad b = 0 \tag{A.9}$$

Substituting these values into (A.2) produces the simple harmonic equation of motion of the oscillating electrons:

$$m \frac{d^2 x}{dt^2} + \frac{n_0 e^2}{\varepsilon_0} x = 0 \tag{A.10}$$

The general solution to an equation such as (A.2) can be written in the form

$$x(t) = C \cos \omega t + D \sin \omega t + h$$

where C and D can take any real values determined by the position and velocity of the particles at $t = 0$, $h = b/a$ and

$$\omega = \sqrt{\frac{a}{m}} \tag{A.11}$$

Substituting our result (A.9) into (A.11) gives the frequency of oscillation — the plasma frequency:

$$\omega_p = \left(\frac{ne^2}{\varepsilon_0 m} \right)^{\frac{1}{2}} \tag{A.12}$$

Appendix B

Solution of circularly-polarised wave cut-offs

The dispersion relation for right-circularly polarised waves is given by equation (6.26). Ignoring the contribution from the ions:

$$k_R = \frac{\omega}{c}\left(1 - \frac{\omega^2_{pe}}{\omega(\omega - \omega_{ce})}\right)^{\frac{1}{2}} \tag{B.1}$$

The cut-off frequency, ω_R, is obtained by setting equation (B.1) equal to zero and labelling ω as ω_R:

$$\frac{\omega_R}{c}\left(1 - \frac{\omega^2_{pe}}{\omega_R(\omega_R - \omega_{ce})}\right)^{\frac{1}{2}} = 0$$

Since c and ω_R cannot be zero:

$$1 - \frac{\omega^2_{pe}}{\omega_R(\omega_R - \omega_{ce})} = 0$$

$$1 = \frac{\omega^2_{pe}}{\omega_R(\omega_R - \omega_{ce})}$$

$$\omega_R(\omega_R - \omega_{ce}) = \omega^2_{pe}$$

$$\omega^2_R - \omega_R\omega_{ce} - \omega^2_{pe} = 0 \tag{B.2}$$

Using the quadratic formula, $ax^2 + bx + c = 0$, which has roots:

$$x = \frac{-b \pm \sqrt{b^2 - 4ac}}{2a}$$

and substituting $a = 1$, $b = -\omega_{ce}$, and $c = -\omega^2_{pe}$ gives roots for equation (B.1).

$$\omega_R = \frac{1}{2}\left[\omega_{ce} \pm (\omega^2_{ce} + 4\omega^2_{pe})^{\frac{1}{2}}\right] \tag{B.3}$$

By convention, ω_R is positive; a negative ω_R is therefore meaningless. Thus

$$\omega_R = \frac{1}{2}\left[\omega_{ce} + (\omega^2_{ce} + 4\omega^2_{pe})^{\frac{1}{2}}\right] \tag{B.4}$$

(and similarly for ω_L).

Solution of X-wave cut-offs

The dispersion relation is given by equation (6.33):

$$\frac{c^2 k^2}{\omega^2} = 1 - \frac{\omega^2_{pe}(\omega^2 - \omega^2_{pe})}{\omega^2(\omega^2 - \omega^2_{UH})}$$

where

$$\omega^2_{UH} = (\omega^2_{ce} + \omega^2_{pe})$$

Substituting for ω^2_{UH} and setting k equal to zero produces

$$1 - \frac{\omega^2_{pe}(\omega^2 - \omega^2_{pe})}{\omega^2(\omega^2 - (\omega^2_{ce} + \omega^2_{pe}))} = 0$$

$$1 = \frac{\omega^2_{pe}(\omega^2 - \omega^2_{pe})}{\omega^2(\omega^2 - (\omega^2_{ce} + \omega^2_{pe}))}$$

$$(\omega^2 - \omega^2_{ce} - \omega^2_{pe}) = \frac{\omega^2_{pe}}{\omega^2}(\omega^2 - \omega^2_{pe})$$

$$\frac{(\omega^2 - \omega^2_{ce} - \omega^2_{pe})}{(\omega^2 - \omega^2_{pe})} = \frac{\omega^2_{pe}}{\omega^2}$$

Dividing the left-hand side by $(\omega^2 - \omega^2_{pe})$:-

$$1 - \frac{\omega^2_{ce}}{(\omega^2 - \omega^2_{pe})} = \frac{\omega^2_{pe}}{\omega^2}$$

Rearranging:

$$\left(1 - \frac{\omega^2_{pe}}{\omega^2}\right) = \frac{\omega^2_{ce}}{(\omega^2 - \omega^2_{pe})} = \frac{\omega^2_{ce}}{\omega^2\left(1 - \frac{\omega^2_{pe}}{\omega^2}\right)}$$

Multiply both sides by $(1 - \omega^2_{pe}/\omega^2)$:-

$$\left(1 - \frac{\omega^2_{pe}}{\omega^2}\right)^2 = \frac{\omega^2_{ce}}{\omega^2}$$

Taking the square root of both sides:-

$$1 - \frac{\omega^2_{pe}}{\omega^2} = \pm\frac{\omega_{ce}}{\omega}$$

Multiply by ω^2 and we have recovered equation (B.2), with ω labelled as ω_R:

$$\omega^2 - \omega^2_{pe} = \pm\omega\omega_{ce}$$

Historical development of Plasma Physics and related fields

1610-11 First telescope observations of sunspots.

1643 Evangelista Torricelli (1608-47) invented the mercury barometer.

1650 Otto von Guericke (1602-86) built the first air-extraction pump.

1662 Robert Boyle (1627-91) published his law relating gas pressure to volume ($PV =$ constant). Discovered independently, in 1676, by Edme Mariotte (1620-84).

1665 Robert Hooke (1635-1703) proposed a simple wave theory of light.

c. 1672 Otto von Guericke invented the electric friction machine.

1675 Jean Picard (1620-82) observed a flickering light in his mercury barometer.
Danish astronomer Olaf Rømer (1644-1710), working in Paris with Picard, made the first estimate of the speed of light: 141,000 miles (226,870 km) per second. It was viewed as a scientific curiosity.

1687 Sir Isaac Newton (1642-1727) published his theory of gravitation and laws of motion (*Principia Mathematica*) and derived the relationship $v = f\lambda$.

1716 Edmund Halley (1656-1742) suggested that geomagnetic effects caused the aurora.

1729 Stephen Gray (1696-1736) found there are good and bad conductors of electricity.

1731 J.J. de Mairan suggested a link between the aurora and the solar atmosphere.

1733 Charles François de Cisternay du Fay (1698-1739) identified two kinds of electric "fluid".

1741 O.P. Hjorter and George Graham observed a connection between auroral and geomagnetic activities.

1745 Leyden jar invented by Ewald Georg von Kleist (d.1748) in Poland, and Pieter van Musschenbroek (1692-1761) in Holland.

1747 Benjamin Franklin (1706-90) proposed single-fluid theory of electricity with positive charge carriers.
Sir William Watson (1715-87) began studying electric discharges.

1752 Franklin showed that lightning is an electrical discharge.
Thomas Melvill (1726-53) made the first observation of spectral lines.
Watson compared electric discharge to the aurora.

1766 English chemist and physicist Henry Cavendish (1731-1810) identified hydrogen.

1774 Alexander Wilson (1714-86) said sunspots are depressions in the Sun's atmosphere.
Wm. Henly thought "particles of electricity" travelled through wires at great speed.

1784 Cavendish discovered the composition of the atmosphere.

1785 Coulomb (1736-1806) published inverse square law for the force between electric charges.

1787 Jacques-Alexandre Cézar Charles (1746-1823) defined law relating gas volume and temperature.

1790 Cavendish measured the height of the aurora.

1800 Alessandro Volta (1745-1827) developed the first electric battery (voltaic cell).
Electrolysis discovered by Wm. Nicholson (1753-1815) and Sir Anthony Carlisle.

1801 Thomas Young (1773-1829) presented his wave theory of light (double-slit experiment).

1802	Joseph Louis Gay-Lussac (1778-1850) defined law relating gas pressure and temperature.
1802-3	John Dalton (1766-1844) proposed his atomic theory of matter.
1808	Sir Humphrey Davy (1778-1829) demonstrated the electric arc.
1820	Hans Christian Oersted (1777-1851) showed that electric current deflects a magnetic compass needle.
1826	Georg Simon Ohm (1789-1854) established the relation $V = IR$ (Ohm's Law). Samuel Christie showed that solar rays possess magnetic properties.
1831	Michael Faraday (1791-1867) discovered electromagnetic induction (discovered in 1830 by Joseph Henry (1799-1878) but not published.)
1838	Faraday classified electric discharges.
1842	Christian Johann Doppler (1803-53) predicted the Doppler shift for sound (confirmed in 1845) and suggested it would apply to light. Dominique François Jean Arago (1786-1853) said that the Sun is wholly gaseous.
1843	Abria observed striations in a discharge tube.
1845	Faraday showed that a magnetic field will rotate the plane of polarisation of light.
1848	Lord Kelvin (1824-1907) proposed absolute zero and absolute temperature scale.
1851	Heinrich Schwabe (1789-1875) reported periodicity in sunspots of about 10 years.
1851-2	Edward Sabine suggested that magnetic storms were linked to sunspot variations.
1852	First good photographs of Moon, and in 1857, first photographs of Sun, taken by Warren de la Rue.
1857	Werner von Siemens (1816-92) produced ozone from oxygen using a silent electric discharge. Rudolph Clausius (1822-88) founded the kinetic theory of matter. Julius Plücker (1801-68) found the light stream in a discharge tube was affected by a magnet.
1858	Richard Carrington (1826-75) noticed that sunspot latitudes change with the solar cycle.
1859	First recorded observation of a solar flare, followed by strong magnetic storm and auroral displays.
1860	James Clerk Maxwell (1831-79) published his *Dynamical Theory of Gases*, developing the kinetic theory of gases and introducing statistical mechanics.
1862	Jean Bernard Léon Foucault (1819-68) used a revolving mirror to determine the absolute velocity of light: $2.98 \times 10^8 \, \text{m s}^{-1}$.
1864	Maxwell published his theory of electromagnetism, identifying light as electromagnetic waves. The theory was extended in 1873.
1868	Sir Joseph Norman Lockyer (1836-1920) and Pierre Jules Cézar Janssen (1824-1907) found helium in the Sun's spectrum 27 years before it was found on Earth. In Sweden, Anders Jonas Ångström (1714-74) published his table of solar spectrum wavelengths.
1869	Russian chemist Dimitri Mendeleyev (1834-1907) produced the periodic table of chemical elements. Johann Wilhelm Hittorf (1824-1914) observed shadows cast by objects placed in

discharge tubes.

1874   George Johnstone Stoney (1826-1911) defined a fundamental unit quantity of electricity.

1876   Eugen Goldstein introduced the term "cathode rays".

1878   Sir William Crookes (1832-1919) established that cathode rays are streams of tiny charged particles and suggested the term "fourth state of matter" for the gases in the discharge tubes.

1879   Albert Abraham Michelson (1852-1931) determined the speed of light to be 2.999 x 10^8 m s^{-1}.

            Thomas Alva Edison (1847-1931) and Sir Joseph Wilson Swan (1828-1914), independently invented the electric light bulb (incandescent lamp).

1880   Balfour Stewart (1828-87) suggested there was an electrically-conducting gas layer in the atmosphere.

1881   Goldstein suggested the aurora was due to particles from the Sun.

1888   Heinrich Rudolph Hertz (1857-94) discovered radio waves, confirming Maxwell's theory.

1889   Arthur Schuster defined atmospheric dynamo theory.

c. 1890   First reports of "whistlers".

1890   J. J. Thomson (1856-1940) showed that cathode rays travel slower than light.

1891   George Johnstone Stoney gave name "electron" to the fundamental electric charge.

1891   Thomson produced the first "electrodeless" ring discharge.

1892   Oliver Lodge (1851-1940) and George Francis FitzGerald (1851-1901) postulated a solar wind.

1895   Wilhelm Conrad Röntgen (1845-1923) discovered X-rays while studying cathode ray luminescence.

            Hendrick Antoon Lorentz (1853-1928) produced an equation of motion for charged particles in an electromagnetic field.

            Jean Baptiste Perrin (1870-1942) found that metal objects in a beam of cathode rays acquired a negative charge.

1896   Pieter Zeeman (1865-1943) found that spectral lines split in a magnetic field.

1897   J.J. Thomson measured the deflection of cathode rays by magnetic fields and discovered the electron. He estimated the amount of charge carried by one electron from the charge to mass ratio, e/m_e.

1901   Townsend developed theory of ionisation.

1902   Sir Oliver Heaviside (1850-1925) and Arthur Edwin Kennelly (1861-1939) independently explained Marconi's radio transmission across the Atlantic Ocean using Stewart's idea of an electrically-conducting layer in the upper atmosphere.

1905   Albert Einstein (1879-1955) explained the photo-electric effect, developed the principle that mass and energy are equivalent ($E = mc^2$) and published the Special Theory of Relativity.

1906   Lord Rayleigh (1842-1919) observed electron oscillations (plasma frequency).

1907   E.F. Northrup described the pinch effect.

1908   George Ellery Hale (1868-1938) found magnetic fields associated with sunspots.

1909	Kristian Birkeland (1867-1917) published his theory of electric currents flowing along magnetic field lines during auroral displays.
1910	American physicist Robert Andrews Millikan (1868-1953) measured the charge on the electron.
1911	Ernest Rutherford (1871-1937) proposed the existence of the nuclear atom.
	Carl Størmer postulated a ring current around the Earth in the upper atmosphere.
1913	Niels Henrik David Bohr (1885-1962) combined Rutherford's nuclear atom with Planck's quantum theory in his model for the structure of the hydrogen atom.
	Frederick Soddy (1877-1956) suggested the existence of isotopes.
	J. Franck and G. Hertz observed energy levels in atoms using electrical discharges.
	Johannes Stark (1874-1957) observed electric field equivalent of the Zeeman effect.
1918	Sidney Chapman began to develop his theory of magnetic storms.
1919	Rutherford demonstrated the existence of protons.
1920	Rutherford proved Bohr's theory experimentally and proposed the existence of the neutron, discovered in 1932 by Sir James Chadwick (1891-1974).
1920-23	Development of Saha equation.
1923	Peter Joseph Wilhelm Debye (1884-1966) and E. Hückel developed the theory of electrolytes, providing a picture of ions in solution.
	Rutherford proposed synthesis of helium from hydrogen as a source of solar energy.
1925	Appleton and Barnett demonstrated reflection of radio waves by upper atmosphere.
1926	Gilbert Lewis introduced the term "photon" for the particle of light.
1928	Irving Langmuir (1881-1957) introduced the term "plasma".
	George Gamow and Edward Condon independently explained α-tunnelling.
1929	Tonks and Langmuir laid the foundations of plasma physics.
	Atkinson and Houtermans said solar energy is due to thermonuclear fusion reactions of hydrogen.
1930	Chapman and Ferraro outlined theory of magnetic storms.
1931	Chapman and Ferraro produced theories of ionosphere and of magnetosphere formation.
	Discovery of deuterium.
1932	Sir John Douglas Cockcroft (1897-1967) and Ernest Walton split the atom.
	Radio waves found in space by Karl Jansky (1905-50).
1934	M.L.E. Oliphant, P. Harteck & Ernest Rutherford published paper on D-D and D-T fusion reactions.
1936	Lev Davidovich Landau (1908-68) produced a kinetic equation for a system of charged particles.
1937	V.C.A. Ferraro introduced image of charged particles "frozen" to the magnetic field.
	Uranium atom split (nuclear fission).
1938	Hans Albrecht Bethe described the CNO cycle, which produces energy in stars.
1942	Hannes Olof Göst Alfvén discovered hydromagnetic waves in the ionised gas of the Sun and devised the set of MHD equations which define the motion of electromagnetic fluids — plasmas.
1945	(6 Aug.) First atom bomb exploded over Hiroshima.

c. 1946	Development of microwave discharges.
1946	US classified fusion research.
	Landau described the interaction between particles and plasma waves.
1947	A.C.B. Lovell, J.A. Clegg and C.D. Ellyett bounced radio signals off the aurora, showing that aurorae have plasma characteristics.
1949	Stig Lundquist demonstrated experimentally the existence of MHD waves.
1949-50	Theory of plasma oscillations established by D. Bohm and E.P. Gross.
1951	Cousins and Ware observed pinch effect in a gas discharge.
1952	First fusion (hydrogen) bomb exploded by US on Eniwetok Atoll in Pacific Ocean.
1953	Model-A stellarator built at Princeton, USA.
	J.W. Dungey proposed magnetic reconnection theory.
	USSR exploded hydrogen bomb in Kazakhstan.
1955	First tokamak (TMB) built in USSR.
1956	US-UK and UK-USSR exchange of classified information on fusion research.
1957	ZETA operational at Harwell, UK.
	Lawson's criterion formulated.
1958	Atoms for Peace Conference: start of international cooperation on fusion research.
	Eugene Parker showed that the corona is constantly expanding.
	Van Allen radiation "belts" discovered by *Explorer I*.
1959	Tom Gold (1920-2004) proposed the term "magnetosphere".
1960	First (ruby) laser developed by Theodore Maiman.
	First ion engine built by N.A.S.A.
	Interplanetary magnetic field discovered by *Pioneer V*.
	The S.I. units (Système International d'Unités) proposed as a system of scientific units.
1961	H.W. Babcock developed the modern theory of sunspots.
1962	Laser fusion research programmes started.
1963	ZETA spontaneously produced RFP configuration.
1968-9	UK team make measurements on Soviet tokamak, confirming temperature claims.
1972	First use of neutral beam heating on CLEO tokamak (Culham, UK).
	USA declassified the laser fusion programme.
1973	Oil crisis led to increased fusion research to reduce West's reliance on fossil fuels.
1974	J.B. Taylor produced a theory of relaxed states; ideal MHD theory extended to include resistive effects.
1978	Construction of JET began — biggest machine in the world.
1982	TFTR operational (Princeton). (Shut down 1997.)
1983	JET operational (Culham).
1989	Fleischmann and Pons claim cold fusion.
1991	First spherical tokamak, START, became operational. (Shut down 1998.)
	JET produced 1.7 MW of power for 2 sec. from a 10% admixture of tritium.
1993	TFTR produced 10.7 MW of fusion power, with a peak of 3 MW, from a 50:50 D-T mixture.
1999	First successful ion-drive space-propulsion system: Deep Space 1.

Glossary

ambipolar diffusion: Charge separation, resulting from electron diffusion being faster than positive ion diffusion, produces an electric field which increases the drift velocity of the ions and slows that of the electrons. Eventually, ions and electrons diffuse with the same velocity, for example: the radial field in the positive column of a glow discharge results in ambipolar diffusion to the walls.

arc: arched electrical discharge. A low voltage, high current electrical discharge through gas, maintained between electrodes.

aspect ratio: ratio of major to minor radius. Indicates how compact a torus is.

atmospheric pressure at sea level: $1.01325 \times 10^5 \, \text{Nm}^{-2}$ (760 mm Hg or 760 torr)

AU: astronomical unit. 1 AU is the distance of the Earth from the Sun.

Boltzmann's distribution: A statistical distribution of large numbers of small particles when subjected to thermal agitation and acted upon by electric, magnetic or gravitational fields.

brush discharge: A discharge, usually between a conductor and air, which is insufficient to cause a spark or arc. Often called "corona".

corona: Air breakdown when electric stress at the surface of a conductor may result in a luminous discharge.

critical or *cut-off density*: The highest plasma density that allows passage of an electromagnetic wave.

diode (early 20th C.): Simple discharge tube, with a heated cathode and anode, used for one-way, and hence rectification properties.

discharge (sometimes *field discharge*): Flow of electric charge through gas or air due to ionisation, e.g. lightning, or at reduced pressure, as in fluorescent tubes.

discharge tube (or *electron tube*): Any system of two electrodes placed in a gas between which a discharge may pass.

"frozen-in" magnetic field: If electrical conductivity is sufficiently large, the magnetic lines of force can be visualised as being carried by the fluid and travelling with it, despite there being no scientific basis for the idea.

isotropic: having the same properties irrespective of direction

magnetic flux density or *magnetic induction*: the magnetic flux passing through a unit area perpendicular to the magnetic induction vector **B**. It is measured in tesla (T), and is related to the *magnetic field strength* **H** via $\mathbf{B} = \mu \mathbf{H}$ (T), where μ is the magnetic permeability (μ in plasma is equal to that of free space, μ_0). In plasma physics, the magnetic induction, **B**, is often referred to as the "magnetic field" or "magnetic field strength".

mean free path: Average distance a particle travels between collisions.

mm Hg: The degree of vacuum was originally measured with mercury manometers which related the pressure in the vacuum system to that of the atmosphere. The common unit of pressure was the millimetre of mercury, mm Hg, in which one standard atmosphere (atm) at 0° C was equal to 760 mm Hg. In the mid 20th century the unit of pressure was changed from mm Hg to the torr (1 torr = 1 mm Hg). The SI unit of pressure is the pascal (Pa). 1 std. atm. = 760 torr =1.01325 x 10^5 Pa. 1 mm Hg $\equiv$ 133.3 Nm^{-2}.

permeability, μ: (absolute permeability) — the ratio of flux density produced to the magnetic field strength producing it: $\mu = B/H$ henry m^{-1}. Indicates by how much the material is permeated by the magnetic field in which it is immersed.

permittivity, ε: (absolute permittivity) — the ratio of electric displacement in a medium to the electric field intensity producing it: $\varepsilon = D/E$ farad m^{-1}. Indicates by how much the material is permeated by the electric field in which it is immersed.

plasma diagnostics: equipment used to study the plasma and measure parameters such as temperature and density.

potential: the relative electrical state of an object

rectifier: component for converting a.c. into d.c.

relaxation time: the time required for a plasma to reach near-equilibrium

singly-ionised atom: an atom with one electron removed

space charge: local net charge distributed through a finite volume (as distinct from a surface). Gradients in electrostatic fields are associated with regions of space charge.

space charge limitation: condition in a thermionic valve when electron current leaving a cathode is limited by the balance between attractive electric forces from other electrodes and repulsion within space charge.

substrate: surface onto which films are deposited or integrated circuits etched

tension: obsolete term used to designate a potential difference

thermionic emission: Charged particles (usually electrons) are evaporated from the surface of a metal at high temperatures.

torr: unit of pressure used mainly in high-vacuum technology. 1 torr = 1 mm Hg = 133.32 N m^{-2} = 1/760 atm

waveguide: hollow metal tube often containing a dielectric. Used to guide electromagnetic waves in the microwave region of the spectrum. The wave is reflected from the internal surfaces of the guide.

SI Units

Quantity	Name of unit	Symbol	(alternative)
length	metre	m	
mass	kilogram	kg	
time	second	s	
electric current	ampere	A	
thermodynamic temperature	Kelvin	K	

derived:

Quantity	Name of unit	Symbol	(alternative)
frequency	hertz	Hz	
speed, velocity	metre per second	$m\ s^{-1}$	
angular velocity	radian per second	$rad\ s^{-1}$	
force	newton	N	$kg\ m\ s^{-2}$
work, energy	joule	J	$N\ m$
power	watt	W	$J\ s^{-1}$
charge, quantity of electricity	coulomb	C	$A\ s$
charge density	coulomb per m^3	$C\ m^{-3}$	
current density	ampere per m^2	$A\ m^{-2}$	
potential difference, electric potential	volt	V	$W\ A^{-1}, J\ C^{-1}$
electric field strength	volt per metre	$V\ m^{-1}\ N\ C^{-1}$	
resistivity	ohm-metre	$\Omega\ m$	$V\ m\ A^{-1}$
magnetic field	tesla	T	$V\ s\ m^{-2}, N/A\ m, N\ s/C\ m$
wave number	radian per metre	$rad\ m^{-1}$	

Index

Alfvén waves   114
alternating current discharge   162-4
Aston dark space   25, 159
aurora   223-8
behavioural parameters   264-5
black-body   129
Boltzmann equation   147-50
Boltzmann factor   53
bounce period   250
breakdown   24, 37-8, 152, 155-6
breakdown voltage   24
bremsstrahlung   122, 123
brush discharge   157
cathode rays   13, 15
cathode spot   161
cathode yield   38, 154
chromosphere   241
collision cross-section   90
collision frequency   92
collisionless shock   246
contamination   120-2
continuity equations   136-7
corona discharge   157, 174-5
coronal domain   131
Coulomb collision   89
Coulomb logarithm   91
Coulomb potential energy   268
coupling parameter   268
criteria for plasma   264
Crookes' (cathode) dark space   14-5, 25-6
current sheet   247
cut-off frequency   105
cyclotron frequency   65
cyclotron radiation   124-5
damping   106-7
dark discharge   29, 153, 157
Debye distance   34
Debye length   83-5
Debye sphere   85
deuterium   184, 188
diagnostics   119, 129-30

diamagnetic drift   70
diamagnetic effect   65
diamagnetism   9, 140
dielectric constant   102
diffusion   91
dispersion   100
distribution function   48-50
Doppler effect   128
double layer   31, 87
drawn arc   162
dusty plasma   238
dynamo layer   253
E x B drift   66-8
electrical discharge definitions   8-9
electron avalanche   63
electron drift velocity   63
electron multiplication factor   154
electron retardation radiation   123
electron runaway   63, 96
electron temperature   52
electron volt   52
electrostatic precipitator   175
electrostatic waves   101
energy confinement time   201
equation of motion   60
extraordinary waves   115
Faraday dark space   9, 159
Faraday rotation   10, 113
Fermi energy   266-8
floating potential   29
flux surface   206
Fokker-Planck equation   149
force-free field   244
fourth state of matter   15
fractional abundance   130-1
fusion cross-section   191
fusion triple product   201
gas laws   43
grad B drift   73-5
gradient-curvature drift   76
granules   241

gravitational drift   69-70
group velocity   100, 238
gyrofrequency   65
ideal gas law   44
ignition temperature   200, 219
impact parameter   89
inertial confinement   199-200, 217-20
instabilities   140-5
ion Landau damping   109
ion propulsion   180-1
ion sound speed   108
ion temperature   52
ionisation coefficient   153
ionisation cross-section   120
ionosphere   230-2, 245-56
ITER   217
JET   199, 209
kelvin   44
kinetic model   145-50
L- and R-waves   110-3
Langmuir oscillations   103
Langmuir probe   29
Larmor orbit   64
Lawson's criterion   200-1, 218
Lenard rays   18
line radiation   126
local thermal equilibrium   51
Lorentz force   60
magnetic bottle   72-3, 143
magnetic field pressure   140
magnetic reconnection   144-5, 244
magnetic helicity   206
magnetic shear   208
magnetic storm   224-30
magnetosheath   247
magnetosonic waves   116
Maxwellian distribution   48
MHD   39, 133-4, 137-8, 142-4
mirror point   72
mode number   208
momentum transfer   134-6
ohmic heating   96, 215
optical mean free path   129

ozone production   170
partition function   54
phase space distribution function   147
phase velocity   100, 238
pinch effect   37
pinch point   203
pinch ratio   212
plasma temperature   51
plasma beta   140, 211
plasma: definition   81
plasma frequency   33, 102-4
plasma, naming of   32-3
plasma nitriding   173
plasma oxidation/anoxidation   173
plasma parameter   88
plasma, perfectly-conducting   94
plasma resistivity   94
plasmasphere   249
polarisation drift   78
polarised waves   110
positive column   159-60
pressure balance condition   139
Project Matterhorn   185
prominence   225, 243
propagation constant   100
proton-proton chain   240
quantum effects   266, 268-9
quasi-neutrality   81
r.f. discharge   162-4, 172
r.f. heating   215
recombination radiation   122, 125-6
relaxation time   53
ring discharge   17
Saha equation   54-7
saturation current density   153
separatrix   145, 207
sheath   30, 85-7
shielding distance   84
skin depth   106
solar flare   225, 242, 244
solar wind   229-30, 242, 244-8, 251
sparking criterion   155
sparking potential   23

spectral line broadening   128
Spitzer resistivity   95
sputtering   120, 171
sputtering yield   171
stable discharge   156
Stark effect   127, 128
START   214-5
stellarator   198, 210
sterilisation   178-9
striations   10, 16, 160
substrate   162, 171
sunspots   223-9, 243
synchrotron radiation   125
TFTR   199
thin-film coatings   172-3
tokamak   198-9, 209
toroidal geometry   204
Townsend discharge   153-4, 157
transparency   129
tritium   185, 188, 190-3
Vlasov equation   148
wave number   100
whistler   113, 256
Zeeman effect   127
ZETA   159, 161, 169